COSMIC EXPLOSIONS IN THREE DIMENSIONS

Supernovae and gamma-ray bursts are the strongest explosions in the Universe. Recent observations have shown that, rather than being symmetrical, they are driven by strong jets of energy and other asymmetrical effects that reveal previously unknown physical properties. These observations have demanded new theories and computations that challenge the biggest computers. This volume marks the transition to a new paradigm in the study of stellar explosions. It highlights the burgeoning era of routine supernova polarimetry and the new insights into core collapse and thermonuclear explosions. With chapters by leading scientists, the book summarizes the status of a rapidly developing new perspective on stellar explosions and should be a valuable resource for graduate students and research scientists.

PETER HÖFLICH received his Ph.D. from the University of Heidelberg in 1986. He then held research posts at the Max Planck Institute for Astrophysics, the University of Munich, and Harvard University. He is currently a Senior Research Scientist at the University of Texas in Austin, where he works mainly on problems of radiation transport, non-LTE atmospheres, polarization, hydrodynamics, and radiation hydrodynamics. Besides analyzing young stellar objects, B stars, and red supergiants, his main field of interest is supernova explosions, both core-collapse and thermonuclear explosions. He works on the use of Type Ia as cosmological probes and the connection of supernovae and gamma-ray bursts.

PAWAN KUMAR received his Ph.D. from Caltech in 1988. He subsequently held positions at the High Altitude Observatory and at MIT, and in 1996 he took up a long-term visiting professorship at the Institute for Advanced Study. In 2002 he became Professor of Astronomy at the University of Texas at Austin where he is a key member of the innovative UTeach science education program. He is an expert on helioseismology and more recently he has actively worked on the physics of gamma-ray bursts, especially models for the collimation of and radiation from the associated energy flow.

J. CRAIG WHEELER received a Ph.D. in physics from the University of Colorado in 1969. He is currently the Samuel T. and Fern Yanagisawa Regents Professor of Astronomy at the University of Texas at Austin, and was Chair of the Astronomy Department from 1986 to 1990. He is a member of the Academy of Distinguished Teachers at the University of Texas. He is currently serving on the Space Studies Board of the National Research Council, and is co-Chair of the NRC Committee on the Origin and Evolution of Life. His research interests are supernovae, black holes, gamma-ray bursts, and astrobiology.

COSMIC EXPLOSIONS IN THREE DIMENSIONS

Asymmetries in Supernovae and Gamma-Ray Bursts

Edited by:

PETER HÖFLICH
PAWAN KUMAR
J. CRAIG WHEELER
University of Texas at Austin

CAMBRIDGE
UNIVERSITY PRESS

PUBLISHED BY THE PRESS SYNDICATE OF THE UNIVERSITY OF CAMBRIDGE
The Pitt Building, Trumpington Street, Cambridge, United Kingdom

CAMBRIDGE UNIVERSITY PRESS
The Edinburgh Building, Cambridge, CB2 2RU, UK
40 West 20th Street, New York, NY 10011–4211, USA
477 Williamstown Road, Port Melbourne, VIC 3207, Australia
Ruiz de Alarcón 13, 28014 Madrid, Spain
Dock House, The Waterfront, Cape Town 8001, South Africa

http://www.cambridge.org

First published 2004

Printed in the United Kingdom at the University Press, Cambridge

Typeface Times 11/14 pt. *System* LaTeX 2ε [TB]

A catalogue record for this book is available from the British Library
Library of Congress Cataloging in Publication Data
Cosmic explosions in three dimensions: asymmetries in supernovae and gamma-ray bursts/edited by Peter Höflich, Pawan Kumar, J. Craig Wheeler.
p. cm. – (Cambridge contemporary astrophysics)
Includes bibliographical references and index.
ISBN 0 521 84286 7
1. Supernovae. 2. Gamma ray bursts. I. Höflich, Peter. II. Kumar, Pawan.
III. Wheeler, J. Craig. IV. Series.
QB843.S95C67 2004
523.8′4465 – dc22 2004045689

ISBN 0 521 84286 7 hardback

Contents

Part I

Introduction

1

3-D explosions: a meditation on rotation (and magnetic fields)

J. C. Wheeler

Department of Astronomy University of Texas at Austin

1.1 Introduction: a brief time for history

There has been a great deal of progress in the thirty-five years or so that I have been working on supernovae and related topics. Two of the classical problems have been with us the whole time: what makes core collapse explode, and what are the progenitors of Type Ia supernovae? This workshop, indeed, the perspectives of three-dimensional astrophysics applied to these problems, gave encouraging evidence that breakthroughs may be made in both of these venerable areas.

On the other hand, what a marvelous array of progress has rolled forth with ever increasing speed. We have an expanded botany of supernovae classification: Type Ia, Ib, Ic, Type IIP, IIL IIb, IIn; but, of course, more than mere classification, a growing understanding of the physical implications of these categories. Neutron stars were discovered as rotating, magnetized pulsars when I was a graduate student, and the extreme form, magnetars, has now been revealed (Duncan & Thompson 1992). The evidence that we are seeing black holes in binary systems and the centers of galaxies has grown from suspicion to virtual certainty, awaiting only the final nail of detecting the black spot in a swirl of high-gravity effects. Supernova 1987A erupted upon us over 16 years ago and is still teaching us important lessons as it reveals its distorted ejecta and converts to a young supernova remnant before our eyes.

There have also been immense theoretical developments. Focus on core collapse has stimulated so much great work on neutrino transport: the invocation of weak neutral currents and neutrino-nucleon scattering; the understanding that neutrinos can and will become degenerate at the highest densities and the concomitant implications for the dynamics and the formation of the homologous core. More recently we have come to general understanding that a prompt shock is unlikely to make an explosion, but that significant layers of the proto-neutron star will be convective with important implications for the neutrino transport. Techniques of neutrino transport have evolved from simple diffusion to full Boltzmann transport. SN 1987A showed dramatically that we are on the right track, even if the details, even important physics, may be missing: core collapse with the predicted production of neutrinos does occur! In terms of the "other mechanism," our understanding has evolved from detonations to deflagrations, to the current paradigm of delayed detonation models. The recent understanding of the associated combustion physics has blossomed with the computational ability to do the required three-dimensional modeling.

Finally, the last few years have seen the birth and maturation of a field that was hinted at long ago, but came to fruition only recently, the systematic study of the polarization of supernovae. This technique has substantially altered our view of core collapse. It was only a few years ago that polarization was still regarded as an oddity, perhaps limited to a few peculiar events. In the last year, the idea that core collapse is asymmetric has become sufficiently accepted that papers are now written saying "as is well known, core collapse is asymmetric" without providing any reference to the hard labor required to establish that! Overnight, it seems, the wonders of the three-dimensional world have become revealed wisdom. The revelations of polarized core collapse have been the most distinct so far, but their implications are far from understood. The application of polarization to Type Ia supernovae had lagged somewhat in drama because the polarization is generally smaller, but this workshop served to provide evidence that important three-dimensional distortions are ubiquitous, and important, in Type Ia as well.

Besides all these developments that have been so central to the development of supernova science, the last few years have seen two outstanding developments that have cast supernovae research, already one of the most central and important in astrophysics, onto broader stages. What a time was 1997/1998! Careful studies of Type Ia supernovae revealed the acceleration of the Universe with the implication of the pervading dark energy. In virtually the same time frame, the discovery of optical transients associated with gamma-ray bursts and then SN 1998bw led to the connection of gamma-ray bursts with supernovae, probably some variety of Type Ic. For a mature field, the study of supernovae had a great deal of life left! Since Type Ia and Type Ic have been especially near and dear to me, this was about more excitement than my mature heart could stand.

I cannot do justice to all the great work on supernovae that has been done over my career, but I would like to touch on one other bit of history, a development that was critical for so much else that followed. I distinctly recall that when I was in graduate school there was a raging debate concerning the nature of the spectra of Type Ia, then called just Type I, supernovae. Some people argued that the spectrum near maximum consisted only of absorption lines and provided the interpretation of the absorption minima in terms of atomic features. Others insisted that the spectrum consisted purely of emission lines and provided an interpretation of the flux peaks, totally incongruent with the first interpretation, of course. David Branch provided the insight that we were looking at P-Cygni lines, hence a blended mix of emission and blue-shifted absorption. That was the insight needed to convince the world that the key feature in the spectrum of a Type Ia was Si II. From that it followed that the presence of silicon and other intermediate mass elements ruled out pure detonation models. This was the base on which so much subsequent analysis of supernovae of all types was built. More work, especially from Bob Kirshner and colleagues revealed that, with patience, the spectrum does evolve to be dominated by emission lines. Type Ia, like all supernovae, eventually evolve to a "supernebular" phase.

1.2 Type Ia

The combination of ever more thorough searches both by people at the eyepiece and by computer-driven telescopes, subsequent multi-wavelength follow-up, and theoretical and computational study has brought the study of Type Ia supernova to an impressive level of maturity. After a spirited debate, the conclusion that Type Ia are not merely thermonuclear explosions in white dwarfs, but specifically explosions in carbon/oxygen white dwarfs of mass very nearly the Chandrasekhar mass is now essentially universally accepted (Höflich & Khokhlov 1996; Nugent *et al.* 1997; Lentz *et al.* 2001). Even more precisely, the paradigm of a slow initial subsonic deflagration phase followed by a rapid supersonic, shock-mediated detonation phase (Khokhlov 1991) has been richly successful in accounting for the observed properties of Type Ia (Höflich 1995). It accounts for the existence of iron-peak elements in the center of the explosion and layers of intermediate mass elements in the outer layers, essentially by design. It also gives a framework in which to understand the variety of light curve shapes with lower transition densities leading to less nickel, and dimmer, cooler, faster light curves (Höflich *et al.* 1996), and it has successfully made predictions about infrared spectra (Höflich *et al.* 2002) and polarization properties (Wang, Wheeler & Höflich 1997; Howell *et al.* 2001). *Delayed detonation works!*

This success has put focus on a wonderful physics problem, the deflagration to detonation transition, or DDT, that astrophysics shares with a host of terrestrial combustion issues. This is a hard problem on Earth or off! One of the most interesting developments in recent years has been the resonance of terrestrial and astrophysical combustion studies. There has been dramatic progress in understanding DDT in laboratory, shock-tube environments by means of sophisticated computational studies of shock-flame interactions (Khokhlov & Oran 1999) and DDT in enclosed environments where boundaries and reflected shocks play a key role (Khokhlov, Oran & Thomas 1999). Still, the astrophysical problem, one of unconfined DDT, remains elusive. This is a quintessential multi-dimensional problem, one for which several promising lines of attack are underway.

The wealth of knowledge of Type Ia revealed by optical studies is too large to summarize here, but it has been amplified and complemented in recent years by studies in the near infra-red. The NIR is an especially powerful spectral range to study because lines are less blended and the continuum is nearly transparent so one sees all the way through the ejecta with a single spectrum probing all the important layers simultaneously. This technique was pioneered for all supernovae by Peter Miekle and his collaborators and is rapidly coming to the fore as a major tool in the study of Type Ia. SN 1999by was a subluminous Type Ia that was, not incidentally, significantly polarized (Howell *et al.* 2001). Höflich *et al.* (2002) showed that a delayed detonation model selected to match the light curve provided a good agreement with the NIR spectra and revealed the products of explosive carbon burning in the outer layers and products of incomplete silicon burning in deeper layers. The results were inconsistent with pure deflagration models or merger models that leave substantial unburned matter on the outside. The data also seemed incompatible with the mixing of unburned elements into the center as predicted by pure deflagration 3-D models. Three-dimensional models in which the inner unburned matter undergoes a detonation, the current most realistic manifestation of the delayed detonation paradigm as presented here by Gamezo *et al.* alleviate that problem. Marion *et al.* (2003) have presented NIR spectra of "normal" Type Ia (see also Hamuy 2002) and shown that the outer layers of intermediate mass elements are not mixed, that very little unburned carbon remains in the outer layers, and perhaps revealed Mn, a sensitive probe of burning conditions.

Another important development concerns work on the quasi-static phase of carbon burning that follows carbon ignition and precedes dynamic runaway. This important "smoldering" phase had not been critically re-examined since the initial study of Arnett (1969). Höflich & Stein (2002) showed that the convective velocities in this phase can exceed the initial speeds of the subsequent deflagration front. This means that the "pre-processing" of the white dwarf by this smoldering phase and the resulting velocity field, rather than the pure Rayleigh-Taylor driven deflagration, will dominate the early propagation of the burning front. This is a crucial, multidimensional, insight that will foment much work in the near future to understand all the implications.

Finally, it is necessary to repeat that polarization studies have revealed that Type Ia are polarized and hence asymmetric (Wang, Wheeler & Höflich 1997). It may be that the subluminous variety are more highly polarized and perhaps more rapidlly rotating than the "normal" type (Howell *et al.* 2001). It may also be that, although the polarization is generally low, all Type Ia are polarized at an interesting level if appropriate, sufficiently accurate observations are made (Wang *et al.* 2003a). This has clear implications for the quest to answer the old problem of whether Type Ia arise in binary systems and, if so, as we all believe, what sort? The asymmetries might also be teaching us lessons yet ungleaned about the combustion process which is undoubtedly complex and three dimensional. The asymmetries must be understood in order to use Type Ia with great confidence as we move to the next phase of cosmological studies where exceptionally precise photometry and tight control of systematic effects will be necessary to probe the equation of state of the dark energy.

In any case, the lesson of recent history and of this workshop is that Type Ia supernovae are three dimensional!

1.3 Asymmetric core collapse

If anything, the polarization studies have had even more dramatic impact on core collapse supernovae. All core collapse supernovae adequately observed are found to be polarized and hence asymmetric in some way (Wang *et al.* 1996; Wang *et al.* 2001, 2002, 2003b; Leonard *et al.* 2000; Leonard & Filippenko 2001; Leonard *et al.* 2001, 2002). Many of these events are substantially bi-polar (Wang *et al.* 2001). The fact that the polarization is higher as one sees deeper in and is higher when the hydrogen envelope is less, strongly indicates that the very machine of the explosion deep in the stellar core is asymmetric and probably predominantly bi-polar. SN 1987A reveals similar evidence (Wang *et al.* 2002). Other famous "spherically symmetric" supernovae are those that gave rise to the Crab Nebula and to Cas A.

Complementary computational work has shown that jet-induced explosions can produce the qualitative asymmetries that are observed (Khokhlov *et al.* 1999, see also MacFadyen *et al.* 2001 Zhang, *et al.* 2003). Khokhlov & Höflich (2001) and Höflich, Khokhlov & Wang (2001) have shown that asymmetric nickel deposition by a jet-like flow can produce polarization by asymmetric heating and ionization even in an otherwise spherically-symmetric density distribution. This very plausibly accounts for the early low polarization in Type II supernovae that grows as the underlying asymmetry is revealed.

The large question remains as to what causes the jet-like flow. My bet is that this involves rotation and magnetic fields at the deepest level. Rotation alone can affect neutrino deposition, but the case can be made that rotation without magnetic fields is highly unlikely. Akiyama *et al.* (2003) have presented a proof of principle that the physics of the magneto-rotational instability (MRI: Balbus & Hawley 1991, 1998) is inevitable in the context of the differentially-rotating environment of protoneutron stars. The magnetic fields can in turn affect the neutrino transport. The ultimate problem of core collapse is intrinsically three-dimensional involving rotation, magnetic fields, and neutrino transport. We have known this all along (despite, not because of, cheap shots after core collapse talks in which some joker always asks "but what about rotation?" or "but what about magnetic fields?"), but the new polarization observations demand a new, integrated view. This makes a devilishly hard problem even harder. Progress will come by isolating and understanding pieces of the problem and eventually sticking them together.

1.4 The magneto-rotational instability and core collapse

The advantage of the MRI to generate magnetic field is that while it works on the rotation time scale of Ω^{-1} (as does field-line wrapping), the strength of the field grows exponentially. This means that from a plausible seed field of 10^{10} to 10^{12} G that might result from field compression during collapse, only $\sim$7–12 e-folds are necessary to grow to a field of 10^{15} G. That is only $(7\text{–}12)/2\pi \sim 1\text{–}2$ full rotations or $\sim$10–20 ms for expected initial rotation periods of order 10 ms. Furthermore, while the growth time may depend on the seed field, the final saturation field is independent of the seed field (unlike a linear wrapping model that ignores the complications of reconnection, see Wheeler *et al.* 2000, 2002, for examples and other references).

Core collapse will lead to strong differential rotation near the surface of the protoneutron star even for initial solid-body rotation of the iron core (Kotake, Yamada & Sato 2003; Ott *et al.* 2004). The criterion for instability to the MRI is a negative gradient in angular velocity, as opposed to a negative gradient in angular momentum for the Rayleigh dynamical instability. This condition is broadly satisfied at the surface of a newly formed neutron star during core collapse and so the growth of magnetic field by the action of the MRI is inevitable. More quantitatively, when the magnetic field is small and/or the wavelength is long ($kv_a < \Omega$) the instability condition can be written (Balbus & Hawley 1991, 1998):

$$N^2 + \frac{\partial \Omega^2}{\partial \ln r} < 0, \tag{1.1}$$

where N is the Brunt-Väisälä frequency. Convective stability will tend to stabilize the MRI, and convective instability to reinforce the MRI. The saturation field given by general considerations and simulations is approximately given by the condition: $v_a \sim \lambda\Omega$ where $\lambda \lesssim r$ or $B^2 \sim 4\pi\rho r^2 \Omega^2$ where v_a is the Alfvén velocity.

These physical properties were illustrated in the calculations of Akiyama *et al.* (2003) who used a spherically-symmetric collapse code to compute the expected conditions, instability, field growth and saturation. Akiyama *et al.* assumed initial rotation profiles, solid body or differential, invoked conservation of angular momentum on shells that should, at least, give some idea of conditions in the equatorial

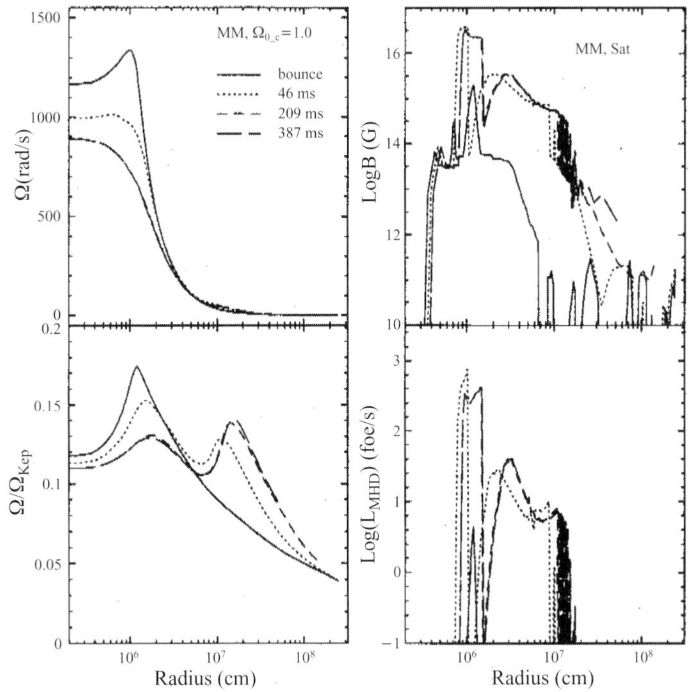

Fig. 1.1. Angular velocity, field strength and MHD luminosity (in units of 10^{51} erg s^{-1}) for a representative initial differential rotation of the iron core as a function of time from Akiyama *et al.* (2003)

plane and computed regions of MRI instability. They assumed exponential growth to saturation. For sub-Keplerian post-collapse rotation, Akiyama *et al.* found that fields can be expected to grow to 10^{15} to 10^{16} G in a few tens of milliseconds. The resulting characteristic MHD luminosity (cf. Blandford & Payne 1982) is:

$$L_{MHD} \sim B^2 r^3 \Omega/2 \sim 3 \times 10^{52}\ \text{erg s}^{-1} B_{16}^2 R_{NS,6}^3 \left(\frac{P_{NS}}{10\ \text{ms}}\right)^{-1} \sim 10^{51} - 10^{52}\ \text{erg s}^{-1}. \tag{1.2}$$

If this power can last for a significant fraction of a second, a supernova could result. Figure 1.1 shows the results for a model in which the iron core began with a smoothly decreasing distribution of angular velocity and a central value of $\Omega = 1$ s^{-1}.

The implication of the work of Akiyama *et al.* (2003) is that the MRI is unavoidable in the core collapse ambience, as pertains to either supernovae or γ-ray bursts. The field generated by the MRI must be included in any self-consistent calculation. These implications need to be explored in much greater depth, but there is at least some possibility that the MRI may lead to strong MHD jets by the magneto-rotational (Meier, Koide & Uchida 2001) or other mechanisms. A key point is that the relevant dynamics will be dictated by large, predominantly toroidal fields that are generated internally, not the product of twisting of external field lines that is the basis for so much work on MHD jet and wind mechanisms. Understanding the role of these internal toroidal fields in producing jets, in providing the ultimate dipole field strength for both ordinary pulsars and magnetars (Duncan & Thompson 1992), in setting the "initial" pulsar spin rate after the supernova dissipates (that is, the "final" spin rate from the supernova dynamicists point of view), and any connection to γ-ray bursts is in its infancy.

1.5 Gamma-ray bursts

We have learned so much about gamma-ray bursts since the revolution of the discovery of afterglows it is impossible to do it justice. Briefly, we now know that the energy is not spread isotropically, but is collimated into jets that seem to have a canonical energy of a few $\times 10^{50}$ ergs (Panaitescu & Kumar 2001; Frail *et al.* 2001). There is growing circumstantial evidence for a connection to massive stars, yet there is evidence that some gamma-ray bursts explode into a rather low density ISM, and little evidence in many cases for the winds that should characterize massive stars (Panaitescu & Kumar 2002).

Now we have the dramatic evidence from GRB030329/SN 2003dh (Stanek *et al.* 2003; Hjorth *et al.* 2003; Kawabata *et al.* 2003) of a definite connection between this burst and a Type Ic-like supernova. We also have the startling result of the observation by Coburn & Boggs (2003) of a large polarization, $80 \pm 20\%$ in GRB021206. One interpretation of this is that the Alfvén speed considerably exceeds the sound speed, implying a dynamically dominant field (Lyutikov, Blandford & Pariev 2003).

Despite these dramatic developments, there is much to be done. In the context of gravitational collapse models we must consider Keplerian shear, nearly equipartition fields, magnetic neutrino cross sections, strong magnetic helicity currents and viscoelastic effects (see my concluding remarks), and a host of other effects that will pertain to this rapidly rotating and inevitably magnetic environment.

One of the issues currently facing the supernova community as we grapple with the supernova/gamma-ray burst connection is whether or not there is a new class of explosions ("hypernovae," see, e.g. Maeda *et al.* 2002 and in these proceedings) or if the events we see with large photospheric velocities are just an extension of a population with a continuum of properties. There are several complications that must be borne in mind. First, the velocity at the photosphere is a very sensitive function of time for Type Ic supernovae. Just to say a velocity is "high" is not a terribly useful statement. It is the whole velocity evolution that must be compared

Table 1.1.

EVENT	Peak M_v	v at Peak (1000 km s^{-1})	Ref
SN 1983V	18.1	15	(1)
SN 1983N	17.4	10	(2)
SN 1987K	16.9	10	(3)
SN 1987M	18.5	10	(4)
SN 1992ar	19.3	15	(5)
SN 1994I	18.1	14	(6, 7)
SN 1993J	17.7	10	(8)
SN 1997ef	17.2	11	(9)
SN 1998bw	19.4	15	(10, 11)
SN 2002ap	17.7	15	(12)
SN 2003dh	20.5	23	(13)

REFERENCES – (1) Clocchiatti *et al.* (1997); (2) Clocchiatti, Wheeler, Benetti & Frueh (1996); (3) Filippenko (1988); (4) Filippenko, Porter & Sargent (1990); (5) Clocchiatti *et al.* (2000); (6) Richmond *et al.* (1996); (7) Millard *et al.* (1999); (8) Wheeler & Filippenko (1996); (9) Mazzali, Iwamoto & Nomoto (2000); (10) Galama *et al.* (1998); (11) Patat *et al.* (2001); (12) Gal-Yam, Ofek & Shemmer (2002); (13) Hjorth *et al.* (2003)

to make a valid contrast of one event with another. Even then different velocity evolution can and will result from different envelope masses without substantial differences in explosion energy. Another key factor is that we now have substantial evidence, some of it quite direct, that Type Ic supernovae are strongly asymmetric. This means that we might see different photospheric velocities in different directions (Höflich, Wheeler & Wang 1999). We can also see different luminosities in different directions. This is then related to the deduction of nickel masses, a key factor in the definition, at least in some cases, of "hypernovae." In addition, non-spherical explosions can affect the resulting density distribution and hence the gamma-ray deposition and even the late time luminosity. Since explosion energies and nickel masses are quantities derived from spherical models, great care must be taken in their interpretation when strong asymmetries are suspected.

To illustrate the empirical case, Table 1.1 gives photospheric velocity at maximum light and peak brightness for a sample of Type Ic and related supernovae for which such data were available. Whether or not a comparison at maximum light is valid or the best way to do this is not clear. The very fact that the data are sparse is a cautionary note to both advocates and critics of "hypernovae." Nevertheless, this table illustrates that Type Ic come with a considerable dispersion in both peak brightness and photospheric velocity. While it may be that SN 1998bw and SN 2003dh are especially bright, there are others as bright; and while those events may have shown high velocities, so have other Type Ic with modest peak brightness. Further data may reveal differently, but this table reveals no special pattern nor obvious bifurcation in properties between normal Type Ic supernovae and "hypernovae."

Even the great triumph of SN 2003dh has brought some new issues to the fore. The spectral evolution of SN 2003dh looks remarkably like that of SN 1998bw. How could that be since SN 2003dh was associated with a classic gamma-ray burst and must have been observed nearly down the jet axis and SN 1998bw was either associated with an odd, very subluminous gamma-ray burst, or it was seen substantially off axis. The recent report of a supernova-like spectrum in GRB 021211 by Della Valle *et al.* (2003) also adds a twist. Here again, the supernova must be seen "down the pipe," but the velocities seem to be modest. Clearly there is still much to learn about the supernova gamma-ray burst connection.

1.6 Conclusions

As we enter this workshop, we can point to several area of critical interest. Type Ia supernovae sometimes have significant polarization and hence asymmetry. This may yield clues to their binary origin. Perhaps we are seeing evidence of how the combustion physics proceeds. Perhaps there are hints, specifically, to the mechanism of the crucial deflagration/detonation transition.

Much hard work has also shown that all core collapse explosions are significantly polarized and hence asymmetric. This means that both the dynamics and the radiative processes (photons and neutrinos!) are asymmetric. An account of this asymmetry must be made in the analysis of core collapse.

In particular, core collapse is an intrinsically shearing environment, That makes it subject to the MRI, the resulting turbulence, and hence to strong dynamo action and the exponential growth of magnetic fields. The implication is that rotation and magnetic fields are intrinsic to the process of core collapse for either neutron stars or black holes, for supernovae or gamma-ray bursts.

Welcome to the brave new world of three-dimensional explosions!

Acknowledgements

I am most grateful to Peter Höflich and Pawan Kumar for hosting this wonderful meeting and putting in the hard work before and after to make it scientifically successful and stimulating. I learned so much from so many colleagues over the years, but would like to especially acknowledge those who have recently led me into three-dimensional perspectives – Shizuka Akiyama, Dietrich Baade, Vadim Gamezo, Andy Howell, Alexei Khokhlov, Itamar Lichtenstadt, Dave Meier, Elaine Oran,

Peter Williams and especially Lifan Wang and Peter Höflich. This work was supported in part by NSF AST-0098644 and by NASA NAG5–10766.

References

Akiyama, S. Wheeler, J. C., Meier, D. & Lichtenstadt, I. 2003, *Astrophysical Journal*, **584**, 954

Arnett, D. W. 1969, *Astrophysics & Space Science*, 5, 180

Balbus, S. A. & Hawley, J. F. 1991, *Astrophysical Journal*, **376**, 214

Balbus, S. A. & Hawley, J. F. 1998, *Review of Modern Physics*, **70**, 1

Blandford, R. D. & Payne, D. G. 1982, *Monthly Notices of the Royal Astronomical Society*, **199**, 833

Clocchiatti, A. *et al.* 1997, *ApJ*, **483**, 675

Clocchiatti, A., Wheeler, J. C., Benetti, S., & Frueh, M. 1996, *ApJ*, **459**, 547

Clocchiatti, A. *et al.* 2000, *ApJ*, **529**, 661

Coburn, W. & Boggs, S. E. 2003, *Nature*, **423**, 415

Della Valle, M. *et al.* 2003, *Astronomy & Astrophysics*, **406**, L33

Duncan, R. C. & Thompson, C. 1992, *ApJ*, **392**, L9

Filippenko, A. V. 1988, *AJ*, **96**, 1941

Filippenko, A. V., Porter, A. C., & Sargent, W. L. W. 1990, *AJ*, **100**, 1575

Frail, D. A. *et al.* 2001, *ApJ*, **562**, L55

Galama, T. J. *et al.* 1998, *Nature*, **395**, 670

Gal-Yam, A., Ofek, E. O., & Shemmer, O. 2002, *MNRAS*, **332**, L73

Hamuy, M. *et al.* 2002, *AJ*, **124**, 417

Hjorth, J. *et al.* 2003, *Nature*, **423**, 847

Höflich, P. 1995, *ApJ*, **443**, 89

Höflich, P., Gerardy, C. L., Fesen, R. A., & Sakai, S. 2002, *ApJ*, **568**, 791

Höflich, P. & Khokhlov, A. 1996, *ApJ*, **457**, 500

Höflich, P., Khokhlov, A., Wheeler, J. C., Phillips, M. M. Suntzeff, N. B. & Hamuy, M. 1996, *ApJ*, **472**, L81

Höflich, P., Khokhlov, A. & Wang, L. 2001, in *Proc. of the 20th Texas Symposium on Relativistic Astrophysics*, eds. J. C. Wheeler & H. Martel (New York: AIP), 459

Höflich, P. & Stein, J. 2002, *ApJ*, **568**, 779

Höflich, P., Wheeler, J. C., & Wang, L. 1999, *ApJ*, **521**, 179

Howell, D. A., Höflich, P., Wang, L., & Wheeler, J. C. 2001, *ApJ*, **556**, 302

Kawabata, K. S. *et al.* 2003, *ApJ Lett.*, **593**, L19

Khokhlov, A. M. 1991, *Astronomy & Astrophysics*, 245, 114

Khokhlov, A. & Höflich, P. 2001, in *Explosive Phenomena in Astrophysical Compact Objects*, eds. H.-Y, Chang, C.-H. Lee & M. Rho, AIP Conf. Proc. No. 556, (New York: AIP), p. 301

Khokhlov, A. M. & Oran, E. S. 1999, *Combustion & Flame*, **119**, 400

Khokhlov, A. M., Oran, E. S. & Thomas, G. O. 1999, *Combustion & Flame*, **117**, 323

Khokhlov A. M., Höflich P. A., Oran E. S., Wheeler J. C. Wang, L., & Chtchelkanova, A. Yu. 1999, *ApJ*, **524**, L107

Kotake, K., Yamada, S. & Sato, K. 2003, *ApJ*, **595**, 304

Lentz, E. J., Baron, E., Branch, D., & Hauschildt, P. H. 2001, *ApJ*, **547**, 402

Leonard, D. C., Filippenko, A. V., Barth, A. J., & Matheson, T. 2000, *ApJ*, **536**, 239

Leonard, D. C. & Filippenko, A. V. 2001, *Publications of the Astronomical Society of the Pacific*, **113**, 920

Leonard, D. C., Filippenko, A. V., Ardila, D. R., & Brotherton, M. S. 2001, *ApJ*, **553**, 861
Leonard, D. C., Filippenko, A. V., Chornock, R. & Foley, R. J. 2002, *PASP*, **114**, 1333
Lyutikov, M., Pariev, V. I. & Blandford, R. D. 2003, *ApJ*, **597**, 998
MacFadyen, A., Woosley, S. E. & Heger, A. 2001, *ApJ*, **550**, 410
Maeda, K., Nakamura, T., Nomoto, K., Mazzali, P., Patat, F. & Hachisu, I. 2002, *ApJ*, **565**, 405
Marion, G. H., Höflich, P., Vacca, W. D., & Wheeler, J. C. 2003, *ApJ*, **591**, 316
Mazzali, P. A., Iwamoto, K., & Nomoto, K. 2000, *ApJ*, **545**, 407
Meier, D. L., Koide, S. & Uchida, Y. 2001, *Science*, **291**, 84
Millard, J. *et al.* 1999, *ApJ*, **527**, 746
Nugent, P., Baron, E., Branch, D., Fisher, A., & Hauschildt, P. H. 1997, *ApJ*, **485**, 8
Ott, C. D., Burrows, A., Livne, E. & Walder, R. 2004, *ApJ*, **600**, 834
Panaitescu. A. & Kumar, P. 2001, *ApJ*, **560**, L49
Panaitescu, A. & Kumar, P. 2002, *ApJ*, **571**, 779
Patat, F. *et al.* 2001, *ApJ*, **555**, 900
Richmond, M. W. *et al.* 1996, *AJ*, **111**, 327
Stanek, K. Z. *et al.* 2003, *ApJ*, **591**, L17
Wang, L., Howell, D. A., Höflich, P., & Wheeler, J. C. 2001, *ApJ*, **550**, 1030
Wang, L., Wheeler, J. C., & Höflich, P. 1997, *ApJ Lett.*, **476**, L27
Wang, L., Wheeler, J. C., Li, Z., & Clocchiatti, A. 1996, *ApJ*, **467**, 435
Wang, L. *et al.* 2002, *ApJ*, **579**, 671
Wang, L. *et al.* 2003a, *ApJ*, **591**, 1110
Wang, L. *et al.* 2003b, *ApJ*, **592**, 457
Wheeler, J. C. & Filippenko, A. V. 1996, *IAU Colloq. 145: Supernovae and Supernova Remnants*, 241
Wheeler, J. C., Meier, D. L. & Wilson, J. R. 2002, *ApJ*, **568**, 807
Wheeler, J. C., Yi, I., Höflich, P. & Wang, L. 2000, *ApJ*, **537**, 810
Zhang, W., Woosley, S. E. & MacFadyen, A. I. 2003, *ApJ*, **586**, 356

Part II

Supernovae: Observations Today

2

Supernova explosions: lessons from spectropolarimetry

L. Wang

Lawrence Berkeley National Laboratory

Abstract

Supernovae can be polarized by an asymmetry in the explosion process, an off-center source of illumination, scattering in an envelope distorted by rotation or a binary companion, or scattering by the circumstellar dust. Careful polarimetry should thus provide insights to the nature of supernovae. Spectropolarimetry is the most powerful tool to study the 3-D geometry of supernovae. A deep understanding of the 3-D geometry of SNe is critical in using them for calibrated distance indicators.

2.1 Introduction

Polarimetry of supernovae (SNe) reveals the intrinsic ejecta asymmetries (Shapiro & Sutherland 1982, McCall 1984, Höflich 1991, 1996). SN 1987A represented a breakthrough in this area, by providing the first detailed record of the spectropolarimetric evolution (e.g. Mendez *et al.* 1988; Cropper *et al.* 1988). SN 1993J also provided a wealth of data (Trammell, Hines, & Wheeler 1993; Tran *et al.* 1997). Most of the theoretical interpretations of the polarimetry data are based on oblate or prolate spheroid geometries. A very different picture of SN polarization is discussed in Wang & Wheeler (1996) where time-dependent dust scattering is shown to be a potential mechanism. New attempts are being made with more complicated geometrical structures (Kasen *et al.* 2003; 2004).

We started a systematic program of supernova spectropolarimetry in 1995 using the 2.1 meter telescope of the McDonald Observatory which nearly doubled the number of SNe with polarimetry measurements in the first year of the program. The early qualitative conclusion was that core-collapse SNe are in general polarized at about the 1% level and that Type Ia are much less polarized ($\leq$0.3%) (Wang *et al.* 1996). We have also observed several SNe for their spectropolarimetry using the

Image Grism Polarimeter (IGP) at the 2.1 meter telescope and a polarimeter at the 2.7 meter telescope. The spectropolarimetry data show that normal SN Ia are not highly polarized, but a hint of intrinsic polarization is detected at levels around 0.3% for SN 1996X at around optical maximum. We found also that a peculiar, subluminous SN Ia 1999by was clearly polarized at a much higher level (~0.7%, Howell *et al.* 2001) than normal SN Ia. The polarization of core-collapse SNe evolve with time (Wang *et al.* 2001; Leonard *et al.* 2001), with the general trend being that the polarization is larger post optical maximum than before or around optical maximum. Our observations indicate further that for core-collapse SNe the degree of polarization is larger for those with less massive envelopes (Wang *et al.* 2001).

The ESO-VLT, with its extremely flexible ToO capability, makes it possible to study in great detail SN polarimetry. Not only can the SNe be follow to late nebular stages, but also they can be observed early enough so that the most energetic outer layers can be probed. We have obtained high quality multi-epoch spectropolarimetry data of a normal Type Ia, SN 2001el. The data allow us to probe both the mechanisms of SN Ia explosions, and the effect of asymmetry on the use of SN Ia as distance indicators (Wang *et al.* 2003a).

For core-collapse events, the inferred degree of asymmetry is in general of the order of 10–30% if modeled in terms of oblate/prolate spheroids. The asphericity of the photosphere of SN Ia is perhaps not larger than 10–15%.

SN polarimetry is still a rapidly growing field in terms of both observations and theories. Spectropolarimetry of SNe has been a driving force for 3-D models of SN Ia and in turn the advance of 3-D models of SN explosion will help us to better understand the observed phenomena. Here I will concentrate on the observational efforts of our SN polarimetry program.

2.2 Core-collapse SNe

2.2.1 SN 1987A

The best observed core-collapse SN is, of course SN 1987A. Extensive spectropolarimetry was obtained soon after the SN explosion. The polarimetry data were recently revisited by Wang *et al.* (2002) and compared with recent HST images and HST+STIS spectroscopy (see Fig. 2.1). By combining the early spectropolarimetry observations with the HST data we found a prolate geometry for the radioactive material at the innermost core of the ejecta, and an oblate spheroid geometry for the oxygen shell and the hydrogen envelope. The entire structure can be approximated by an axially symmetric structure with the symmetry axis close to, but noticeably offset from, the minor axis of the circumstellar rings. These and more recent images

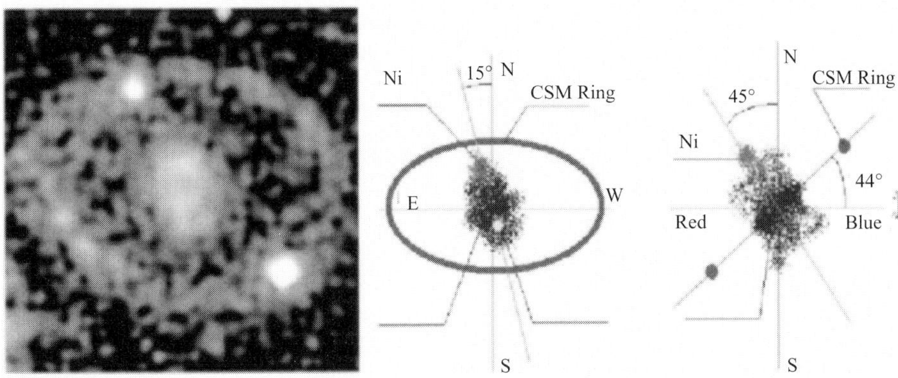

Fig. 2.1. Left: HST image of SN 1987A taken on 2000 June 11 with filter F439W. The outer ellipse shows the CSM matter around the SN. The elongated central patch shows the SN 1987A ejecta. The apparent asymmetry of the SN ejecta represents the distribution of ^{56}Fe. Middle and right: Schematic drawing of the inner CSM ring – ejecta structure projected onto the plane of the sky (middle), and the plane of the line of sight and the direction to the North (right). The symmetry axis of the ejecta is about 14° off to the east of the symmetry axis of the ring. The ^{56}Ni clump can be consistently placed at about 45° from the line of sight. Oxygen is predominantly distributed in the plane of the CSM ring. There may also be an enhancement of dust opacity in the plane of the ring that makes the image of the ejecta fainter near the center.

from HST show an emission minimum at the center of the SN 1987A ejecta that could be explained by an enhanced dust distribution on the plane defined by the CSM ring.

2.2.2 SN 1996cb

SN 1996cb is a Type IIb SN whose photometric and spectroscopic behavior bear strong resemblance to that of SN 1993J (Fig. 2.2). These SNe show weak hydrogen features but strong helium lines. They are produced by stars that have lost most of their hydrogen envelope. Surprisingly, the spectropolarimetry data obtained at the McDonald Observatory show that they are also very similar to each other in spectropolarimetry. Detailed theoretical models were done for SN 1993J (Höflich *et al.* 1996). Based on those models, the SN photosphere could be highly aspherical with minor to major axis ratios around 0.6 if modeled in terms of oblate spheroids.

A natural question to ask is why SN 1993J and SN 1996cb are so similar despite the fact that they are so highly aspherical and clumpy (Wang, & Hu 1994; Spyromilio 1994; Wang *et al.* 2001)? The complementary question is whether these SNe are

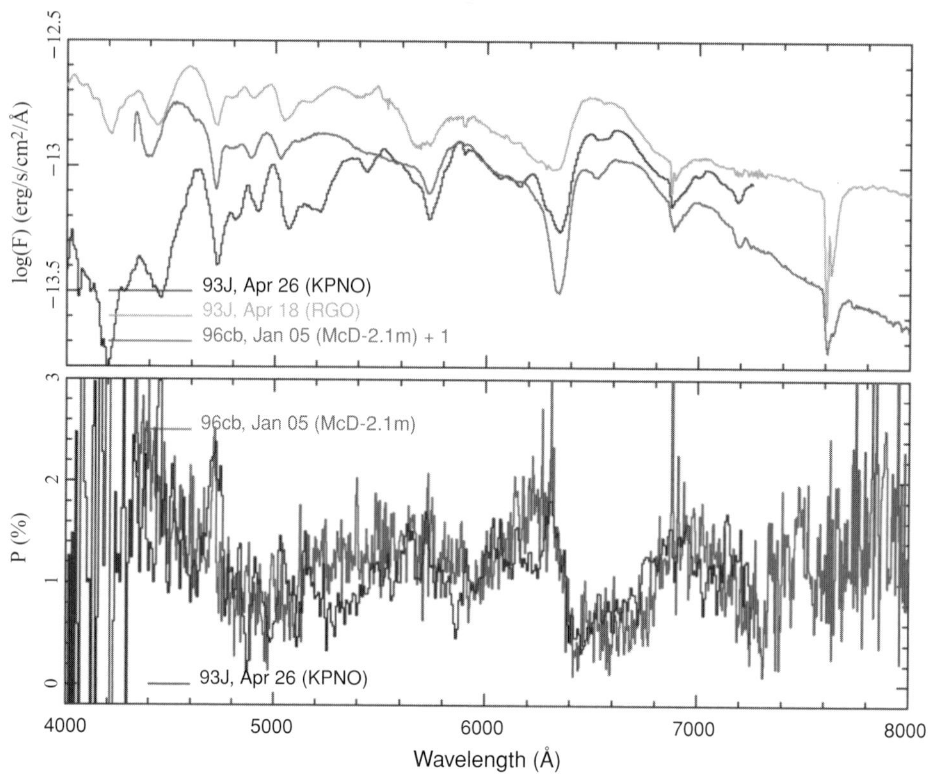

Fig. 2.2. Spectropolarimetry of SN 1996cb compared to that of SN 1993J obtained at comparable stages. The upper panel shows the spectroscopic data. The lower panel shows the degree of polarization. The similarity of the polarization of the two SNe is striking.

still Type IIb when viewed at different viewing angles. Our guess is that they are likely a select subgroup of a more common phenomena. More data, especially spectropolarimetry, will help to create a unified picture of SN IIb.

2.2.3 SN 1997X

SN 1997X is a typical Type Ic SN with no hydrogen or helium features. This is an especially exciting observation because SN 1997X, like all SN Ib/c, must have occurred in a nearly naked, non-degenerate, carbon/oxygen core. There is little mass to dilute an asymmetrical explosion or to drive Rayleigh-Taylor instabilities, and with rapid expansion, little time to propagate transverse shocks. There is the tantalizing expectation that in SN 1997X we may be seeing the direct effects of asymmetric explosion (Wang *et al.* 2001). The degree of intrinsic polarization is the highest for any observed SNe which points to a disk-like geometry or a

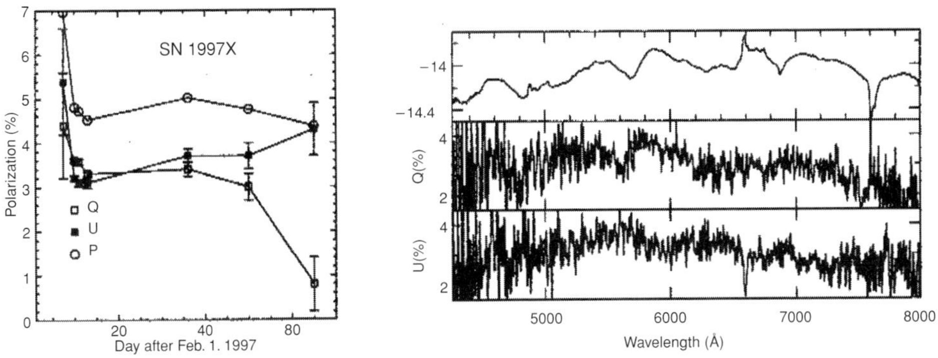

Fig. 2.3. Left: Evolution of the continuum polarization of SN 1997X synthesized from the spectropolarimetry data. Right: Flux spectrum on a logarithmic scale and the observed degree of polarization obtained on Feb. 10, 1997 at McDonald Observatory. Data at other epochs have similar quality. The degree of polarization evolves with time. Conspicuous spectral features are detected in both Q and U.

prolate spheroid with large major to minor axis ratios (see Fig. 2.3). Models of core-collapse that, while convective, are spherical in the mean may be inadequate to induce the observed polarization unless instabilities at large scale can grow substantially (Blondin, Mezzacappa, & DeMarino 2003). The large polarization of SN 1997X may mean that the effects of rotation and perhaps magnetic fields have to be included *ab initio* in the collapse calculations.

2.2.4 SN 1998S

SN 1998S is a Type IIn SN showing strong narrow Hα emission lines. We have secured two epochs of polarimetry data with the 2.1 meter telescope at McDonald Observatory (Fig. 2.4; Wang *et al.* 2001) but more data are reported by Leonard *et al.* (2000). Emission of Type IIn arises from ejecta-CSM interaction. SN 1998S showed a very well-defined symmetry axis that can be associated with the geometry of the CSM matter around the SN. The degree of polarization approaches 3%. In the extreme case that the polarization is modeled in terms of a central source that is scattered by a CSM disk, the opening angle of the disk would be on the order of a few degrees (McDavid 2001; Melgarejo *et al.* 2001; See also Wang *et al.* 2004 and the discussion of SN 2002ic below).

2.2.5 SN 1999em

SN 1999em is a well-observed Type II plateau SN. The polarization data show an extremely well-defined symmetry axis (Fig. 2.5). The degree of polarization increased sharply from before optical maximum to the plateau phase.

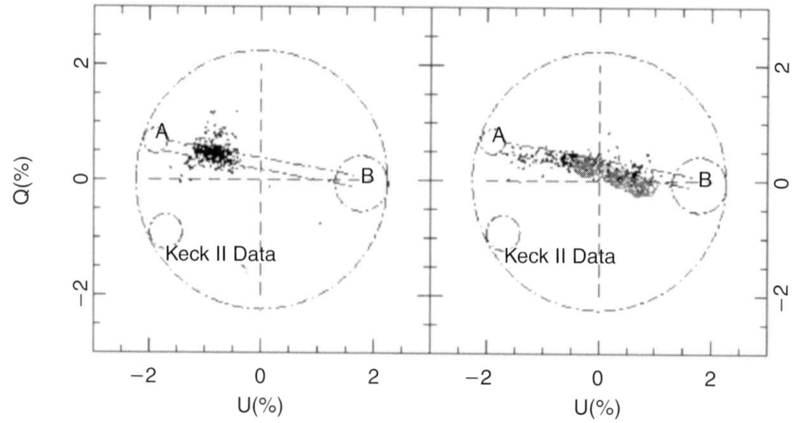

Fig. 2.4. Spectropolarimetry of the Type II plateau SN 1998S on the Q-U plane. The dot-dashed circles represent the limit of polarization caused by interstellar dust. Point A gives the actual location of the interstellar polarization. The data points at the later epoch are dispersed along a well-defined line that indicates that the degree of polarization increased significantly during the two epochs of observations.

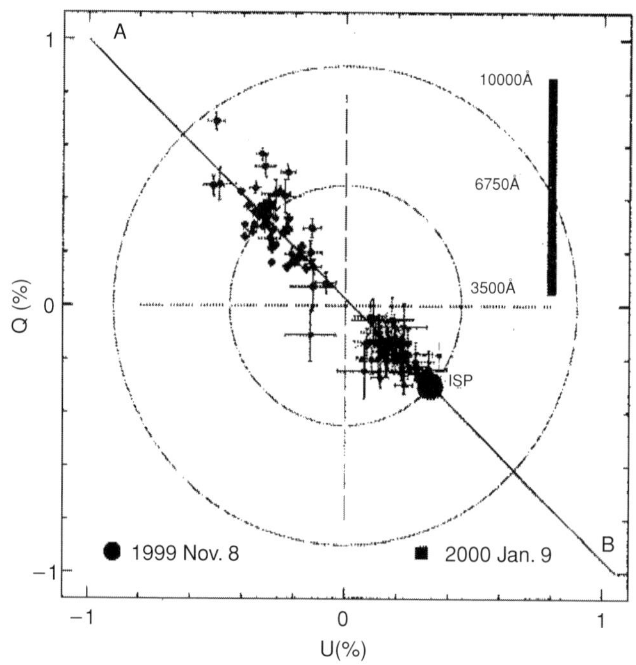

Fig. 2.5. Spectropolarimetry of the Type II plateau SN 1999em. The pre-max data cluster at the lower-right quadrant, while the data in the plateau phase are dispersed along a very well-defined line – an indication of a very well-defined geometry with no significant small scale fluctuations.

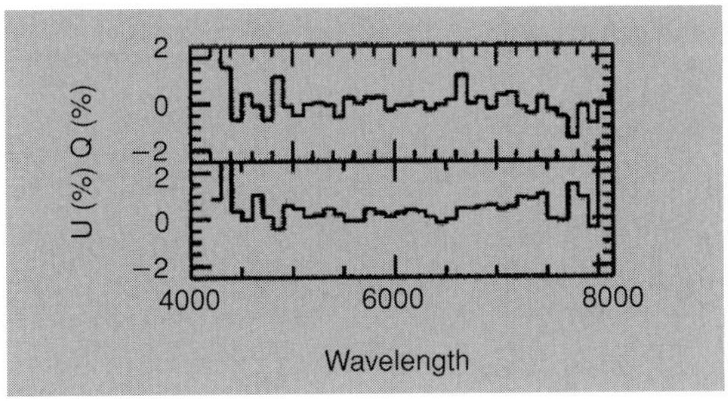

Fig. 2.6. Spectropolarimetry of the peculiar Type Ic SN 1997ef. The level of intrinsic polarization is $\leq$0.5%

2.2.6 *Hypernovae*

The association of peculiar Type Ic SN 1998bw with a faint GRB has led to the suggestion that SN Ib/c may be responsible for some GRBs. The discovery of SN 2003dh/GRB 030329 proves these suggestions (Stanek *et al.* 2003; Matheson, 2003, these proceedings). Spectropolarimetry is reported for SN 1998bw (Patat *et al.* 2001) and SN 2003dh (Kawabata *et al.* 2003).

Two other SNe with similar spectral properties were observed with polarimetry. SN 1997ef was observed at McDonald Observatory and the data show that the degree of polarization cannot be significantly larger than 0.5% (see Fig. 2.6), which is consistent with what was found for SN 1998bw and SN 2003dh.

The nearby SN 2002ap is of similar spectral behavior but with no associated GRB. Extensive spectropolarimetry was obtained at the VLT (Wang *et al.* 2003b), Keck (Leonard *et al.* 2002), and the Subaru (Kawabata *et al.* 2002). In contrast to the low levels of observed polarization for SN 2003dh/GRB 030329, SN 1998bw and SN 1997ef, SN 2002ap was found to be rather highly polarized. The earliest data were from the ESO VLT and highly polarized features from OI 7773 and Ca II IR triplet were detected (Fig. 2.7). The polarized features show sharp evolution with time, with the dominant axis of asymmetry changing with time. Post maximum observations show increases of the degree of polarization. This behavior suggests that the ejecta are highly distorted in the central region, and thus point to a highly aspherical central engine for the explosion. This behavior could be understood if the ejecta are "jet-like" with the jet pointing away from the observer for SN 2002ap, but head-on for SN 2003dh and SN 1998bw, making the projected geometry more symmetrical for the SN Ib/c with GRB counterparts.

2.3 Thermonuclear supernovae

Broad band polarimetry study showed that Type Ia SNe are in general less polarized than core-collapse SNe, with typical degree of polarizations less than 0.3% (Wang

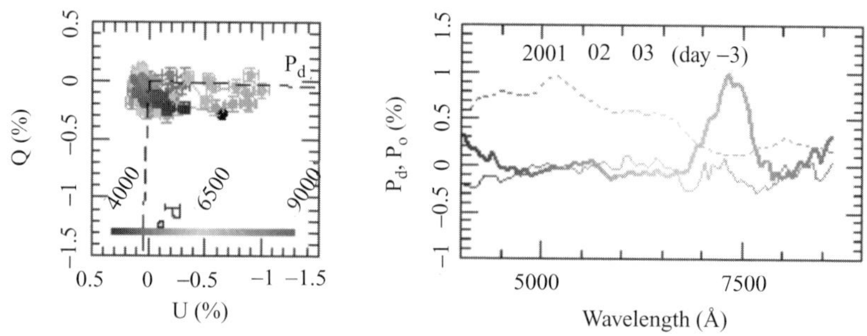

Fig. 2.7. Spectropolarimetry of the peculiar Type Ic SN 2002ap obtained on Feb. 03, about −3 days from *B* maximum (the date of maximum is take from Foley *et al.* 2003). The data on the Q-U plane are shown on the left with the principal axis and the axis orthogonal to it (the secondary axis) drawn in dashed lines. The flux, and the Stokes parameters projected onto the principal and secondary axes are shown on the right hand side. The largest asymmetry is seen in the OI 7773 line and the Ca II IR triplet (Wang et al. 2003b).

et al. 1996). The polarization is thus hard to detect. SN 1992A was the only SN Ia with spectropolarimetry observations before our program. Spectropolarimetry data of SN 1992A two and seven weeks past optical maximum were reported by Spyromilio & Bailey (1992) and no significant polarization was detected.

2.3.1 SN 1996X

SN 1996X was observed close to optical maximum, and the data are suggestive of intrinsic polarization at a level of ~0.3% (Fig. 2.8). It is realized that the interpretations of SN Ia polarization can be difficult due to strong wavelength modulations of opacity. An asphericity of 11% was derived from SN 1996X (Wang, Wheeler, Höflich 1997).

2.3.2 SN 1999by

SN 1999by is the first Type Ia with clear evidence of asphericity (Howell *et al.* 2001; Fig. 2.9). The degree of polarization is 0.7% – comparable to those of core-collapse SNe. However, SN 1997cy is a subluminous Type Ia and the observed polarization does not probe the chemical layers in thermonuclear equilibrium. Nonetheless, it suggests that some SN Ia may be aspherical despite the rather low level of polarization found for other spectroscopically normal SN Ia.

2.3.3 SN 2001el

SN 2001el is the first *normal* SN Ia with multiple epoch observations covering the first two months after explosion (Wang *et al.* 2003a; Fig. 2.10). Thanks to the

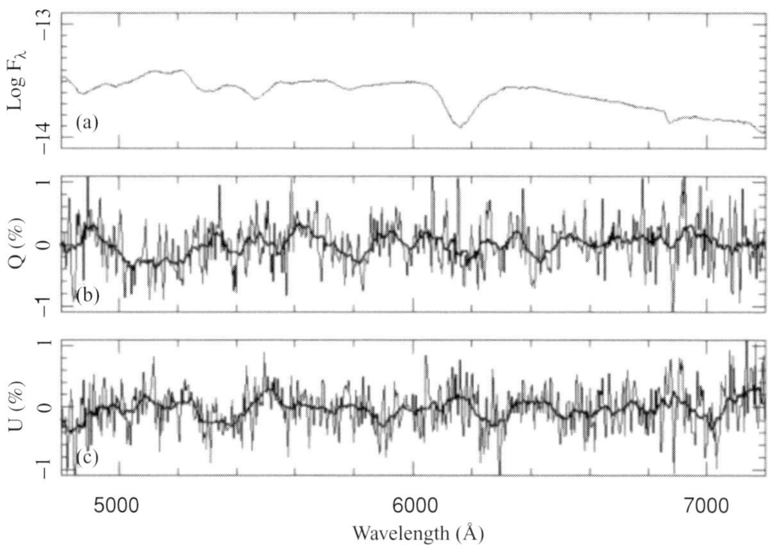

Fig. 2.8. Flux spectrum (a), Stokes Q (b), and U (c) of the normal Type Ia SN 1996X obtained at optical maximum. A low level of polarization of about 0.3% is tentatively detected.

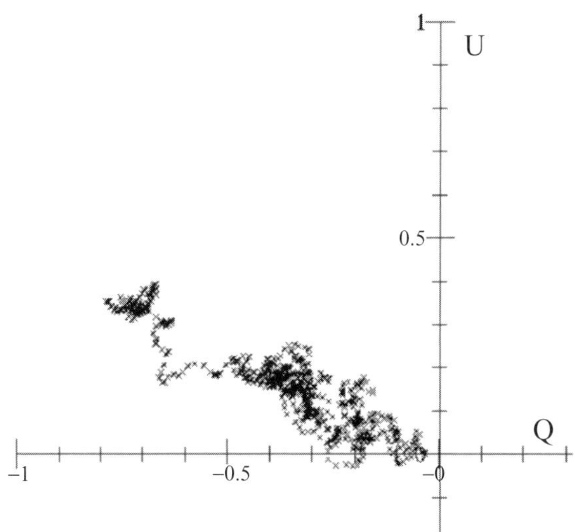

Fig. 2.9. Spectropolarimetry of the subluminous Type Ia SN 1999 by showing that it is significantly polarized, and thus highly aspherical. A well-defined symmetry axis is observed. Note that the data points are smoothed and the errors are highly correlated.

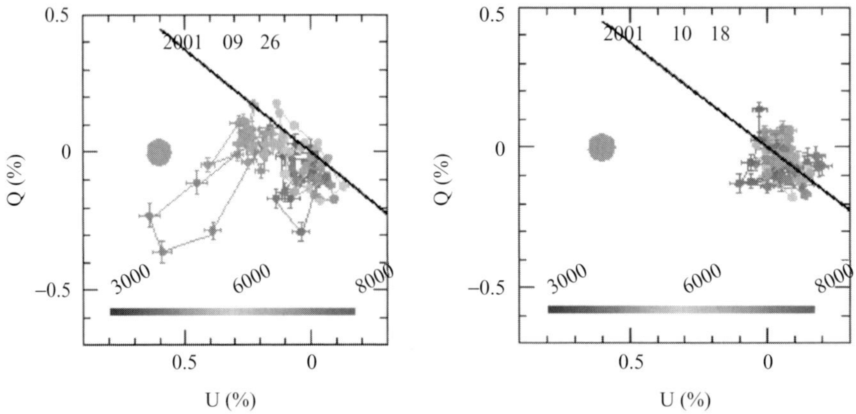

Fig. 2.10. Spectropolarimetry of the normal Type Ia SN 2001el. The polarization −4 days from optical maximum is shown on the left hand side. The distribution of the data points on the Q-U plane indicates intrinsic polarization of about 0.3% for the photosphere and 0.7% for the high velocity Ca II line. About two weeks past optical maximum, the polarization decreased substantially and became consistent with observational noise as shown on the right hand side panel.

spectacular performance of the VLT, for the first time we obtained high quality data which can provide a clear picture of the geometrical structure of an SN Ia in different chemical layers. In the outermost region, a shell, a ring, or clumps of high velocity (∼20,000–26,000 km/sec) material, perhaps enriched in calcium, is identified. This high velocity matter was found to be causing an absorption feature at 8000Å and makes that feature polarized at 0.7%. At around 10,000 km/sec, the Ca II and Si II lines show polarizations around 0.3%. Deeper inside, the material is found to be spherical as evidenced by a sharp decrease of the polarization degree past optical maximum. This indicates that there is no significant chemical mixing in regions that were burned to thermonuclear equilibrium. Such a structure is consistent with what one expects from delayed detonation models but is in conflict with pure deflagration models. The observed degree of polarimetry suggest that the photosphere could be aspherical at the 10% level and lead to magnitude dispersions of ∼ 0.1–0.2 mag.

2.3.4 SN 2002ic

SN 2002ic is the first SN Ia showing evidence of ejecta-CSM interaction (Hamuy *et al.* 2003). We have obtained late time spectropolarimetry data at Subaru and the ESO VLT. The required hydrogen mass is on the order of several solar masses with densities higher than $10^8/\mathrm{cm}^3$. Such a dense CSM must have a small volume filling factor. The Hα line shows a depolarization of nearly 1%. If the polarization is caused by scattering of the supernova light by CSM matter, the dense CSM must

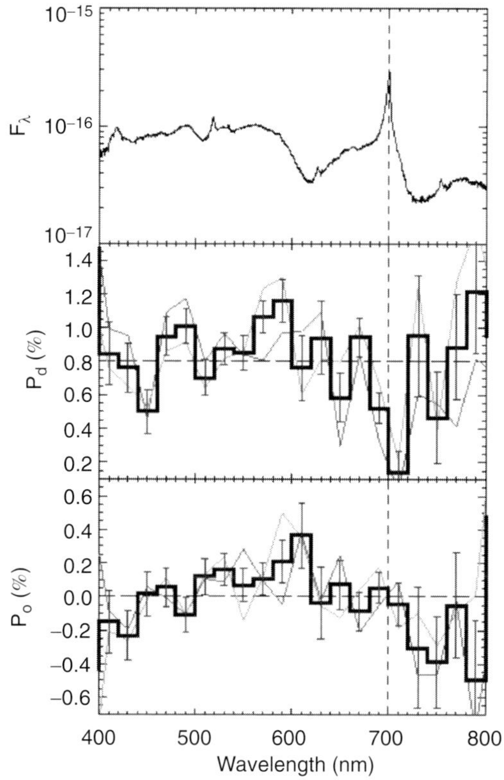

Fig. 2.11. Flux (top), and the Stokes parameter projected onto the principal axis of asymmetry (middle) and the axis orthogonal to the principal axis (lower) of the hybrid SN 2002ic. The Q and U for two different epochs are shown as thin lines and their weighted averages are shown as thick histograms

be distributed in a dense CSM disk (Wang *et al.* 2004; McDavid 2001; Melgarejo *et al.* 2001). The CSM matter bears striking resemblance to some well-observed proto-planetary nebulae. This suggests that thermonuclear explosions may occur at all stages of the formation of a white-dwarf, with most of them, of course, at stages when the systems contain no significant amount of CSM.

2.4 Summary

Spectropolarimetry is still a rapidly growing field. It is the best method to probe the geometrical structures of SNe.

Our studies have shown that a systematic polarimetry program of supernovae can provide detailed maps of the geometrical structures of the various chemical layers of supernova ejecta. These studies are important ingredients in the studies of the nature of the supernova phenomena, and they are crucial in making SN Ia better calibrated standard candles.

The intrinsic luminosity dispersion introduced by the observed asphericity of SN Ia is around 0.1–0.2 magnitudes which is of the same order of the residual luminosity dispersion of SN Ia after correction for light curve shapes. The fact that SN Ia are aspherical before and at optical maximum implies that geometrical orientation alone can introduce a significant amount of the luminosity dispersion. Since the luminosity-light curve shape relation corrects mostly the effect of the amount of ^{56}Ni, and is insensitive to the geometrical effect, it is then clear that SN Ia are indeed an extremely homogeneous group of events in their own right. Unless there is significant evolution of the geometrical structure of SN Ia, geometrical effects can be corrected with a large enough data sample, and this is extremely simple compared to all the other possible complications such as progenitor chemical or mass evolution. Furthermore, SN 2001el also indicates that the geometrical effect is much less prominent after optical maximum which makes these epochs extremely useful for deriving distance scales. It also makes the CMAGIC approach (Wang *et al.* 2003c) attractive as CMAGIC exploits exactly these later stages when there are no substantial departures from spherical symmetry.

Acknowledgment

I am grateful to D. Baade, P. Höflich, and especially J. C. Wheeler for collaborations on the polarimetry program. This work is supported by the Director, Office of Science, Office of High Energy and Nuclear Physics, of the U. S. Department of Energy under Contract No. DE-AC03-76SF000098.

References

Blondin, J. M., Mezzacappa, A., DeMarino, C. 2003, ApJ, 584, 971
Cropper, M., Bailey, J., McCowage, J., Cannon, R. D., Couch, Warrick J. 1988, MNRaS, 231, 695
Foley *et al.* 2003, PASP, 115, 1220
Hamuy *et al.* 2003, Nature, 424, 651
Höflich, P. A 1991, A&A, 246, 481
Höflich, P., Wheeler, J. C., Hines, D. C., & Trammell, S. R. 1996, ApJ, 459, 307
Howell, D. A., Höflich, P., Wang, L., Wheeler, J. C. 2001, 556, 302
Kasen, D. *et al.* 2003, ApJ. 593, 788
Kasen, D., Nugent, P., Thomas, R. C., Wang, L. 2004, ApJ, submitted (astro-ph/0311009)
Kawabata, K. S. *et al.* 2002, ApJ, 580, 39
Kawabata, K. S., Deng, J., Wang, L., Mazzali, P. at al. 2003, ApJ, 593, L19
Leonard, D. C., Filippenko, A. V., Ardila, D. R., Brotherton, M. S. 2001, ApJ, 553, 861
Leonard, D. C., Filippenko, A. V., Barth, A. J., Matheson, T. 2000, ApJ, 536, 239
Leonard, D. C., Filippenko, A. V., Chornock, R., Foley, R. J. 2002, PASP, 114, 1333
McCall, M. L. 1984, MNRaS, 210, 829

McDavid, D. 2001, ApJ, 553, 1027
Melgarejo, R. Magalhaes, A. M., Carcofi, A. C. & Rodrigues, C. V. 2001, A&A, 377, 581
Mendez, M., Clocchiatti, A., Benvenuto, O. G., Feinstein, C., Marraco, H. G. 1988, ApJ, 334, 295
Patat, F. *et al.* 2001, ApJ, 555, 900
Shapiro, P. R., Sutherland, P. G. 1982, ApJ, 263, 902
Spyromilio, J. 1994, MNRaS, 266, L61
Spyromilio, J., Bailey, J. 1993, PASAu, 10, 293
Stanek, K. Z. *et al.* 2003, ApJ, 591, 17S
Trammell, S. R., Hines, D. C., Wheeler, J. C. 1993, ApJ, 414, L21
Tran, H. D., Filippenko, A. V., Schmidt, G. D., Bjorkman, K. S., Jannuzi, B. T., Smith, P. S. 1997, PASP, 109, 489
Wang, L., Baade, D. *et al.* 2003a, ApJ, 591, 1110
Wang, L., Baade, D., Höflich, P., Wheeler, J. C. 2003b, ApJ, 592, 457
Wang, L., Baade, D., Höflich, P., Wheeler, J. C. *et al.* 2004, ApJ, submitted
Wang, L., Goldhaber, G., Aldering, G., Perlmutter, S. 2003c, ApJ, 590, 944
Wang, L., Howell, D. A., Höflich, P., Wheeler, J. C. 2001, ApJ, 550, 1030
Wang, L., & Hu, J. 1994, Nature, 369, 380
Wang, L., & Wheeler, J. C. 1996, ApJ, 462, L27
Wang, L., Wheeler, J. C., Höflich, P. A. 1997, ApJ, 476, L27
Wang, L., Wheeler, J. C., Höflich, P. A., Baade, D. *et al.* 2002, ApJ, 579, 671
Wang, L., Wheeler, J. C., Li, Z., Clocchiatti, A. 1996, ApJ, 467, 435

3

Spectropolarimetric observations of supernovae

Alexei V. Filippenko

Department of Astronomy, University of California, Berkeley, CA 94720–3411

D. C. Leonard

Five College Astronomy Department, University of Massachusetts, Amherst, MA 01003–9305

Abstract

We briefly review the existing database of supernova spectropolarimetry, concentrating on recent data and on results from our group's research. Spectropolarimetry provides the only direct known probe of early-time supernova geometry. To obtain reliable conclusions, however, it is very important to correctly account for interstellar polarization. We find that Type IIn supernovae (SNe IIn) tend to be highly polarized, perhaps in part because of the interaction of the ejecta with an asymmetric circumstellar medium. In contrast, SNe II-P are not polarized much, at least shortly after the explosion. At later times, however, there is evidence for increasing polarization, as one views deeper into the expanding ejecta. Moreover, core-collapse SNe that have lost part (SN IIb) or all (SN Ib) of their hydrogen (or even helium; SN Ic) layers prior to the explosion tend to show substantial polarization; thus, the deeper we probe into core-collapse events, the greater the asphericity. There is now conclusive evidence that at least some SNe Ia are intrinsically polarized, although only by a small amount. Finally, SN spectropolarimetry provides the opportunity to study the fundamental properties of the interstellar dust in external galaxies. For example, we have found evidence for extremely high polarization efficiency for the dust along the line-of-sight to SN 1999gi in NGC 3184.

3.1 Introduction

Since extragalactic supernovae (SNe) are spatially unresolvable during the very early phases of their evolution, explosion geometry has been a difficult question to approach observationally. Although SNe are traditionally assumed to be spherically symmetric, several pieces of *indirect* evidence have cast doubt on this fundamental assumption in the past decade, especially for core-collapse events. On the observational front, high-velocity "bullets" of matter in SN remnants (e.g., Taylor *et al.* 1993), the Galactic distribution and high velocities of pulsars (e.g., Kaspi &

Helfand 2002, and references therein), the aspherical morphology of many young SN remnants (Manchester 1987; see, however, Gaensler 1998), and the asymmetric distribution of material inferred from direct speckle imaging of young SNe (e.g., SN 1987A, Papaliolios *et al.* 1989; see, however, Höflich 1990) collectively argue for asymmetry in the explosion mechanism and/or distribution of SN ejecta. Moreover, recent advances in the understanding of the hydrodynamics and distribution of material in the pre-explosion core (Bazan & Arnett 1994; Lai & Goldreich 2000), coupled with results obtained through multidimensional numerical explosion models (Burrows, Hayes, & Fryxell 1995), imply that asphericity may be a generic feature of the explosion process (Burrows 2000).

Sparking even more interest in SN morphology is the strong spatial and temporal association between some "hypernovae" (SNe with early-time spectra characterized by unusually broad line features; see K. Maeda's contribution to these Proceedings) and gamma-ray bursts (GRBs; e.g., Galama *et al.* 1998; Iwamoto *et al.* 1998; Woosley, Eastman, & Schmidt 1999; Stanek *et al.* 2003; Hjorth *et al.* 2003). These associations have fueled the proposition that some (or, perhaps all) core-collapse SNe explode due to the action of a "bipolar" jet of material (Wheeler, Meier, & Wilson 2002; Khokhlov *et al.* 1999; MacFadyen & Woosley 1999), as opposed to the conventional neutrino-driven mechanism (Colgate & White 1966; Burrows *et al.* 2000, and references therein). Under this paradigm, a GRB is only produced by those few events in which the progenitor has lost most or all of its outer envelope material (i.e., it is a "bare core" collapsing), and is only observed if the jet is closely aligned with our line of sight. Such an explosion mechanism predicts severe distortions from spherical symmetry in the ejecta.

An exciting, emerging field is SN spectropolarimetry, an observational technique that allows the only *direct* probe of early-time SN geometry. As first pointed out by Shapiro & Sutherland (1982; see also McCall 1984), polarimetry of a young SN is a powerful tool for probing its geometry. The idea is simple: A hot young SN atmosphere is dominated by electron scattering, which by its nature is highly polarizing. Indeed, if we could resolve such an atmosphere, we would measure changes in both the position angle and strength of the polarization as a function of position in the atmosphere. For a spherical source that is unresolved, however, the directional components of the electric vectors cancel exactly, yielding zero net linear polarization. If the source is aspherical, incomplete cancellation occurs, and a net polarization results (Fig. 3.1). In general, linear polarizations of $\sim 1\%$ are expected for moderate ($\sim 20\%$) SN asphericity. The exact polarization amount varies with the degree of asphericity, as well as with the viewing angle and the extension and density profile of the electron-scattering atmosphere; through comparison with theoretical models (e.g., Höflich 1991), the early-time geometry of the expanding ejecta may be derived. In addition to bulk asymmetry, a wealth of information about the specific

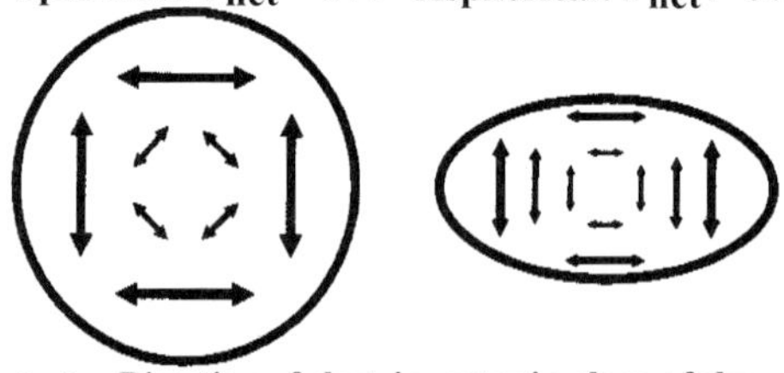

Fig. 3.1. Polarization magnitude and direction (in the plane of the sky) for a resolved electron-scattering atmosphere; for an unresolved source (i.e., a supernova), only the *net* magnitude and direction can be measured.

nature of the implied asphericity can be gleaned from a detailed analysis of line features in the spectropolarimetry (e.g., Leonard & Filippenko 2001; Leonard *et al.* 2000a, 2001, 2002a; Kasen *et al.* 2003).

Largely due to the difficulty of obtaining the requisite signal for all but the brightest objects, the field of SN spectropolarimetry remained in its infancy until quite recently. Indeed, prior to our recent efforts and those of a few other groups, spectropolarimetry existed only for SN 1987A in the LMC (see Jeffery 1991 and references therein) and SN 1993J in M81 (Tran *et al.* 1997). The situation has changed dramatically in the last 5 years. Detailed spectropolarimetric analysis now exists for more than a dozen SNe (Leonard *et al.* 2000a,b, 2001, 2002a,b; Leonard & Filippenko 2001; Howell *et al.* 2001; Kasen *et al.* 2003; Wang *et al.* 2001, 2003), and the basic landscape of the young field is becoming established. The fundamental result is that asphericity is a ubiquitous feature of young SNe of all types, although the nature and degree of the asphericity vary considerably. Here we review the spectropolarimetric characteristics of SNe, concentrating on observations made by our team at UC Berkeley; data obtained by the Texas (Austin) group are discussed by Lifan Wang and others in these Proceedings.

3.2 Interstellar polarization

It is important to first note that a difficult problem in the interpretation of all SN polarization measurements is proper removal of interstellar polarization (ISP), which is produced by directional extinction resulting from aspherical dust grains along the line of sight that are aligned by some mechanism such that their optic axes have a preferred direction. The ISP can contribute a large polarization to the observed signal. Fortunately, ISP has been well studied in the Galaxy and shown to be a smoothly varying function of wavelength and constant with time (e.g.,

Serkowski, Mathewson, & Ford 1975; Whittet & van Breda 1978), two properties that are not characteristics of SN polarization. This has allowed us to develop a number of techniques that confidently eliminate ISP from observed SN spectropolarimetry. For example, the Galactic component can be quantified with observations of distant, intrinsically unpolarized Galactic "probe stars" along the line of sight to the SN, while limits on the host-galaxy component can be estimated from the observed reddening of the SN. Alternatively, one may assume specific emission lines or spectral regions to be intrinsically unpolarized, and derive the ISP from the observed polarization at these wavelengths.

Improperly removed, ISP can increase or decrease the derived intrinsic polarization, and it can change "valleys" into "peaks" (or vice versa) in the polarization spectrum. Since it is so difficult to be certain of accurate removal of ISP, it is generally (but not always) safest to focus on (a) temporal changes in the polarization with multiple-epoch data, (b) distinct line features in spectropolarimetry having high signal-to-noise ratios, and (c) continuum polarization unlike that produced by dust. We will show near the end of this paper, however, that technique (c) can be risky, since dust with unusual properties may be the true cause of observed polarization.

3.3 Type IIn supernovae

We have found that Type IIn supernovae (SNe IIn; i.e., dominated by relatively narrow hydrogen lines – see Filippenko 1997, and references therein) tend to be highly polarized. One of our earliest and best-studied examples is SN 1998S, a SN IIn in which the progenitor star had lost most, but not all, of its hydrogen envelope prior to exploding. Immediately after exploding, an intense interaction ensued between the expanding ejecta and a dense circumstellar medium (CSM; $n_e \approx 10^7$ cm^{-3}). We combined one early-time (3 days after discovery) Keck spectropolarimetric observation with total-flux spectra spanning nearly 500 days to produce a detailed study of the SN explosion and its CSM (Leonard *et al.* 2000a). The high S/N polarization spectrum is characterized by a flat continuum (at $p \approx 2\%$) with distinct changes in polarization associated with both the broad ($\gtrsim$10,000 km s^{-1}) and narrow ($<$300 km s^{-1}) line emission seen in the total-flux spectrum. The very high intrinsic polarization implies a global asphericity of $\gtrsim$45%; the line profiles favor a ring-like geometry for the circumstellar gas, generically similar to what is seen directly in SN 1987A, but much denser and closer to the progenitor in SN 1998S (Fig. 3.2a).

Another SN IIn with high polarization is SN 1997eg, for which we obtained three epochs of spectropolarimetry over a three-month period shortly after discovery (Nakano & Aoki 1997). Although different in detail, the basic spectral

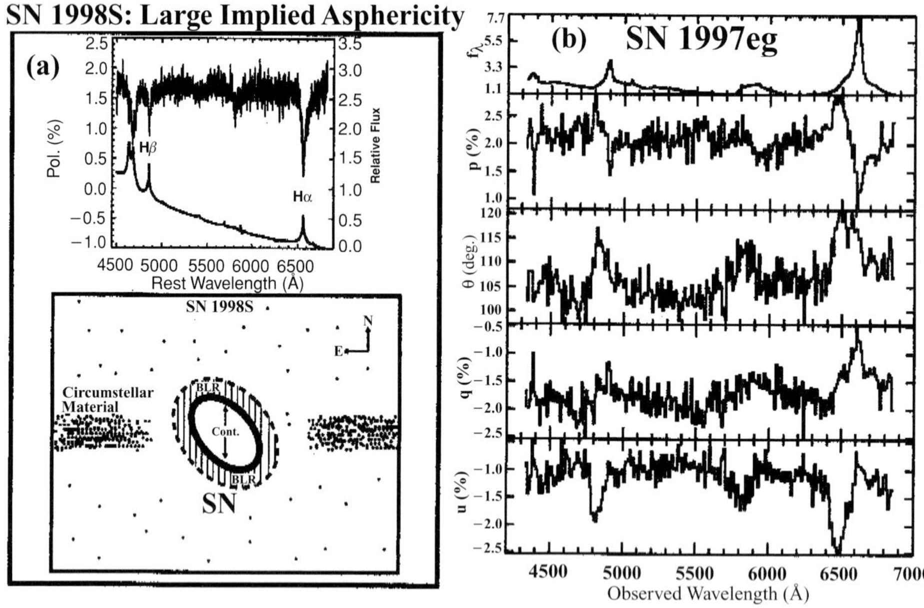

Fig. 3.2. (*a*) Spectropolarimetry and flux data (*top, noisy* and *smooth lines*, respectively), and inferred geometry (*bottom*) for SN 1998S, a peculiar type IIn event. (*b*) Polarization data for the Type IIn SN 1997eg obtained 1998 January 17, about 55 days after discovery. From top to bottom are shown the total flux, in units of 10^{-15} ergs s^{-1} cm^{-2} Å^{-1}, observed degree of polarization, polarization angle in the plane of the sky, and the normalized q and u Stokes parameters.

characteristics of this SN IIn most closely resemble those of SN 1988Z (Stathakis & Sadler 1991), with narrow (unresolved, FWHM $\lesssim$200 km s^{-1}), intermediate (FWHM $\approx$2000 km s^{-1}), and broad (FWHM $\approx$15000 km s^{-1}) line emission lacking the P-Cygni absorption seen in more normal type II events dominating the spectrum. There are clear changes in both the magnitude (by $\sim$1%) and position angle of the polarization across strong spectral lines (e.g., Hα; Fig. 3.2b) in all three epochs, and the overall continuum level of polarization changes by $\sim$1% over the three months of observation. Both of these results argue for at least a 1% polarization intrinsic to the SN, although the polarization could be produced in part by an interaction of SN ejecta with an asymmetric CSM.

3.4 Type II-P supernovae

It appears that SNe II-P are not polarized much, at least shortly after the explosion. Examples include SNe 1997ds and 1998A, which show little if any evidence for intrinsic polarization; the weak line and continuum features are probably partly (or

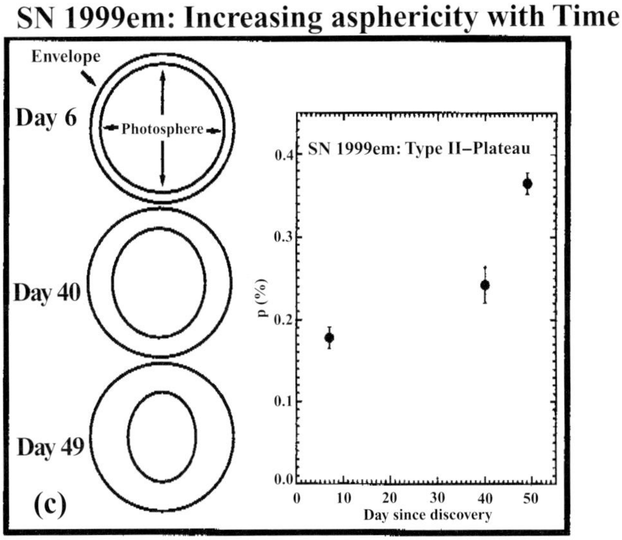

Fig. 3.3. The temporal increase in the polarization of the Type II-P SN 1999em suggests greater asphericity deeper into the ejecta.

mostly) due to ISP (Leonard & Filippenko 2001). Thus, the massive, largely intact hydrogen envelopes of the progenitors of SNe II-P are essentially spherical at the time of the explosion.

However, for SN 1999em, an extremely well-observed SN II-P for which rare, multi-epoch spectropolarimetry exists, we find that the polarization increased with time (Fig. 3.3), implying a substantially spherical geometry at early times that perhaps becomes more aspherical at late times when the deepest layers of the ejecta are revealed. As we will see in the next sections, there is also evidence for large polarizations in SNe that have lost much of their envelope prior to exploding. Both lines of evidence suggest that the deeper we peer into core-collapse events, the greater the asphericity.

The lack of evidence for early-time large-scale departures from spherical symmetry of SNe II-P is encouraging for the use of this class of objects as extragalactic distance indicators through the expanding photosphere method (EPM; Kirshner & Kwan 1974; Eastman, Schmidt, & Kirshner 1996; Leonard *et al.* 2002c,d; see also Leonard *et al.* 2003b), since this technique relies on the assumption of a spherically symmetric flux distribution during the early stages of development (i.e., the plateau phase). Spectropolarimetry of SNe II-P therefore provides a critical test of the cosmological utility of these events, which the small number of SNe II-P observed thus far have passed.

3.5 Other core-collapse supernovae

3.5.1 SNe IIb

SNe IIb are believed to be core-collapse SNe in which the progenitor lost much of its hydrogen envelope through winds or mass transfer to a companion star (e.g., Filippenko 1997, and references therein). The best-studied case is SN 1993J: Tran *et al.* (1997) found a continuum polarization of $\sim 1\%$ a month after the explosion. More importantly, they showed that its polarized-flux spectrum resembles the total-flux spectra of SNe Ib, with prominent He I lines! The data are consistent with models in which the polarization is produced by an asymmetric He core configuration of material. Trammell, Hines, & Wheeler (1993; see also Höflich 1995, Höflich *et al.* 1996) also found that SN 1993J was polarized and intrinsically asymmetric, although their derived level of interstellar polarization differed from that derived with the more extensive data set of Tran *et al.* (1997).

A subsequent SN IIb, SN 1996cb, showed substantially similar polarization of its spectra (Wang *et al.* 2001). This was unanticipated; the polarized flux spectra are expected to depend on the viewing angle, which should be random. Most recently, preliminary analysis of Keck spectropolarimetry of SN IIb 2003ed about 1 month after explosion yields similar results: Leonard, Chornock, & Filippenko (2003a) find an average continuum polarization of about 1%, with strong modulations across the Hα and Ca II near-IR triplet features of up to $\sim 1\%$ in the individual Stokes parameters. While some of the continuum polarization may be due to interstellar dust, the changes across the line features suggest substantial polarization due to scattering by aspherical SN ejecta, as had been inferred for SN 1993J and SN 1996cb.

3.5.2 SNe Ib and Ic

The Texas group identified a trend, largely through broad-band polarimetry, that the percent polarization increases along the sequence SN II-P to IIb to Ib to Ic (Reddy, Höflich, & Wheeler 1999; Wheeler 2000; Wang *et al.* 2001). Our group (Leonard & Filippenko 2001; Leonard *et al.* 2001, 2002a,b) confirmed this with more objects using spectropolarimetry. Certainly SNe IIb exhibit higher polarization than SNe II-P, as discussed above.

A very exciting recent object is the peculiar, "SN 1998bw-like" SN Ic 2002ap (Gal-Yam, Ofek, & Shemmer 2002; Mazzali *et al.* 2002; Filippenko 2003; Foley *et al.* 2003), which permitted analysis of the collapse of a "bare core" and provided several critical tests of the "jet-induced" SN explosion model. Our two epochs of Keck spectropolarimetry (Leonard *et al.* 2002a) both show strong evidence for intrinsic polarization, whose character changed dramatically with time (Fig. 3.4).

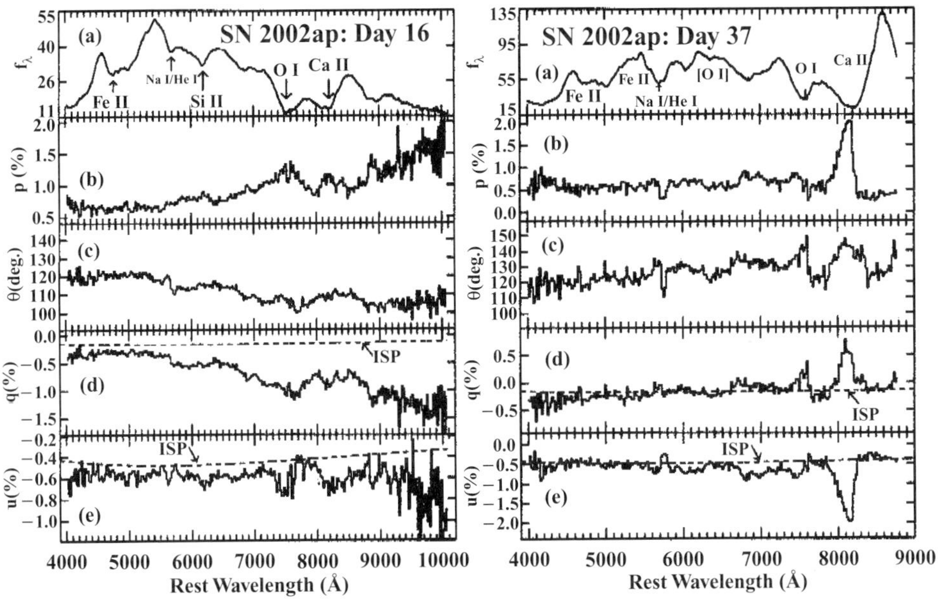

Fig. 3.4. Polarization data (Leonard *et al.* 2002a) for SN 2002ap obtained (*left*) 2002 February 14 and (*right*) 2002 March 7, about 16 and 37 days after explosion (Mazzali *et al.* 2002). The NASA/IPAC Extragalactic Database (NED) recession velocity of 657 km s^{-1} for M74 has been removed. (*a*) Total flux, in units of 10^{-15} ergs s^{-1} cm^{-2} $Å^{-1}$, with prominent absorption features identified by Mazzali *et al.* (2002) indicated. (*b*) Observed degree of polarization. (*c*) Polarization angle in the plane of the sky. (*d, e*) The normalized q and u Stokes parameters, with the level of ISP estimated by Leonard *et al.* (2002a) indicated.

Remarkably, as first noted by Kawabata *et al.* (2002), the intrinsic polarized-flux optical spectrum is similar to the total-flux optical spectrum redshifted by $0.23c$ during our first epoch, which may indicate that much of the polarized continuum at early times results from scattering off of electrons in a relativistic jet of material emitted from the SN during the explosion. Another prediction of jet-induced SN explosion models is that intermediate-mass and heavy elements such as iron are ejected (at high velocity) primarily along the poles whereas elements synthesized in the progenitor (e.g., He, C, Ca, O) are preferentially located at lower velocities near the equatorial plane in the expanding ejecta (Maeda *et al.* 2002; Khokhlov & Höflich 2001). From a careful study of both the temporal and spectral changes of the polarization angle, we find observational evidence supporting a true difference in the distribution of Ca relative to iron-group elements in the expanding ejecta (Leonard *et al.* 2002a).

For core-collapse events, then, it seems that the closer we probe to the heart of the explosion, the greater the polarization and, hence, the asymmetry. The small, but temporally increasing polarization of SNe II-P coupled with the high polarization of stripped-envelope SNe implicate an explosion mechanism that is highly asymmetric for core-collapse events. The current speculation is that the presence of a thick hydrogen envelope dampens the observed asymmetry. Thus, explosion asymmetry, or asymmetry in the collapsing Chandrasekhar core, may play a dominant role in the explanation of pulsar velocities, the mixing of radioactive material seen far out into the ejecta of young SNe (e.g., SN 1987A; Sunyaev *et al.* 1987), and even GRBs.

3.6 Type Ia Supernovae

Although the nearly unanimous consensus is that a mass-accreting white dwarf approaching the Chandrasekhar limit is the progenitor of SNe Ia, observational evidence for this scenario remains elusive. Given a progenitor that is actively accreting material at the time of the explosion, and is likely to be part of a binary system, it seems probable that some distortion of the ejecta should exist, and lend itself to detection through spectropolarimetry. Evidence for intrinsic polarization in SNe Ia has been more difficult to find than for the core-collapse events, and initial investigations found only minimal evidence for polarization, $p \lesssim 0.2\%$ (e.g., Wang, Wheeler, & Höflich 1997; Wheeler 2000).

In the last few years, though, significant advances have been made. Leonard, Filippenko, & Matheson (2000b) reported the first convincing, albeit weak, features in the polarization of an SN Ia, SN 1997dt. The most thorough polarization study of an SN Ia was conducted by Howell *et al.* (2001): the subluminous SN Ia 1999by exhibits a change in p by $\sim$0.8% from 4800 Å to 7100 Å, consistent with an asphericity of about 20% observed equator-on. However, the unusual (subluminous) nature of SN 1999by still left some doubt about intrinsic polarization in *normal* SNe Ia.

That doubt has recently been put to rest, with the work of Wang *et al.* (2003) and Kasen *et al.* (2003) on SN 2001el. Similar to SN 1999by, after subtraction of the ISP, the percent polarization increases from blue to red wavelengths in spectra obtained 1 week before maximum brightness. However, the extraordinary feature here is the existence of distinct high-velocity Ca II near-IR triplet absorption (v = 18,000–25,000 km s^{-1}) in addition to the "normal" lower-velocity Ca II feature. A similar, but much weaker such feature had been previously observed in SN 1994D, and perhaps in other SNe Ia as well; the number of pre-maximum spectra covering the near-IR spectral range is small. The polarization is seen to increase dramatically in this feature, which Kasen *et al.* (2003) interpret in terms of a detached clumpy shell of high-velocity material that partially obscures the underlying photosphere and causes an incomplete cancellation of the polarization

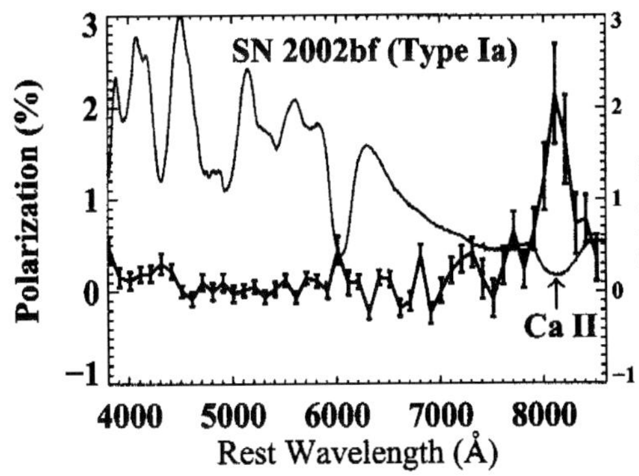

Fig. 3.5. Flux (*thin smooth curve*) and observed polarization (*binned points with error bars*) of the Type Ia SN 2002bf. ISP is not a major concern for this SN, since $E(B - V) = 0.01$ mag for the Milky Way (Schlegel, Finkbeiner, & Davis 1998), and there is no Na I D absorption from the host galaxy.

of the photospheric light, giving rise to the polarization peak. From a detailed study of the continuum polarization of ~0.2% in SN 2001el, Wang *et al.* (2003) conclude that the continuum-producing region itself likely has an axis ratio of about 0.9.

Our group obtained maximum-light spectropolarimetry of the SN Ia 2002bf using the Keck-I telescope. This supernova is characterized by uncharacteristically large photospheric velocities (Filippenko *et al.* 2002), with no obviously detached high-velocity components, although it remains to be seen if it was in any other way peculiar. Figure 3.5 shows the total-flux spectrum and the percent polarization near maximum light. There is a large increase in polarization in the photospheric Ca II near-IR trough, in which the polarization increases from almost zero to about 2%, which is actually similar behavior to what was seen in the SN Ic 2002ap discussed earlier (Fig. 3.4). When interpreted in terms of the simple geometric dilution model used for SN 2002ap, this polarization increase would imply a global asphericity of at least 15%; however, as we have seen from Kasen *et al.* (2003), other interpretations are possible that might not need to resort to global asphericity, including clumpy ejecta. On one of our latest Keck runs, in May 2003, we observed the SN Ia 2003du, and preliminary reductions of the spectropolarimetry show it to be similar to SN 2002bf. The continuum polarization slowly rises from blue to red, increasing from 0% to about 0.2%, and there are modulations across both the Si II 6350 and Ca II near-IR features.

Thus, from these studies, we can conclusively say that at least some SNe Ia are intrinsically polarized. However, this field is still very much in its infancy, and more work, especially of the theoretical sort, needs to be done to gain confidence in the interpretations of the growing database of empirical data. We have not yet seen clear evidence for duplicity of the SN Ia progenitors, which is puzzling.

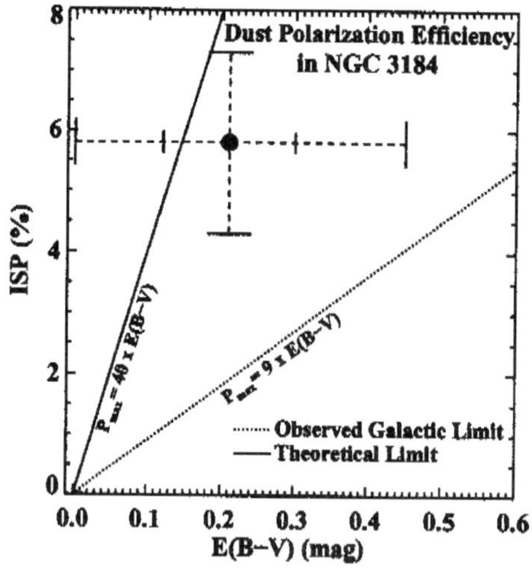

Fig. 3.6. ISP and reddening along the line-of-sight to SN 1999gi in NGC 3184, with the observed polarization and best reddening estimate indicated by the *filled circle*. The 1 σ reddening uncertainty is indicated (*short error bars*), as are the lower and upper limits for both the reddening and ISP (*long error bars*). Also shown is the empirical limit on the polarization efficiency of MW dust (*dotted line*; Serkowski *et al.* 1975), as well as the theoretical upper limit for dust grains consisting of completely aligned infinite dielectric cylinders (*solid line*; Whittet *et al.* 1992).

3.7 SN Spectropolarimetry as a Probe of Interstellar Dust

A rich byproduct of supernova spectropolarimetry is the ability to study fundamental properties of the interstellar dust in external galaxies, a research area with notoriously few direct observational diagnostics. The degree of polarization produced for a given amount of extinction (or reddening) is referred to as the "polarization efficiency" of the intervening dust grains. Through study of the polarization of thousands of Galactic stars, an upper bound on the polarization efficiency of Galactic dust has been derived (Serkowski *et al.* 1975): ISP/$E_{B-V} < 9.0\%$ mag^{-1}. Our observations of the Type II-P SN 1999gi (Leonard & Filippenko 2001) resulted in the discovery of an extraordinarily high polarization efficiency for the dust along the line-of-sight in the host galaxy, NGC 3184: ISP/$E(B-V) = 31^{+22}_{-9}\%$ mag^{-1} (Leonard *et al.* 2002b). This is more than three times the empirical Galactic limit, strains the theoretical Mie scattering limit (ISP/$E(B-V) < 40\%$ mag^{-1}; Whittet *et al.* 1992), and represents the highest polarization efficiency yet confirmed for a single sight line in either the Milky Way or an external galaxy (Fig. 3.6). While the polarization properties of the dust grains along the line-of-sight in NGC 3184 are quite unusual, our analysis also revealed the average size of the grains to be quite similar to the inferred size of typical dust grains in the Milky Way, ~0.14 μm. We speculate that the very high polarization efficiency of the grains may indicate an unusually regular magnetic field in NGC 3184 or even a different dust grain alignment mechanism than has traditionally been assumed.

Acknowledgments

We thank Craig Wheeler, in whose honor this meeting was held, for many stimulating discussions of supernovae and other (mostly explosive) phenomena over the years. A.V. F. is grateful to the workshop organizers for travel support and for their incredible patience while waiting for the written version of his presentation. The research of A.V.F.'s group at UC Berkeley is supported by NSF grant AST-0307894, as well as by NASA through grants GO-9155 and GO-9428 from the Space Telescope Science Institute, which is operated by the Association of Universities for Research in Astronomy, Inc., under NASA contract NAS 5-26555. D.C.L. acknowledges additional support by NASA through the American Astronomical Society's Small Research Grant Program. Most of our spectropolarimetry was obtained at the W. M. Keck Observatory, which is operated as a scientific partnership among the California Institute of Technology, the University of California, and NASA; the observatory was made possible by the generous financial support of the W. M. Keck Foundation.

References

Bazan, G., & Arnett, D. 1994, ApJ, 433, 41
Burrows, A. 2000, Nature, 403, 727
Burrows, A., Hayes, J., & Fryxell, B. A. 1995, ApJ, 450, 830
Burrows, A., Young, T., Pinto, P., Eastman, R., & Thompson, T. A. 2000, ApJ, 539, 865
Colgate, S. A., & White, R. H. 1966, ApJ, 143, 626
Eastman, R. G., Schmidt, B. P., & Kirshner, R. P. 1996, ApJ, 466 911
Filippenko, A. V. 1997, ARA&A, 35, 309
Filippenko, A. V. 2003, in *From Twilight to Highlight: The Physics of Supernovae*, ed. W. Hillebrandt and B. Leibundgut (Berlin: Springer-Verlag), 171
Filippenko, A. V., Chornock, R., Leonard, D. C., Moran, E. C., & Matheson, T. 2002, IAU Circ.7846
Foley, R. J., *et al.*, 2003, PASP, 115, 1220
Gaensler, B. M. 1998, ApJ, 493, 781
Galama, T. J., *et al.* 1998, Nature, 395, 670
Gal-Yam, A., Ofek, E. O., & Shemmer, O. 2002, MNRAS, 332, L73
Hjorth, J., *et al.*, 2003, Nature, 423, 847
Höflich, P. 1990, A&A, 229, 191
Höflich, P. 1991, A&A, 246, 481
Höflich, P. 1995, ApJ, 440, 821
Höflich, P., Wheeler, J. C., Hines, D. C., & Trammell, S. R. 1996, ApJ, 459, 307
Howell, D. A., Höflich, P., Wang, L., & Wheeler, J. C. 2001, ApJ, 556, 302

Iwamoto, K., *et al.* 1998, Nature, 395, 672
Jeffery, D. J. 1991, ApJ, 375, 264
Kasen, D., *et al.* 2003, ApJ, 593, 788
Kaspi, V. M., & Helfand, D. J. 2002, in ASP Conf. Ser. 271: Neutron Stars in Supernova Remnants, ed. P. O. Slane & B. M. Gaensler (San Francisco:ASP), 3
Kawabata, K. S., *et al.* 2002, ApJ, 580, L39
Khokhlov, A. M., & Höflich, P. A. 2001, in *Explosive Phenomena in Astrophysical Compact Objects*, ed. H.-Y. Chang, *et al.* (NY: AIP), 301
Khokhlov, A. M., *et al.* 1999, ApJ, 524, L107
Kirshner, R. P., & Kwan, J. 1974, ApJ, 193, 27
Lai, D., & Goldreich, P. 2000, ApJ, 535, 402
Leonard, D. C., Chornock, R., & Filippenko, A. V. 2003a, IAU Circ.8144
Leonard, D. C., & Filippenko, A. V. 2001, PASP, 113, 920
Leonard, D. C., Filippenko, A. V., Ardila, D. R., & Brotherton, M. S. 2001, ApJ, 553, 861
Leonard, D. C., Filippenko, A. V., Barth, A. J., & Matheson, T. 2000a, ApJ, 536, 239
Leonard, D. C., Filippenko, A. V., Chornock, R., & Foley, R. J. 2002a, PASP, 114, 1333
Leonard, D. C., Filippenko, A. V., Chornock, R., & Li, W. 2002b, AJ, 124, 2506
Leonard, D. C., *et al.* 2002c, PASP, 114, 35
Leonard, D. C., *et al.* 2002d, AJ, 124, 2490
Leonard, D. C., Filippenko, A. V., & Matheson, T. 2000b, in Cosmic Explosions, ed. S. S. Holt & W. W. Zhang (New York: AIP), 165
Leonard, D. C., Kanbur, S. M., Ngeow, C. C., & Tanvir, N. R. 2003b, ApJ, 594, 247
MacFadyen, A. I., & Woosley, S. E. 1999, ApJ, 524, 262
Maeda, K., *et al.* 2002, ApJ, 565, 405
Manchester, R. N. 1987, A&A, 171, 205
Mazzali, P. A., *et al.* 2002, ApJ, 572, L61
McCall, M. L. 1984, MNRAS, 210, 829
Nakano, S. & Aoki, M. 1997, IAU Circ. 6790
Papaliolios, C., *et al.* 1989, Nature, 338, 565
Reddy, N. A., Höflich, P. A., & Wheeler, J. C. 1999, BAAS, 194, 8602
Schlegel, D. J., Finkbeiner, D. P., & Davis, M. 1998, ApJ, 500, 525
Serkowski, K., Mathewson, D. L., & Ford, V. L. 1975, ApJ, 196, 261
Shapiro, P. R., & Sutherland, P. G. 1982, ApJ, 263, 902
Stanek, K., *et al.*, 2003, ApJ, 591, L17
Stathakis, R. A. & Sadler, E. M. 1991, MNRAS, 250, 786
Sunyaev, R., *et al.* 1987, Nature, 330, 227
Taylor, J. H., Manchester, R. N., & Lyne, A. G. 1993, ApJS, 88, 529
Trammell, S. R., Hines, D. C., & Wheeler, J. C. 1993, ApJ, 414, L21
Tran, H. D., Filippenko, A. V., Schmidt, G. D., Bjorkman, K. S., Jannuzi, B. T., & Smith, P. S. 1997, PASP, 109, 489
Wang, L., Baade, D., Höflich, P., & Wheeler, J. C. 2003, ApJ, 592, 457
Wang, L., Howell, D. A., Höflich, P., & Wheeler, J. C. 2001, ApJ, 550, 1030
Wang, L., Wheeler, J. C., & Höflich, P. 1997, ApJ, 476, L27
Wheeler, J. C. 2000, in *Cosmic Explosions*, ed. S. S. Holt & W. W. Zhang (NY: AIP), 445
Wheeler, J. C., Meier, D. L., & Wilson, J. R. 2002, ApJ, 568, 807
Whittet, D. C. B., Martin, P. G., Hough, J. H., Rouse, M. F., Bailey, J. A., & Axon, D. J. 1992, ApJ, 386, 562
Whittet, D. C. B., & van Breda, I. G. 1978, A&A, 66, 57
Woosley, S. E., Eastman, R. G., & Schmidt, B. P. 1999, ApJ, 516, 788

4

Observed and physical properties of Type II plateau supernovae

M. Hamuy

The Observatories of the Carnegie Institution of Washington
813 Santa Barbara St., Pasadena, CA 91101, USA

Abstract

I use photometry and spectroscopy data for 24 Type II plateau supernovae to examine their observed and physical properties. This dataset shows that these objects encompass a wide range in their observed properties (plateau luminosities, tail luminosities, and expansion velocities) and their physical parameters (explosion energies, ejected masses, initial radii, and ^{56}Ni yields). Several regularities emerge within this diversity, which reveal (1) a continuum in the properties of Type II plateau supernovae, (2) a one parameter family (at least to first order), (3) evidence that stellar mass plays a central role in the physics of core collapse and the fate of massive stars.

4.1 Introduction

Type II supernovae (SNe II, hereafter) are exploding stars characterized by strong hydrogen spectral lines and their proximity to star forming regions, presumably resulting from the gravitational collapse of the cores of massive stars ($M_{ZAMS} >$ 8 $M_\odot$). SNe II display great variations in their spectra and lightcurves depending on the properties of their progenitors at the time of core collapse and the density of the medium in which they explode. Nearly 50% of all SNe II belong to the plateau subclass (SNe IIP) which constitutes a well-defined family distinguished by 1) a characteristic "plateau" lightcurve (Barbon *et al.* 1979), 2) Balmer lines exhibiting broad P-Cygni profiles, and 3) low radio emission (Weiler *et al.* 2002). These SNe are thought to have red supergiant progenitors that do not experience significant mass loss and are able to retain most of their H-rich envelopes before explosion. In section 4.2 I summarize the observed properties of SNe IIP based on a sample of 24 objects, and in section 4.3 I use published models to derived physical parameters for a subset of 13 SNe.

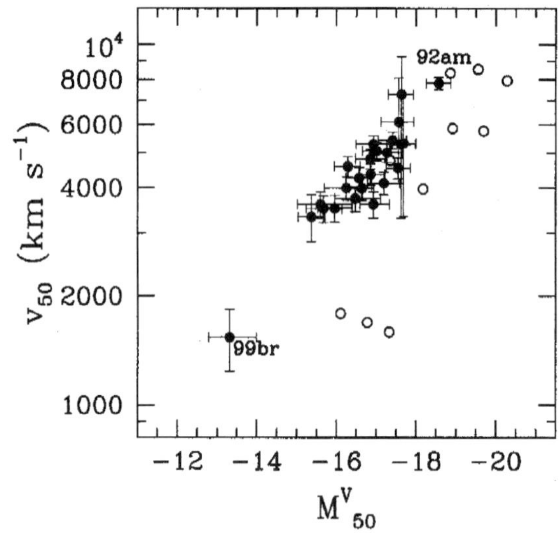

Fig. 4.1. Envelope velocity versus absolute plateau *V* magnitude for 24 SNe IIP, both measured in the middle of the plateau (day 50) (filled circles). The expansion velocities were obtained from the minimum of the Fe II λ5169 lines. The absolute magnitudes were derived from redshift-based distances and observed magnitudes corrected for dust extinction. Open circles correspond to explosion models computed by Litvinova & Nadezhin (1983, 1985) for stars with $M_{ZAMS} \geq 8\ M_{\odot}$.

4.2 Observed properties of Type II plateau supernovae

In Hamuy (2003; H03 hereafter) I compiled photometric and spectroscopic data from my own work and a variety of publications, for a sample of 24 SNe II. In Table 2 of that paper I summarized observed parameters, such as the absolute *V* magnitude near the middle of the plateau (M^V_{50}), the duration of the plateau, the velocity of the expanding envelope measured near the middle of the plateau (v_{50}), and the luminosity of the exponential tail (converted into ^{56}Ni mass ejected in the explosion). The wide range in luminosities and expansion velocities is clear manifestation of the great diversity of SNe IIP.

Figure 4.1 shows that, despite this diversity, the SN plateau luminosities are well correlated with the expansion velocities. Also shown with open circles are the explosion models of Litvinova & Nadezhin (1983, 1985, hereafter LN83 and LN85) for stars with $M_{ZAMS} \geq 8\ M_{\odot}$. It is clear that the luminosity-velocity relation is also present in the theoretical calculations. This comparison suggests that one of the main parameters behind this diversity is the explosion energy, which causes great variation in the kinetic and internal energies. A similar result was recently found by Zampieri *et al.* (2003b).

In figure 4.2 I compare the luminosity during the plateau and exponential phases. The latter is expressed in terms of the mass of ^{56}Ni ejected, M_{Ni}, assuming that the

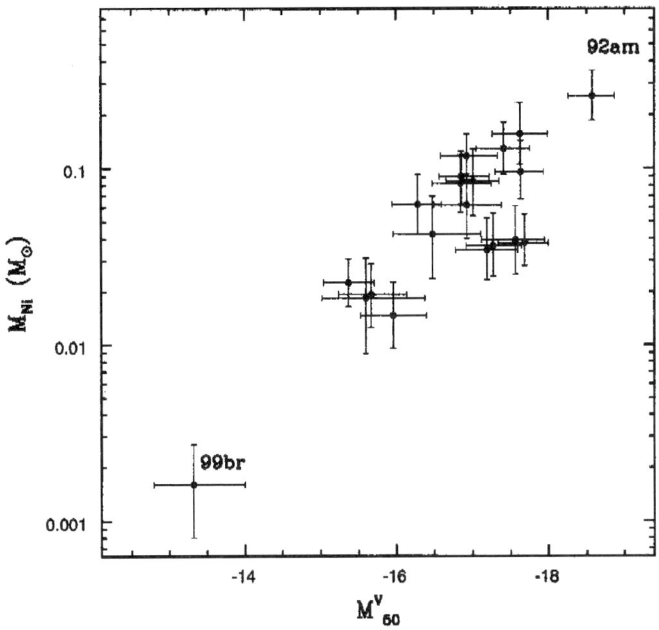

Fig. 4.2. Mass of ^{56}Ni ejected versus plateau luminosity measured 50 days after explosion.

late-time lightcurve is powered by the full trapping and thermalization of the γ rays due to $^{56}Co \rightarrow {}^{56}Fe$ (^{56}Co is the daughter of ^{56}Ni, which has a half life of only 6.1 days). There is clear evidence that SNe with brighter plateaus also have brighter tails. A similar result was recently found by Elmhamdi *et al.* (2003). Note that this correlation is independent of the distance and reddening adopted for each SN.

The previous analysis shows that several regularities emerge among the observed properties of SNe IIP. Within the current uncertainties a single parameter is required to explain the variations in luminosity and expansion velocity.

4.3 Physical properties of Type II plateau supernovae

Using hydrodynamic models, LN83 and LN85 derived approximate relations that connect the explosion energy (E), the mass of the envelope (M), and the progenitor radius (R_0) to three observable quantities, namely, the duration of the plateau, the absolute V magnitude, and the photospheric velocity observed in the middle of the plateau. These formula provide a simple and quick method to derive E, M, and R_0 from observations of SNe II-P, without having to craft specific models for each SN.

Of the 24 SNe II-P considered above only 13 have sufficient data to apply the method of LN85. The light curves for these SNe are shown in Fig. 4.3. The input

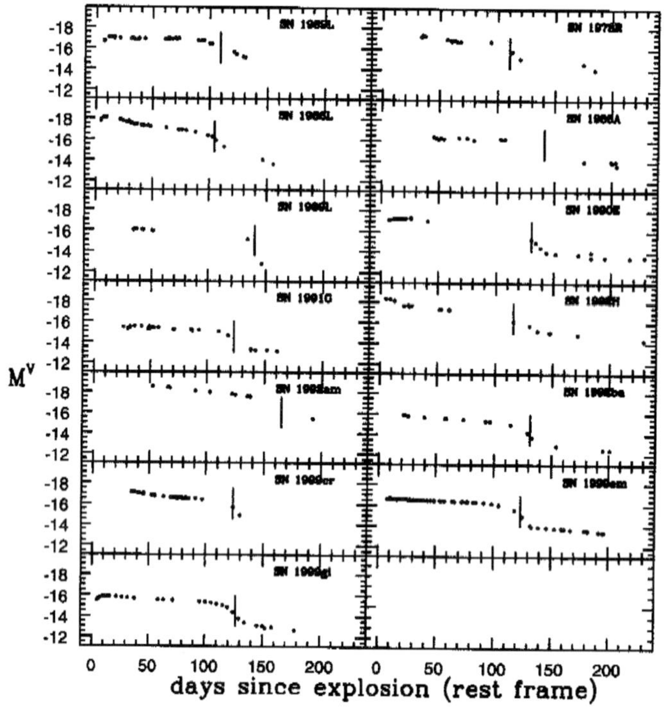

Fig. 4.3. Extinction corrected absolute *V*-band lightcurves of the 13 SNe IIP. The vertical bars indicate the end of the plateau phase for each supernova.

parameters are listed in Table 3 of H03 and the output parameters are summarized in Table 4.1. This table includes physical parameters for 3 additional SNe available in the literature, namely, SN 1987A (Arnett 1996), SN 1997D and SN 1999br (Zampieri *et al.* 2003a). Although SN 1987A showed an atypical lightcurve due to the compact nature of its blue supergiant progenitor, it was not fundamentally different than ordinary SNe II-P in the sense that it had a hydrogen-rich envelope at the time of explosion. For this reason I include it in this analysis. To my knowledge these are the only 16 SNe IIP with available physical parameters.

Among this sample, 9 SNe have explosion energies close to the canonical 1 foe value (1 foe $= 10^{51}$ ergs), 6 objects exceed 2 foes, and one has only 0.6 foes. SN 1992am and SN 1999br show the highest and lowest energies with 5.5 and 0.6 foes, respectively. This reveals that SNe II encompass a wide range in explosion energies. The ejected masses vary between 14 and 56 $M_{\odot}$. Although the uncertainties are large it is interesting to note that, while stars born with more than 8 $M_{\odot}$ can in principle undergo core collapse, they do not show up as SNe II-P. Perhaps they undergo significant mass loss before explosion and are observed as SNe IIn or SNe Ib/c. It proves interesting also that stars as massive as 50 $M_{\odot}$ seem able to

Table 4.1. *Physical Parameters for Type II Supernovae*

SN	Energy (10^{51} ergs)	Ejected Mass ($M_\odot$)	Initial Radius ($R_\odot$)	References
1969L	$2.3^{+0.7}_{-0.6}$	28^{+11}_{-8}	204^{+150}_{-88}	1
1973R	$2.7^{+1.2}_{-0.9}$	31^{+16}_{-12}	197^{+128}_{-78}	1
1986L	$1.3^{+0.5}_{-0.3}$	17^{+7}_{-5}	417^{+304}_{-193}	1
1987A	1.7	15	42.8	2
1988A	$2.2^{+1.7}_{-1.2}$	50^{+46}_{-30}	138^{+80}_{-42}	1
1989L	$1.2^{+0.6}_{-0.5}$	41^{+22}_{-15}	136^{+118}_{-65}	1
1990E	$3.4^{+1.3}_{-1.0}$	48^{+22}_{-15}	162^{+148}_{-78}	1
1991G	$1.3^{+0.9}_{-0.6}$	41^{+19}_{-16}	70^{+73}_{-31}	1
1992H	$3.1^{+1.3}_{-1.0}$	32^{+16}_{-11}	261^{+177}_{-103}	1
1992am	$5.5^{+3.0}_{-2.1}$	56^{+40}_{-24}	586^{+341}_{-212}	1
1992ba	$1.3^{+0.5}_{-0.4}$	42^{+17}_{-13}	96^{+100}_{-45}	1
1997D	0.9	17	128.6	3
1999br	0.6	14	114.3	3
1999cr	$1.9^{+0.8}_{-0.6}$	32^{+14}_{-12}	224^{+136}_{-81}	1
1999em	$1.2^{+0.6}_{-0.3}$	27^{+14}_{-8}	249^{+243}_{-150}	1
1990gi	$1.5^{+0.7}_{-0.5}$	43^{+24}_{-14}	81^{+110}_{-51}	1

Code: (1) Hamuy (2003); (2) Arnett (1996); (3) Zampieri *et al.* (2003a)

retain a significant fraction of their H envelope and explode as SNe II. Objects with $M > 35\ M_\odot$ are supposed to lose their H envelope due to strong winds, and become Wolf-Rayet stars before exploding (Woosley *et al.* 1993). This result suggests that stellar winds in massive stars are not so strong as previously thought, perhaps due to smaller metallicities. Except for four objects, the initial radii vary between 114 and 586 $R_\odot$. Within the error bars these values correspond to those measured for K and M red supergiants (van Belle *et al.* 1999), which lends support to the generally accepted view that the progenitors of SNe II-P have extended atmospheres at the time of explosion (Arnett 1996). Three of the SNe II-P of this sample, however, have $R_0 \sim 80\ R_\odot$ which corresponds to that of G supergiants. This is somewhat odd because in theory such objects cannot have plateau lightcurves but, instead, one like that of SN 1987A. Note, however, that the uncertainties are quite large and it is possible that these objects did explode as red supergiants.

Figure 4.4 shows M and M_{Ni} as a function of E for the 16 SNe II-P. Despite the large error bars, this figure reveals that a couple of correlations emerge from this analysis. The first interesting result (top panel) is that the explosion energy appears

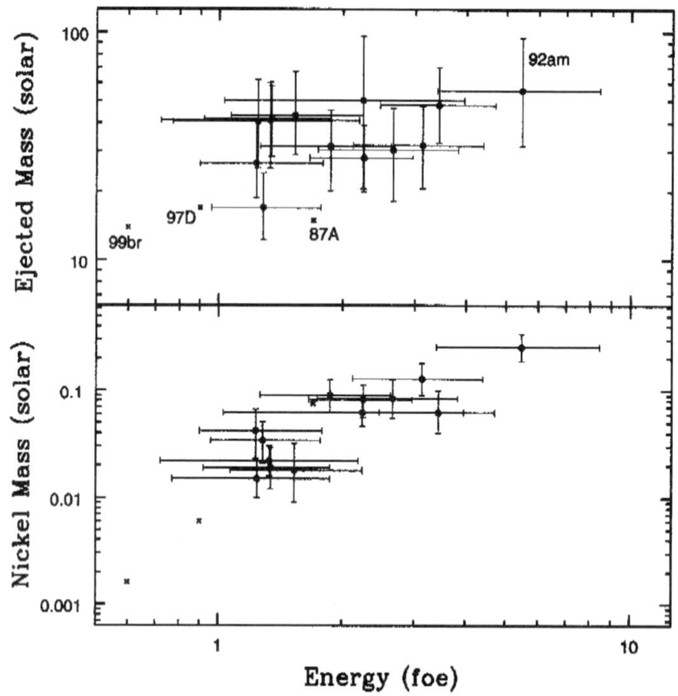

Fig. 4.4. Envelope mass and nickel mass of SNe II, as a function of explosion energy. Solid points represent the 13 SNe II-P for which I was able to apply the technique of LN85. The three crosses correspond to SN 1987A, SN 1997D, and SN 1999br which have been modeled in detail by Arnett (1996) and Zampieri *et al.* (2003a). The nickel yield for SN 1999br comes from H03.

to be correlated with the envelope mass, in the sense that more massive progenitors produce more energetic SNe. This suggests that stellar mass plays a central role in the physics of core collapse. The second remarkable result (bottom panel) is that SNe with greater energies produce more nickel (a result previously suggested by Blanton *et al.* 1995). This could mean that greater temperatures and more nuclear burning are reached in such SNe, and/or that less mass falls back onto the neutron star/black hole in more energetic explosions.

4.4 Conclusions

1) SNe II-P encompass a wide range of ~5 mag in plateau luminosities, a five-fold range in expansion velocities, and a 100-fold range in tail luminosities.
2) Despite this great diversity, SNe II-P show several regularities such as correlations between plateau luminosities, expansion velocities, and tail luminosities, which suggests a one parameter family, at least to first order.

3) There is a continuum in the properties of SNe II-P from faint, low-velocity, nickel-poor events such as SN 1997D and SN 1999br, and bright, high-velocity, nickel-rich objects like SN 1992am.
4) SNe IIP encompass a wide range in explosion energies (0.6–5.5 foes), ejected masses (14–56 $M_\odot$), initial radii (80–600 $R_\odot$), and ^{56}Ni yields (0.002–0.3 $M_\odot$).
5) Despite the large error bars, a couple of correlations emerge from the previous analysis: (1) more ^{56}Ni is ejected in SNe with greater energies; (2) progenitors with greater masses produce more energetic explosions. This suggests that the physics of the core collapse and the fate of massive stars is, to a large extent, determined by the mass of the progenitor.

Acknowledgements

Support for this work was provided by NASA through Hubble Fellowship grant HST-HF-01139.01-A awarded by the Space Telescope Science Institute, which is operated by the Association of Universities for Research in Astronomy, Inc., for NASA, under contract NAS 5-26555.

References

Arnett, D., 1996, Supernovae and Nucleosynthesis (New Jersey: Princeton Univ. Press).
Barbon, R., Ciatti, F., & Rosino, L., 1979, *Astron. Astrophys.,* 72, 287–292.
Blanton, E. L., Schmidt, B. P., Kirshner, R. P., Ford, C. H., Chromey, F. R., & Herbst, W., 1995, *Astron. J.*, 110, 2868–2875.
Elmhamdi, A., Chugai, N. N., & Danziger, I. J., 2003, *Astron. Astrophys.*, 404, 1077–1086.
Hamuy, M., 2003. *Astrophys. J.*, 582, 905–914 (H03).
Litvinova, I. Y., & Nadezhin, D. K., 1983, *Astrophysics and Space Science*, 89, 89–113 (LN83).
Litvinova, I. Y., & Nadezhin, D. K., 1985, *Soviet Astronomy*, 11, L145 (LN85).
Zampieri, L., Pastorello, A., Turatto, M., Cappellaro, E., Benetti, S., Altavilla, G., Mazzali, P., & Hamuy, M., 2003a, *Mon. Not. R. astr. Soc.*, 338, 711–716.
Zampieri, L., *et al.*, 2003b, in *Proceedings of IAU Colloquium 192, "Supernovae (10 years of SN1993J)"*, eds Marcaide, J. M., & Weiler, K. W., Springer Verlag, in press.
van Belle, G. T., *et al.* 1999, *Astron. J.*, 117, 521–533.
Weiler, K. W., Panagia, N., Montes, M. J., & Sramek, R. A., 2002, *Ann. Rev. Astron. Astrophys.*, 40, 387–438.
Woosley, S. E., Langer, N., & Weaver, T. A., 1993, *Astrophys. J.*, 411, 823–839.

5

SN 1997B and the different types of type Ic supernovae

A. Clocchiatti

Pontificia Universidad Católica de Chile

B. Leibundgut, J. Spyromilio

European Southern Observatory

S. Benetti, E. Cappellaro, M. Turatto

Osservatorio Astronomico di Padova

M. M. Phillips

Las Campanas Obervatory

Abstract

We present the *V* light curve and optical/infrared spectra of the Type Ic SN 1997B. We show that (1) this SN displayed lines of the He I series; (2)the expansion velocities were higher than those of SNe with traces of H or large He masses in their envelopes (like SN 1993J); the light curve of SN 1997B decayed slower than that of SN 1993J. The smaller mass to kinetic energy ratio and shallower light curve of SN 1997B are inconsistent with it being a He stripped version of some of the best studied Type Ib or II-transition SNe. We infer that Type Ib/c and II-transition SN progenitors come, at least, with two different types of inner structure.

5.1 Introduction

A few years ago the presence of He in the atmospheres of Type Ic SNe, the nature of their progenitors, and the relation between Type Ib and Type Ic SNe was subject of debate. On the one hand, empirical evidence and theoretical interpretation supported the view that SNe of Type Ib and Ic are different enough to insure that their progenitors result from different paths of stellar evolution. If so, Type Ic SNe originated in bare C+O cores and were expected not to display He I lines in their spectra. On the other hand, it was stressed that Type Ib and Ic SNe could originate in similar stars evolving as interacting binaries. Small differences in orbital parameters would provide different outer layers at the time of explosion, and a gradient of optical and near infrared He I line strengths connecting He–Rich (Type Ib) with He–Poor (Type Ic) SNe was expected in this scenario. The discovery of the infrared line He I λ10830 in SN 1994I (Filippenko *et al.* 1995), as well as lines of the optical series of He I in SN 1994I, SN 1987M and presumably other Type Ic SNe (Clocchiatti *et al.* 1996a, 2001) terminated this discussion and generalized the view that Type Ic SNe may display weak He I lines at high velocity.

The discovery of He in Type Ic SNe also meant a step towards a unifying view of the Type Ib/c and II-transition* (hereafter II-t) SN subclasses. The similarity of the light curves of SN 1983N and 1993J (Nomoto *et al.* 1993; Clocchiatti *et al.* 1996b) and the theoretical models put forth to explain the light curves and spectra of both SN 1993J and SN 1994I (Iwamoto *et al.* 1994; Woosley, Langer & Weaver, 1995; Wheeler *et al.* 1994), generally converged to the view of low mass He stars with differences in the residual outer layers as the common progenitor for Ib/c and II-t SNe. Theoretical interpretation suggest that these SNe explode by gravitational core-collapse, the energy transferred to the ejecta is $\sim 10^{51}$ ergs, and ejected ^{56}Ni is $\sim 0.07\ M_{\odot}$. Different spectra at early times result from varying light element outer residual layers: $\sim 0.15\ M_{\odot}$ of H and $\sim 1.5 M_{\odot}$ of He in the Type II-t SN 1993J (Swartz *et al.* 1993, Woosley *et al.* 1994), $\sim 0.001\ M_{\odot}$ of H and $\sim 1.5 M_{\odot}$ of He in the Type Ib SN 1983N (Wheeler *et al.* 1994), and essentially no H or He in Type Ic SNe, like SN 1994I (Iwamoto *et al.* 1994; Woosley, Langer, & Weaver 1995).

The unified view of low mass He stars as common progenitor for Ib/c and II-t SNe did not completely solve the earlier debate concerning their nature and relations, because detailed analysis of both light curves and spectra was done just for a couple of events. The question can still be asked: Do all Type Ib/c and II-t SNe correspond to a variation of the low mass He star basic structure that successfully explained SNe 1993J and 1994I? We show here that the answer to this question is "no." Some Type Ib/c and II-t SNe correspond to exploding low mass He stars or their naked C+O cores. Some others, however, do not.

5.2 Data of SN 1997B and other SNe

The complete dataset of photometry, optical – near infrared, and infrared spectra of SN 1997B and its evolution in time will be described in a forthcoming paper (Benetti *et al.*, in preparation). Spectra of SN 1997B and photometry were obtained from several runs at different observatories in Chile. We show in Figure 5.1 *J* band spectra of SN 1997B, the Type II-t SN 1993J (Swartz *et al.* 1993), and the Type Ic SN 1990W (Wheeler *et al.* 1994), and, in Figure 5.2, portions of the optical spectra of SN 1997B, and the Type Ic SNe 1994I (Clocchiatti *et al.* 1996a) and 1987M (Filippenko, Porter & Sargent 1990). In Figure 5.3 we show the relative *V* light curves of SN 1997B and 1993J during the first ~ 250 days.

* These are SNe that display H I lines near maximum, but do a transition to a Type Ib/c spectrum at late times, like SN 1993J. For a justification of the name see Clocchiatti & Wheeler (1997).

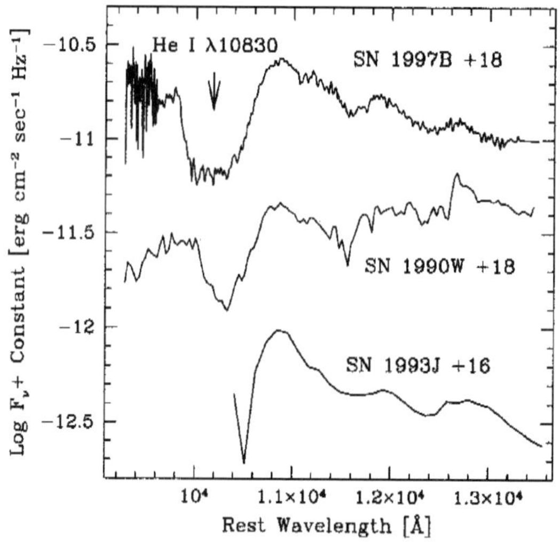

Fig. 5.1. Infrared *J* band spectra of SN 1997B, SN 1990W and SN 1993J.

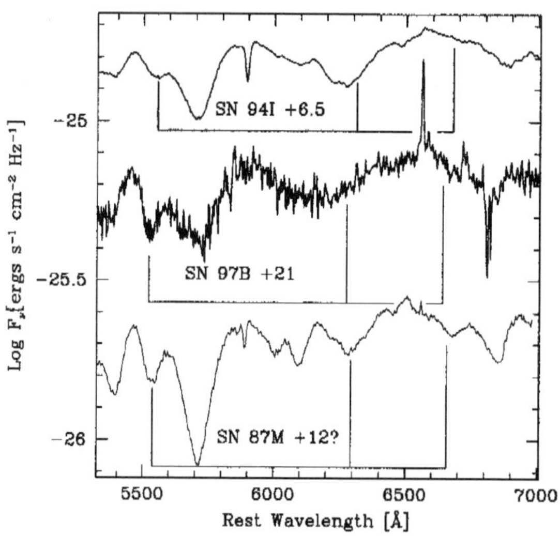

Fig. 5.2. Optical spectra of SN 1997B, SN 1994I and SN 1987M. The vertical marks show the positions where the three lines of the He I series should appear.

Matching the post-maximum decay of SN 1997B (Fig. 5.3) suggests that the spectra of figures 5.1 and 5.2 are ∼18 and ∼21 days after maximum, respectively.

5.3 Discussion

The spectra of SN 1997B shown in figures 5.1 and 5.2 confirm that it was Type Ic, but had a residual He layer. The line He I λ10830 is strong in Fig. 5.1, and He I

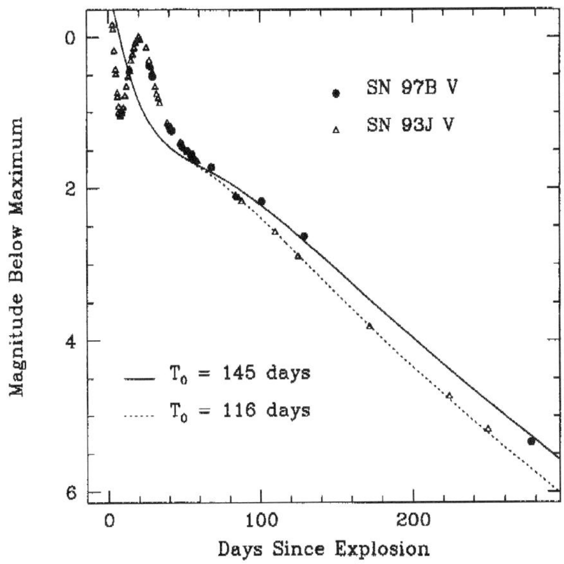

Fig. 5.3. *V* light curve of SN 1997B (Benetti *et al.* 2003), compared with that of SN 1993J (Richmond *et al.* 1994). The solid lines are fits of the bolometric decay of expanding spherical shells with point source γ-ray deposition (see § 1.3). Explosion time for SN 1993J was taken on JD = 2,449,074.6 (Clocchiatti *et al.* 1995) and the light curve of SN 1997B was shifted until its post-maximum decline matched that of SN 1993J.

λ5876 is visible in Fig. 5.2. The expansion velocity of the He I layer measured from the blueshifted minimum of the He I λ10830 line is 17380 km s^{-1}. The bluemost edge of the line at rest frame 9897 Å implies a maximum expansion velocity of ~26900 km s^{-1} for the He I layer. The expansion velocity from the He I λ5876 line is ~18460 km s^{-1}. Due to low S/N ratio the other lines of the He I series that should have been seen are not detected. The fair matching of velocities between the two lines detected, however, confirms that they are He I lines.

The photospheric velocity of SN 1997B at a phase of 21 days, estimated from the minimum of the Fe I λ5169 line (see Fig. 5.2) was ~11000 km s^{-1}. The large difference between the photospheric velocity and the HeI velocity suggests that there is little He and that it was located at high velocity.

The comparison of the IR spectrum of SN 1997B with those of SNe 1990W and 1993J at approximately the same phases is illustrative of a progression of the He layer towards larger velocities as one moves from bottom to top in Fig. 5.1. The velocities of the He I layer suggest that the expansion of SN 1997B ejecta was faster than that of SN 1993J, and detailed study of optical spectra (Benetti *et al.* 2003) shows that Fe I λ5169 is systematically faster as well. The last spectrum, ~80 days after maximum, gives $v_{SN1997B}/v_{SN1993J}$ ~1.5.

SN 1993J is a prototypical event interpreted as the outcome of core-collapse in a low mass He star. One of its successful models (Woosley *et al.* 1994), results from the interacting evolution of a binary system with initial masses of 13 and 9 $M_\odot$, and initial separation of 3 AU. The progenitor mass before explosion is 3.67 $M_\odot$, He mass in the outer ejecta is $\sim 1.5 M_\odot$ and H mass is $\sim 0.15 M_\odot$. The low velocity mass cut that reproduced the light curves implied a collapsed remnant with $\sim 1.37 M_\odot$ and a total mass for the ejecta of $\sim 2.3 M_\odot$. The light curve of SN 1993J has been compared with those of some Ib/c SNe and found to be remarkably similar. In the cases of the Type Ib SN 1983N (Clocchiatti *et al.* 1996b), or the Type Ic SN 1990B (Clocchiatti *et al.* 2001), the similarity of light curves and expansion velocities suggests that they are a variation of the same initial configuration with small differences in explosion and/or progenitor parameters. An extreme version of the configuration of SN 1993J was also used to match the light curve of SN 1994I (Woosley *et al.* 1995), and its maximum light spectrum (Wheeler *et al.* 1994). All of these SNe would fit into the scenario of progenitor stars with similar inner structure and different residual layers of light elements exploding by the same mechanism.

Within the picture of low mass He stars as common progenitors for Type Ib/c and II-t SNe, the progression of velocities in Fig.5.1 is the result of a diminishing He I layer and, hence, total mass of the ejecta. *The smaller mass to energy ratio of SN 1997B would imply a faster light curve, however.* As seen in Fig. 5.3 this is not the case. SN 1997B is a Type Ic SN with a light curve *slower* than that of SN 1993J. The *V* exponential tail of SN 1997B has a slope of ~ 0.015 mag day^{-1}, while that of SN 1993J is ~ 0.0187 mag day^{-1} (Richmond *et al.* 1994).

It is possible to show that the smaller slope of the light curve combined with the larger expansion velocity means that the total mass in the ejecta of SN 1997B was larger than that of SN 1993J. Using the simple model of late time light curves of Clocchiatti & Wheeler (1997), one can write

$$\left(\frac{v_{\rm min,97B}}{v_{\rm min,93J}}\frac{T_{0,97B}}{T_{0,93J}}\right)^2 \sim \frac{M_{97B}}{M_{93J}}, \qquad (5.1)$$

where T_0 is the characteristic time of decay of the γ-ray optical depth in the ejecta (a shallower slope of the light curve means a larger T_0), and it was assumed that the density distributions of the shocked matter in both SNe are similar.

This equation shows that the smaller mass to energy ratio (i.e. larger $v^2{}_{\rm min}$) *and* a shallower slope of the late time light curve (i.e. a larger T_0) of SN 1997B imply that its total ejected mass was significantly larger. The models fitted in Fig.5.3 give $T_{0,1997B} \sim 145$ and $T_{0,1993J} \sim 116$ days, respectively, so that ($T_{0,97B}/T_{0,93J} \sim 1.25$). Using this, together with the latest ratio $v_{\rm min,97B}/v_{\rm min,93J} \sim 1.5$ quoted earlier, one finds $M_{97B}/M_{93J} \sim 3.5$.

In other words, if the density distribution and γ-ray absorption coefficient were similar, the ejecta of SN 1997B must have been more massive than that of SN 1993J. This, in turn, implies that the progenitor of SN 1997B was not a helium stripped variation of that of SN 1993J, and hence must correspond to a different end stage of stellar evolution. SN 1997B needs a progenitor with a much larger C+O core.

Departures of spherical symmetry can provide an alternative explanation. If the matter does not expand at the same rate in all directions the expansion velocities and their time evolution will depend on the viewing angle, and the light curves could depend on the degree of asymmetry. Core-collapse SNe *are expected* to display departures from spherical symmetry, and more detailed theoretical models are required to explore its consequences. We tend to think, however, that departures of spherical symmetry consistent with the extensive spectroscopic and photometric databases of these SNe would fail to transform one of them (SN 1997B) into a He stripped version of the other (SN 1993J).

SN 1997B which could not result from the explosion of a low mass He star and SNe 1994I, which could, are Type Ic SNe with indistinguishable spectra. Although it can be possible to differentiate them using extensive datasets, it is not possible to get a "direct estimation of the physical characteristics" (Keenan 1963) of these SNe based only on spectroscopic types, as is done with normal stars.

We conclude that there are at least two families of stripped envelope SNe, resulting from the explosion of C+O cores of substantially different mass. Both families have the potential to originate Type Ic, Ib, or II-t (IIb) events, according with the residual layers of light elements remaining at the time of explosion. The two families should correspond to different paths of stellar evolution and the possibility exists that there are considerable differences in their explosion mechanisms and asymmetries, as well.

References

Clocchiatti, A., & Wheeler, J. C., 1997, ApJ, 491, 375
Clocchiatti, A., *et al.*, 1995, ApJ, 446, 167
Clocchiatti, A., *et al.*, 1996a, ApJ, 462, 462
Clocchiatti, A., *et al.*, 1996b, ApJ, 459, 547
Clocchiatti, A., *et al.*, 2001, ApJ, 553, 886
Filippenko, A. V., Porter, A. C., & Sargent, W. L. W., 1990, AJ, 100, 1575
Filippenko, A. V., *et al.*, 1995, ApJ, 450, L11
Galama, T. J., *et al.*, 1998, Nature, 395, 670
Hachisu, I., Matsuda, T., Nomoto, K., Shigeyama, T., 1991, Ap.J., 368, L27
Iwamoto, K., *et. al.*, 1994, ApJ, 437, L115
Keenan, P. C., 1963, Stars and Stellar Systems, Vol. 3, Strand, K.AA. ed., Chicago: University of Chicago Press, 78
Nomoto, K., *et al.*, 1993, Nature, 364, 507

Richmond, M. W., *et al.*, 1994, AJ, 107, 1022
Swartz, D. A., *et al.* 1993, 365, 232
Wheeler, J. C., *et al.*, 1993, ApJ, 417, L71
Wheeler, J. C., *et al.*, 1994, ApJ, 436, L135
Woosley, S. E., Eastman, R. G., Weaver, T. A., & Pinto, P. A., 1994, ApJ, 429, 300
Woosley, S. E., Langer, N., & Weaver, T. A., 1995, ApJ, 448, 315

6

Near-infrared spectroscopy of stripped-envelope supernovae

C. L. Gerardy

W. J. McDonald Postdoctoral Fellow, University of Texas at Austin

R. A. Fesen

Dartmouth College

G. H. Marion, P. Höflich, and J. C. Wheeler

University of Taxas at Austin

K. Nomoto and K. Motohara

University of Tokyo

Abstract

Near-infrared (NIR) spectroscopy of several stripped-envelope core-collapse supernovae (SNe) are presented. NIR spectra of these objects are quite rich, exhibiting a large number of emission features. Particularly important are strong lines of He I and C I, which probe the outermost ejecta and constrain the pre-collapse mass-loss. Interestingly, the SN 1998bw-like broad-line Type Ic SN 2002ap does not exhibit the strong C I features seen in other Type Ic SNe. NIR spectra also exhibit strong, relatively isolated lines of Mg I, Si I, Ca II, and O I that provide clues into the kinematics and mixing in the ejecta. Finally, late-time NIR spectra of two Type Ic events: SN 2000ew and SN 2002ap show strong first-overtone carbon monoxide (CO) emission, providing the first observational evidence that molecule formation may not only be common in Type II SNe, but perhaps in all core-collapse events.

6.1 Introduction

Near-infrared (NIR) spectroscopy is a powerful tool for the study of supernovae (SNe), offering new insights into the kinematic, chemical, and evolutionary properties of these events. Here we present applications of NIR spectroscopy for the study of three stripped-envelope supernovae, the Type Ib SN 2001B, the Type Ic SN 2000ew and the broad-line Type Ic SN 2002ap. All of the data presented here were obtained using TIFKAM on the 2.4 m Hiltner telescope at MDM Observatory, except for the SN 2002ap data set which also includes spectra obtained

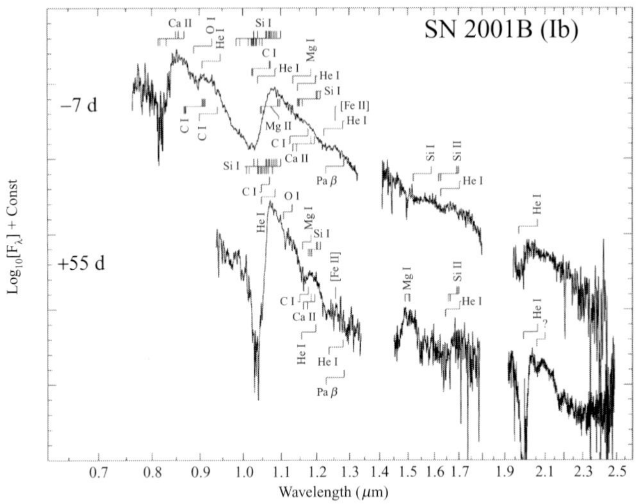

Fig. 6.1. Near-infrared spectra of the Type Ib SN 2001B. Possible line IDs are marked showing both the rest wavelength and the likely location of a P-Cygni minimum. For the day −7 spectrum, all lines assume a photospheric velocity of 13 000 km s^{-1}. For the day 55 spectrum, expansion velocities of 13 000, 10 000, and 6000 km s^{-1} have been assumed for prospective hydrogen, helium, and metal lines respectively.

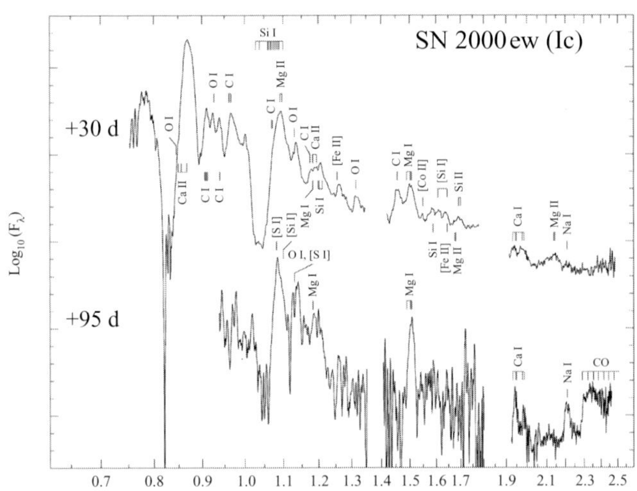

Fig. 6.2. Near-infrared spectra of the Type Ic SN 2000ew. The rest wavelengths of possible line identifications have been marked.

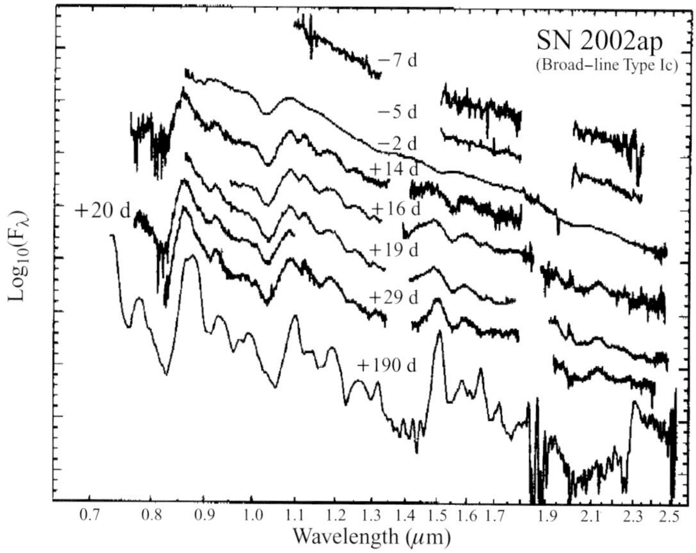

Fig. 6.3. Near-infrared spectra of the broad-line Type Ic SN 2002ap.

at Lick Observatory, IRTF, and Subaru. The reduced spectra are presented in Figures 6.1–6.3.

6.2 He I lines

The presence or absence of helium lines in the spectra of stripped-envelope supernovae has profound implications for the pre-collapse mass-loss and evolution of the progenitor. However, making this determination is difficult in the optical where the He I lines are relatively weak and usually blended with stronger features.

Two strong He I lines are available in the near-IR: the famous 1.083 μm triplet and the somewhat weaker 2.058 μm singlet line. The wavelength region containing the 1.083 μm line has been occasionally observed at the extreme red end of optical SN spectra and attributed to He I 1.083 (e.g., Filippenko *et al.* 1995). Unfortunately, the detection of a feature in this region is a poor diagnostic for He I emission as several other strong lines, including Mg II, C I, and H I (Paγ), could also be the cause of such a feature.

A better diagnostic is to compare the line profiles of both the 1.083 and 2.058 μm lines and look for matching profiles. In Figure 6.4, we show that there is a good match for He I in the Type Ib SN 2001B, but not in the Type Ic SN 2002ap. The Type Ic SN 2000ew is also a poor match for He I, and in both SNe Ic the 2 μm feature is likely Mg II, and the 1.05 μm feature a blend of Mg II, C I, and Si I.

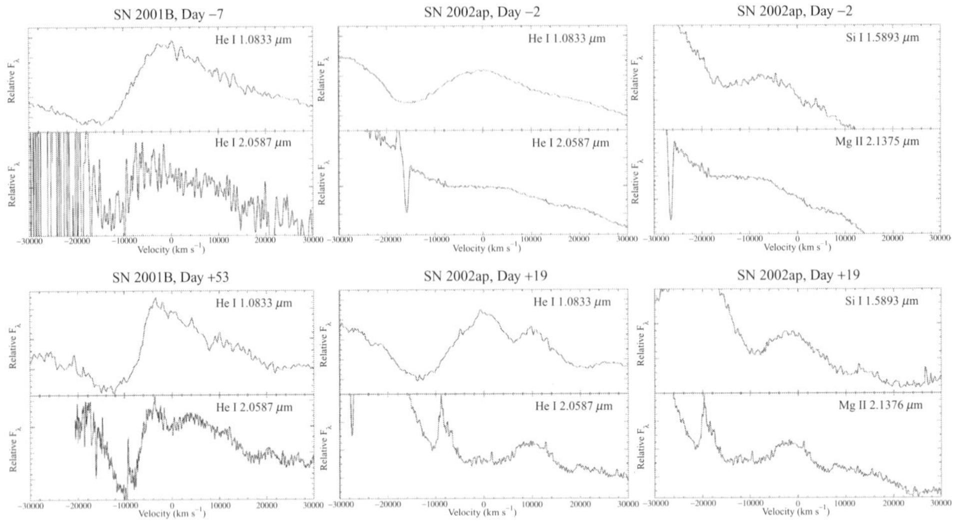

Fig. 6.4. Line profile comparison for the Type Ib SN 2001B and the Type Ic SN 2002ap. In SN 2001B, the line profiles for features at 1.05 and 2.0 μm are a good match for He I 1.08 and 2.06 μm. In SN 2002ap however, the He I identifications result in poorly matching line profiles. The 2.0 μm feature in SN 2002ap is a better match to Mg II 2.14 μm.

6.3 C I lines

Near-infrared spectroscopy also offers the opportunity to study carbon-rich gas in the SN ejecta as several strong lines of C I are observable. Four C I features are clearly seen in SN 2000ew (Fig. 6.2), and can also be seen in the Type IIp SNe SN 1999em and SN 1999gi begining late in the plateau phase (Gerardy *et al.* 2004, in prep). Three of these C I features lie in the 9000–10,000 Å region and appear at the red end of optical spectra of several Type Ic SNe (e.g. SN 1994I; Matheson *et al.* 2001).

However, these C I features are conspicuously absent (or at least very weak) in the spectra of SN 2002ap (see Fig.6.5). The absence of C I lines is not an ionization effect, because other features of similar ionization potential, (Si I, Mg I) are still observed in SN 2002ap. Rather the lack of C I lines is an indication of a lower carbon abundance in the photosphere of SN 2002ap compared to other SNe Ic. This might be related to the unusually strong [O I] and Mg I] emission observed in the nebular phase, and could be an indication of more extreme mass-loss prior to collapse (Filippenko 2002). Alternatively, the carbon depletion might indicate a larger mass and/or lower metallicity progenitor, which would produce less C relative to O (e.g. Woosley, Heger & Weaver 2002). In either case, the ejecta of SN 2002ap cannot be entirely devoid of carbon, as carbon monoxide (CO) formed in the ejecta at late times. (See below).

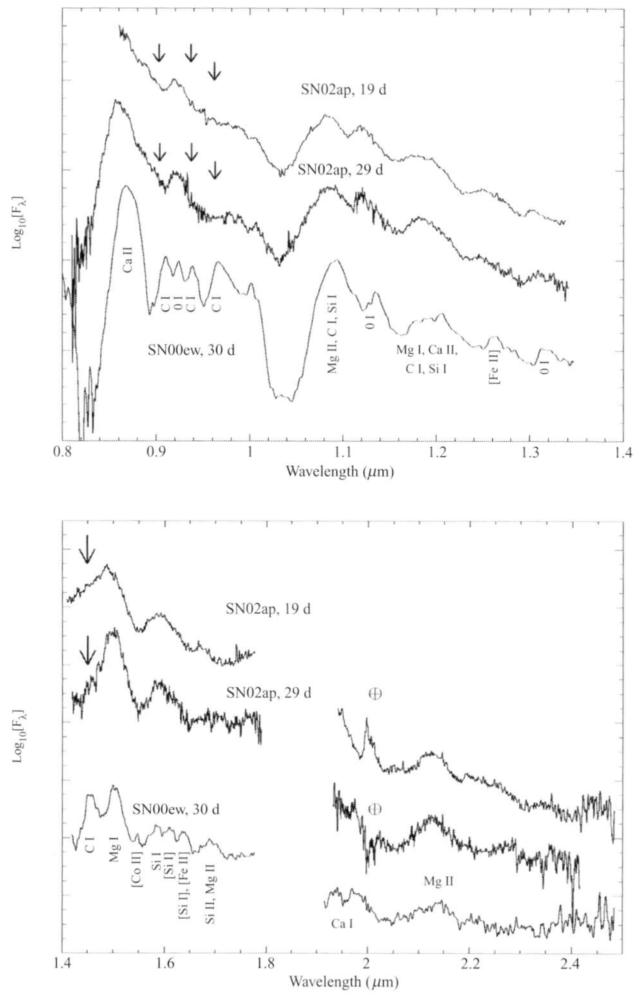

Fig. 6.5. Comparison of SN 2002ap on days 19 & 29 with SN 2000ew on day 30. While most SN 2000ew features can be seen in SN 2002ap, the C I features are conspicuously absent.

6.4 Carbon monoxide emission

Both SN 2000ew and SN 2002ap exhibit strong emission from the first-overtone band of CO near 2.3 μm. CO emission has been detected in a number of Type II supernovae and CO formation appears to be common in these events (Gerardy *et al.* 2000, 2003; Spyromilio, Leibundgut & Glimozzi 2001). With the detection of CO in the two Type Ic events SN 2000ew and SN 2002ap, it appears that CO formation may also be common in SNe Ic. The epoch of CO emission appears to begin at around 1000–200 days, (see Table 6.1).

Table 6.1. *CO detections in NIR spectra of core-collapse SNe*

Event	SN Type	Epoch of last CO non-detection	Epoch of first CO detection	Reference
		CO Detections		
SN 1987A	IIP Pec	112 d	192 d	1
SN 1995ad	IIP		105 d	2
SN 1998S	IIn	44 d	109 d	3
SN 1999dl	IIP		152 d	4
SN 1999em	IIP	118 d	178 d	4,5
SN 1999gi	IIP	74 d	126 d	5
SN 2000ew	Ic	39 d	97 d	5,6
SN 2002ap	Ic Pec	29 d	190 d	5
		Non-Detections		
SN 1990W	Ic	90 d		7
SN 1995V	IIP	84 d		8
SN 2001B	Ib	60 d		5

References: (1) Meikle *et al.* (1989); (2) Spyromilio & Leibundgut (1996); (3) Fassia *et al.* (2001); (4) Spryomilio, Leibundgut & Glimozzi (2001); (5) Gerardy *et al.* 2004, in prep.; (6) Gerardy *et al.* (2003); (7) Wheeler *et al.* (1994); (8) Fassia *et al.* (1998);

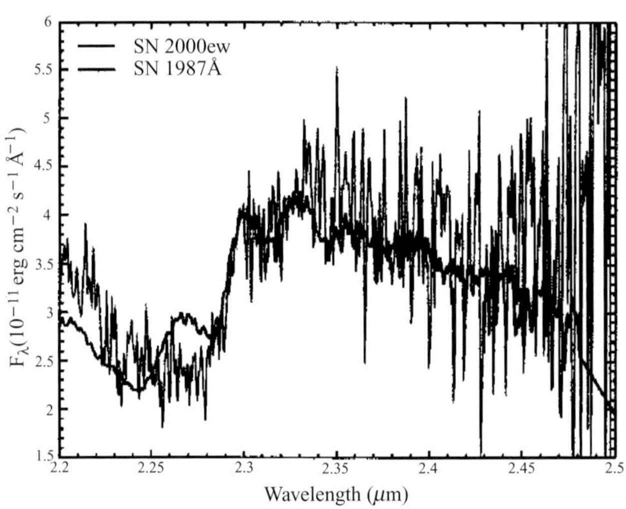

Fig. 6.6. Comparison of the carbon monoxide emission seen in SN 2000ew with that seen in SN 1987A. The blue edge of the CO feature is quite similar in both events, indicative of CO expansion velocity of about 2000 km s^{-1}.

The shape of the CO profile provides some information about the velocity of the CO-rich gas, and in nearly all the observed CO profiles, the velocity is around 2000 km s^{-1} or less. While such a velocity is not unexpected for carbon and oxygen rich gas in a Type II SN, it is surprisingly low for SNe Ic. The fact that we do not see high-velocity CO in Type Ic SNe suggests that it is difficult to form CO in higher velocity gas, where the density drops quickly before molecules have a chance to form. It also suggests that the core is significantly mixed during the collapse and explosion, since the outermost C/O material is mixed into the inner regions of the ejecta. This is also suggested by the lack of velocity stratification in the metal lines of all three SNe, as compared to the strong velocity stratification seen in the NIR spectra of Type Ia SNe (Höflich *et al.* 2002; Marion *et al.* 2003).

References

Fassia, A., Meikle, W. P. S., Geballe, T. R., Walton, N. A., Pollacco, D. L., Rutten, R. G. M., & Tinney, C. 1998, MNRAS, 299, 150
Fassia, A., *et al.* 2001, MNRAS, 318, 1093
Filippenko, A. V. *et al.* 1995, ApJ, 450, L11
Filippenko, A. V. 2002, in From Twilight to Highlight: The Physics of Supernovae, eds. W. Hillebrandt & B. Leibundgut, ESO Astrophysics Symposia (Berlin: Springer), p. 171
Gerardy, C. L., Fesen, R. A., Höflich, P., & Wheeler, J. C. 2000, AJ, 119, 2968
Gerardy, C. L., Fesen, R. A., Nomoto, K., Maeda, K., Höflich, P. & Wheeler, J. C. 2002, PASJ, 54. 905
Höflich, P. Gerardy, C. L., Fesen, R. A., & Sakai, S. 2002, ApJ, 568, 791
Iwamoto, K, Nomoto, K., Höflich, P., Yamaoka, H., Kumagai, S., & Shigeyama, T. 1994, ApJ, 437, L115
Maeda, K., Nakamura, T., Nomoto, K., Mazzali, P. A., Patat, F., & Hachisu, I. 2002, ApJ, 565, 405
Marion, G. H., Höflich, P, Vacca, W. D., & Wheeler, J. C. 2003, ApJ, 591, 316
Matheson, T., Filippenko, A. V., Li, W., Leonard, D. C., & Shields, J. C. 2001, AJ, 121, 1648
Mazalli, P. *et al.* 2002, ApJ, 572, L61
Meikle, W. P. S., Spyromilio, J., Varani, G.-F., & Allen, D. A. 1989, MNRAS, 238, 193
Nomoto, K, Suzuki, T., & Iwamoto, K. 1995, Phys. Rep. 256, 173
Spyromilio, J. & Leibundgut, B. 1996, MNRAS, 283, L89
Spryomilio, J., Liebundgut, B., & Glimozzi, R. 2001, A&A, 376,188
Wheeler, J. C., Harkness, R. P., Clocchiatti, A., Benetti, S., Brotherton, M. S., Depoy, D. L., & Elias, J. 1994, ApJ, 436, L135
Woosley, S. E., Heger, A., & Weaver, T. A. 2002, Rev. Mod. Phys., 74, 1015

7

Morphology of supernova remnants

R. A. Fesen

6127 Wilder Lab, Department of Physics & Astronomy Dartmouth College, Hanover NH 03755 USA

Abstract

Emission morphologies of young, Galactic supernova remnants can be used for investigating SN expansion dynamics, elemental distributions, and progenitor mass loss history and properties at the time of outburst. The remnants of two suspected Galactic Type Ia SNe, Tycho and SN 1006, show spherical morphologies, with Si-rich ejecta near the forward shock front suggestive of significant mixing. Searches for possible surviving binary companions near the centers of these remnants may help clarify the progenitor binary system(s) involved in SNe Ia. On the other hand, high mass, core collapse remnants, such as SNR 1987A and Cas A, exhibit strongly asymmetrical morphologies, with Cas A showing some evidence for bipolar ejecta jets. However, it is currently unclear if such ejecta jets are consistent with any of the recently proposed jet induced SN explosion models.

7.1 Introduction

For a workshop on the 3-D signatures of stellar explosions, it seems worth-while to first explain why one might be interested in the properties of supernova remnants (SNRs). Even the youngest Galactic SN remnants are hundreds and even thousands of years removed from the actual SN events, so SNRs may seem at first to be relatively poor tools for any meaningful testing of SN models or explosion theories. However, young supernova remnants, and especially the nearby Galactic ones, offer chemical and kinematic data on SN ejecta on much finer spatial scales than possible from extragalactic SN/SNR investigations. Specifically, the chemical and expansion properties of a SNR's high-velocity, outermost ejecta can sometimes offer clues regarding the type of progenitor involved and well as exploring asymmetries in the explosion. In addition, young Galactic Type Ia remnants may prove helpful for investigating possible surviving companions and hence help determine the make-up of the progenitor binary systems involved.

Table 7.1. *Young Galactic Supernova Remnants (age < 1000 yr)*

SNR Name	Galactic ID	SN Date	Probable Type
Cassiopeia A	G111.7–2.1	≥1671	Ib
Kepler's SNR	G004.5+6.8	1604	Ia?
Tycho's SNR	G120.1+1.4	1572	Ia
SNR 3C58	G130.7+3.1	1181	IIpec
Crab Nebula	G184.6–5.8	1054	IIpec
SN 1006	G327.6+14.6	1006	Ia

The large scale emission morphology of a young remnant is, to some measure, directly linked to several key aspects of SNe. These include SN expansion dynamics and asymmetries, explosive elemental production and distribution (mixing), progenitor mass loss history via detection of circumstellar material (CSM), and progenitor properties at the time of outburst (e.g., stellar classification and binarity). However, even a relatively young remnant's morphology can be significantly affected by its local ISM/CSM environment as well as the wavelength band by which it is examined. Thus one must be cautious on interpreting results solely from a remnant's overall morphology. We usually only get to see those parts which radiate strongly at some wavelength and this is often a function of many variables, some which are unknown.

For understanding the chemical and dynamical properties of SNe through morphology, the youngest remnants are both the most interesting and meaningful, as the older ones represent more "interstellar impact statements" than windows into SN physics. Unfortunately, there are only a handful of Galactic SNRs with ages less than 1000 yrs or so. These are listed in Table 7.1. The paucity of young, nearby SNRs severely limits using remnants as investigating tools for exploring the physics of SN explosion. Nonetheless, these few objects are still quite valuable.

7.2 Morphology of Type Ia SNRs

If the currently held model of Type Ia SNe originating from explosions of CO white dwarfs in some sort of binary system is correct, then the remnants of SNe Ia might be both: i) fairly homogeneous if one particular binary type progenitor system dominates the rate, and ii) relatively spherical if CSM losses from the system are small and spatially limited. Here we will examine the youngest and oldest Type Ia remnants listed in Table 7.1.

The youngest Galactic SNR generally believed to be the result of a Type Ia explosion is Tycho's SNR (SN 1572). *Chandra* X-ray images of the remnant covering the 0.5–10 keV energy range are shown in Figure 7.1 (adopted from Hwang *et al.* 2002).

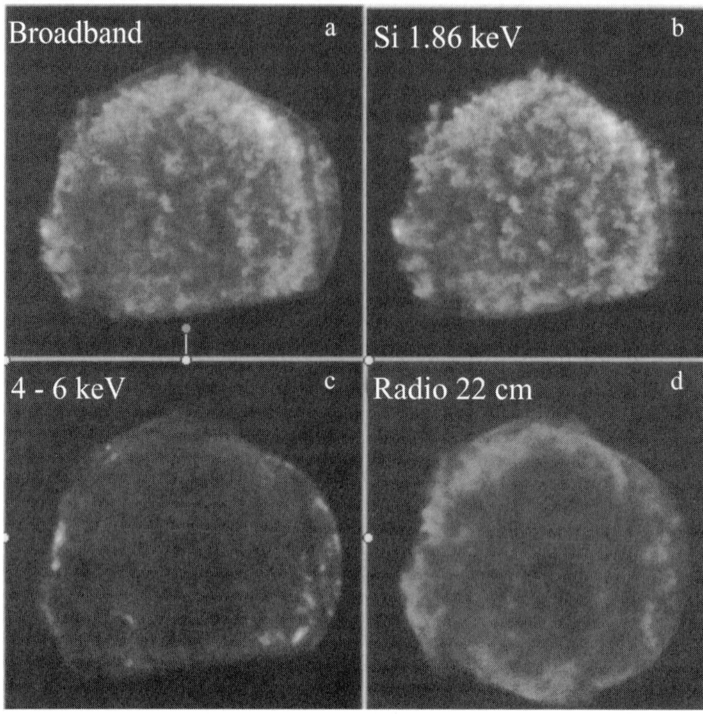

Fig. 7.1. Montage of: a) broadband *Chandra* ACIS image, b) Si 1.86 keV emission image (continuum subtracted), c) 4–6 keV image, and d) a VLA 22 cm image. (Adapted from Hwang *et al.* 2002; radio image courtesy of J. Dickel)

Panel A shows a broadband energy image which reveals both the location of the forward blast wave seen as the thin, faint outer emission filaments along the western and northeastern limbs, and the clumpy Si and S rich ejecta. The ejecta are seen to extend out to locations very close to the forward shock at several places. This can be seen somewhat better in Panel B which is an image centered at the 1.86 keV Si emission features, while Panel C shows only the hard continuum emission (4–6 keV) thereby highlighting just the forward shock emission. Finally, Panel D shows a VLA 22 cm radio image of the remnant for radio/X-ray morphology comparison.

Several things can be learned from these images. Firstly, the Si and S rich ejecta are highly clumped. Secondly, the fact they they lie close to the forward shock suggests a significant amount of mixing – either during or after the explosion (Hwang *et al.* 2002). Thirdly, one can examine the remnant's expansion rate, velocity, and uniformity with azimuthal angle, and using *ROSAT* image data, Hughes (2000) found an outer shock velocity of 4600 ± 400 (D/2.3 kpc), and a large expansion rate parameter m (where $r \sim t^{\mathrm{m}}$) of 0.71 ± 0.06 compared to that estimated from radio data (0.471 ± 0.028; Reynoso *et al.* 1997).

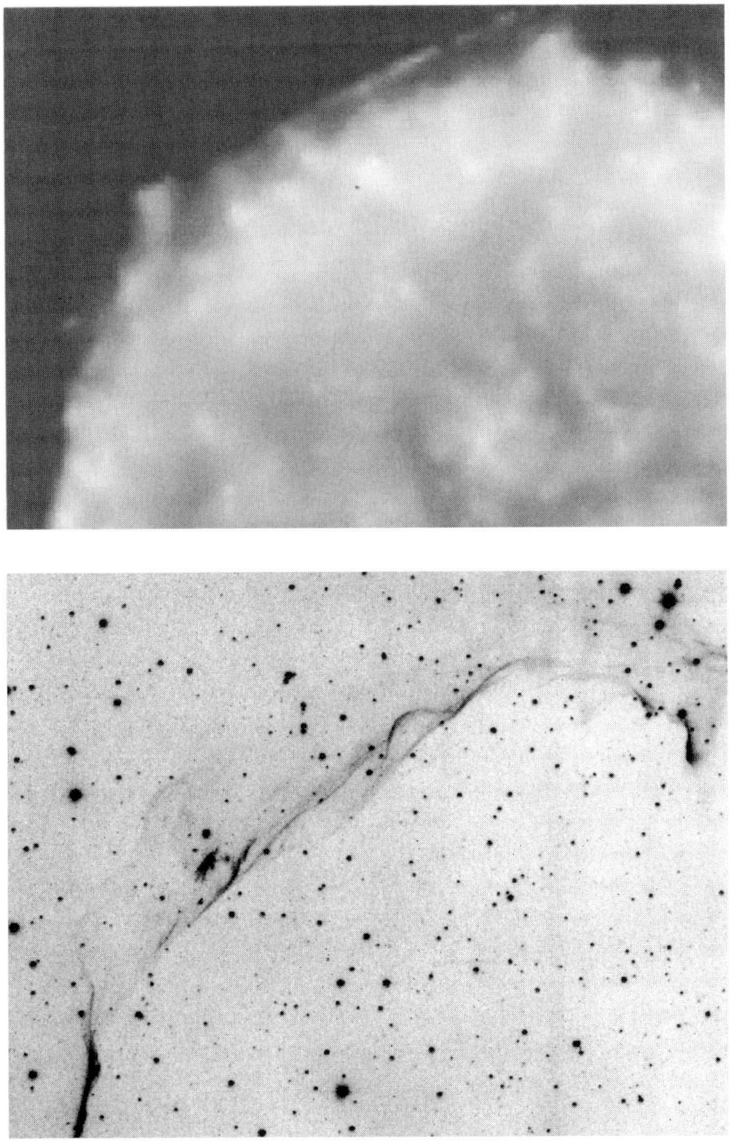

Fig. 7.2. NE region of Tycho's SNR. Upper Panel: Broadband *Chandra* showing expansion excursions of Si-rich ejecta (Hwang *et al.* 2002). Lower Panel: Optical Hα image with filaments showing location of the forward shock.

Although largely spherical, evidence for minor expansion excursions of the shock front and SN ejecta from a smooth spherical symmetry can be seen in the enlarged Si image shown in the upper panel of Figure 7.2. Several similar outer edge excursions are seen in the remnant's optical emission through Hα emission (Fig. 7.2, lower panel) which highlights the location of the remnant's forward shock. Measurements

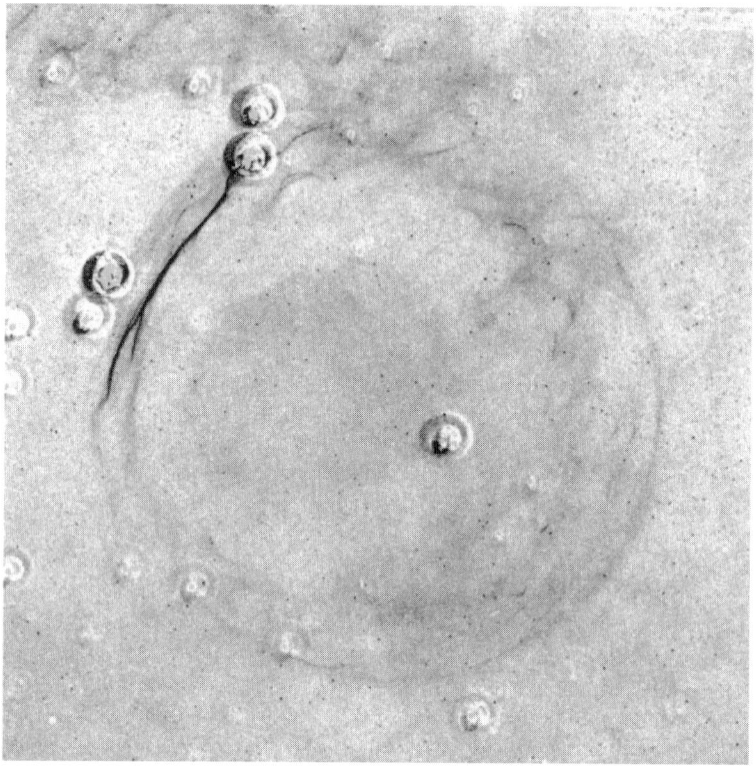

Fig. 7.3. A deep, continuum subtracted Hα image of the SN 1006 remnant showing delicate filaments along most of the remnant's shell. The field of view is 40 arcmin square. (From Winkler, Gupta, & Long 2003)

of the proper motions of these optical filaments may provide a direction comparison between X-ray and optical expansion rates.

On the other side of the age range, the oldest Galactic remnant that is generally thought to be likely connected to a Type Ia event is perhaps the remnant of SN 1006. At a distance of 2.2 kpc (Winkler *et al.* 2003), it is also the closest of the historical SN listed in Table 7.1. A deep optical Hα image of the remnant is shown Figure 7.3. As was the case in Tycho's SNR, one sees a nearly spherically expanding remnant and it is the highly spherical shape of these two Type Ia remnants which prove invaluable in solving a long-standing problem.

One of the most fundamental questions regarding Type Ia SNe is the nature of the progenitor system. Although most likely an explosion of a CO white dwarf with a mass close to the Chandrasekhar limit, the precise progenitor system(s) involved is currently uncertain. The three most often cited progenitor systems are: 1) a double degenerate system where two low-mass WDs merge due to the loss angular momentum (Webbink 1984, Paczyński 1985), 2) mass transfer onto the

WD by a red giant companion, and 3) mass transfer from a close-in main sequence companion such as in a cataclysmic system.

The spherical morphologies of Tycho and SN 1006 may offer a means of directly testing these three binary system scenarios. Early calculations by Colgate (1970) and Cheng (1974) showed that the binary red giant, subgiant, or main sequence companion star would receive a substantial kick, of order 50 to 300 km s^{-1} or more. Such high velocities are large enough to identify the surviving companions via radial velocity or proper motion studies, especially when one takes into account direction and rate of motion away from a remnant's center of expansion and known age. The survivability of such post-SN companions has been investigated further by Wheeler, Lecar, & McKee (1975), and more recently by Marietta, Burrows, & Fryxell (2000). These latter calculations show that all survive to varying degrees, with MS stars losing just 15% of their mass while RGs might lose 98% of their envelopes.

For most SNRs, the coordinates of the projected centers of expansion are rather poorly. However, in the cases of Tycho and SN 1006, the nearly symmetrical morphologies and relative youth of both remnants greatly help limit the uncertainties. Nonetheless, despite several searches of the centers of the Tycho and SN 1006 remnants, no likely companion candidates via proper motions studies have yet been found. Ruiz-Lapuente and collaborators have also obtained radial velocity measurements of several of the brighter stars near these remnant centers (Ruiz-Lapuente *et al.* 2003; Canal, Méndez, & Ruiz-Lapuente 2001), but again no likely surviving companion stars have been identified. A deeper and higher precision proper motion *HST* investigation is now getting underway that may provide stricter limits on SN Ia companions and hence new insights on this long-standing problem.

7.3 Morphology of core collapse SNe

In contrast to the remnants of Type Ia SNe, nicely behaved morphologies are not the rule for the remnants of core collapse supernovae. Figure 7.4 shows a *Chandra* X-ray image of the southern hemisphere remnant, G292.0 + 1.8. This relatively young (~2000 yr) core collapse remnant from a high mass progenitor (Hughes & Singh 1994), shows a complex, asymmetric structure with a X-ray point source, the 135 ms pulsar (PSR J1124-5916), well off from the remnant's nominal emission center (Hughes *et al.* 2003). In the much younger core collapse remnant of SN 1987A, observations show that the SN ejecta are strongly aspherical in overall structure (Wang *et al.* 2002).

One of the best young Galactic SNRs for studying SN expansion asymmetries and ejecta chemistry in a core collapse SN is Cassiopeia A. Because of its relatively

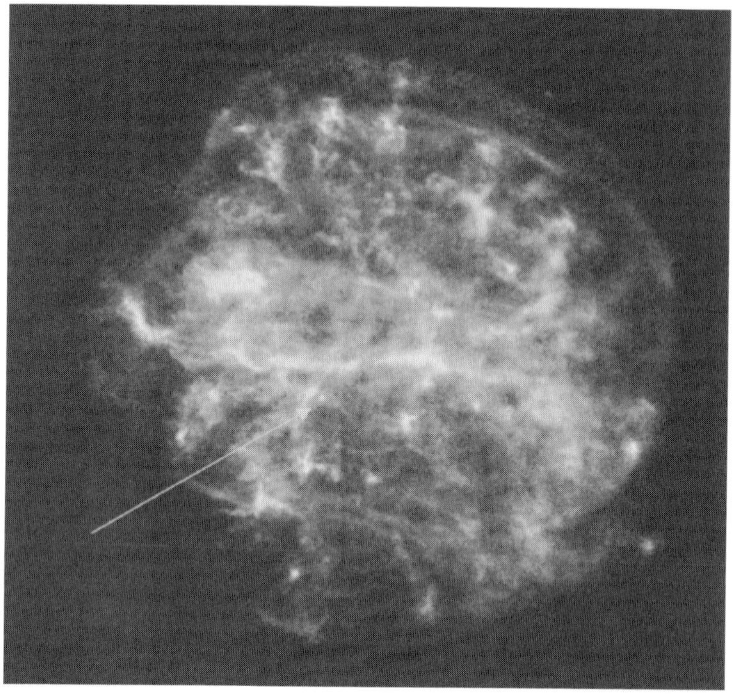

Fig. 7.4. *Chandra* image of SNR G290.0 + 1.8 (Hughes *et al.* 2001). Arrow marks location of PSR J1124–5916.

youth ($\approx$300 yrs), its morphology largely still reflects the dynamics of the SN explosion. The morphology as revealed by *Chandra* images shows three important aspects related to SN explosion physics of a high-mass star: 1) A compact central X-ray point source displace 7 arcsec south of the center of expansion (Thorstensen *et al.* 2001) and roughly perpendicular to a NE-SW jet axis of especially high-velocity ejecta (Fesen 2001), and 2) Fe-rich ejecta seen along the outer SE limb indicating an overturning of the progenitor's initial O, Si, Fe layering (Hughes *et al.* 2000).

Additionally, optical ground-based images and spectra given clear evidence for an asymmetrical morphology consisting of a 5000–6000 km s^{-1} expansion main shell of ejecta (radius = 110″) enclosed within a forward blast wave (radius = 153″; Gotthelf *et al.* 2001) but which is disrupted by a prominent NE jet or plume of ejecta (V_{exp} up to 15 000 km s^{-1}) along with a possible SW counterjet (see Fesen 2001 and references therein). Moreover, there appears to a systematic correlation of elemental abundance with expansion velocity in the jet-counterjet in the sense that ejecta become poorer in O but richer in S with increasing ejecta velocity. Such an inverse relationship between heavy element abundance and ejection velocity in a jet of ejecta might naturally arise from a jet explosion SN model.

Jet-induced explosions of core collapsed SNe was first suggested in the 1970's (LeBlanc & Wilson 1970; Ostriker & Gunn 1971; Bisnovatyi-Kogan 1971) but has

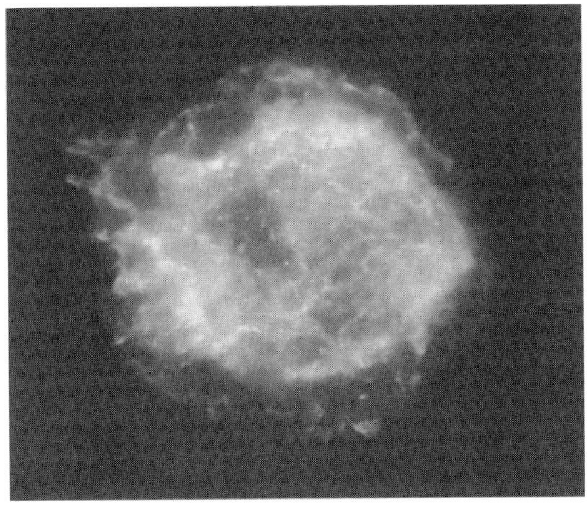

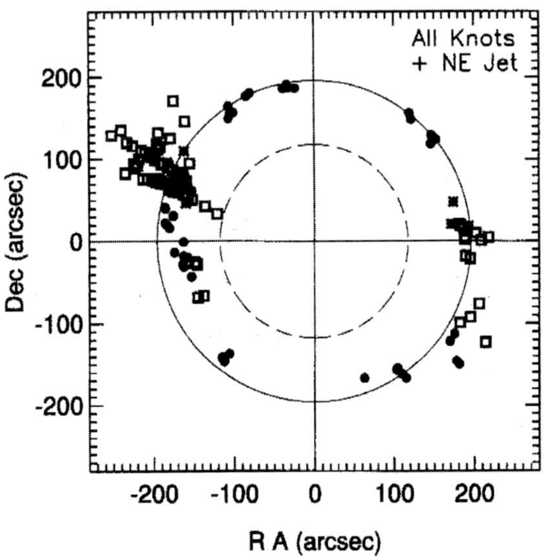

Fig. 7.5. Cas A. Upper Panel: Broadband *Chandra* image (NASA). Lower panel: De-projected expansion velocities of outer ejecta knots (including NE jet), with dashed inner and solid outer circles marking 6000 and 10,000 km s^{-1} respectively (Fesen 2001)

recently experienced a revival of interest. Several jet mechanisms and effects including magnetocentrifugal jets and a link to GRBs have been proposed (Khokhlov *et al.* 1999; Wheeler *et al.* 2000, 2001; MacFadyen *et al.* 2001). The general idea of a jet induced SN explosion is supported by *Chandra* data of Cas A which show Fe and Si ejecta asymmetries not that unlike those predicted for a jet's equatorial Si/Fe enrichment.

However, it is not yet clear if any of the various models and theories for axisymmetric explosions really can explain Cas A, SN 1987A, or any other core collapse

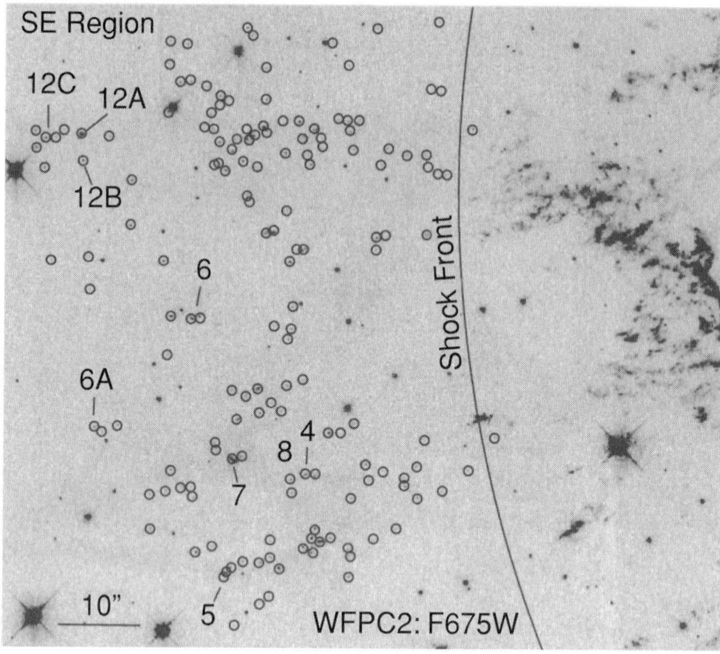

Fig. 7.6. WFPC2 *HST* image of a portion of Cas A's SE region out ahead of the position of the forward shock front as indicated by *Chandra* data. Ejecta knots detected via ground-based images are marked Knots 4–12C, while outer ejecta visible via Hubble images are indicated by the small circles.

SNR since we do not yet have sufficient kinematic and chemical data to make detailed comparisons. SN 1987A is still quite young and 50 kpc away limiting its size and hence the spatial resolution of its ejecta. In the case of the older Cas A remnant, new *HST* images suggest a much larger array of outlying ejecta knots than first suspected which may help clarify its explosion dynamics. Figure 7.6 shows a small portion of the SE limb of Cas A just a bit out ahead of the forward shock front. Although ground-based images showed only a handful of knots (indicated by the knot numbers 4–12C), two epochs of HST images reveal over 100 more knots. If imaged in different filters which isolate various strong line emissions, these outlying, high velocity ejecta might yield more detailed information of the Cas A SN's explosion asymmetries and hence stronger tests of jet-induced explosion models for at least this, the youngest SNR in our galaxy.

References

Bisnovatyi-Kogan, G. S. 1971. *Soviet Astron.-AJ*, **14**, 652.

Canal, R., Méndez, J., & Ruiz-Lapuente 2001. *Astrophys. J.*, **550**, L53–L56.

Cheng, A. 1974. *Astrophys. Sp. Sci.*, **31**, 49–56.

Colgate, S. A. 1970. *Nature*, **225**, 247.

Fesen, R. A. 2001. *Astrophys. J. Suppl.*, **133**, 161–186.
Gotthelf, E. V., Koralesky, B., Rudnick, L., Jones, T. W., Hwang, U., & Petre, R. 2001. *Astrophys. J.*, **552**, L39–L43.
Hughes, J. P. 2000. *Astrophys. J.*, **545**, L53–L56.
Hughes, J. P., Rakowski, C. E., Burrows, D. N., & Slane, P. O. 2000. *Astrophys. J.*, **528**, L109–L113.
Hughes, J. P., & Singh, K. P. 1994. *Astrophys J.*, **422**, 126–135.
Hughes, J. P., Slane, P. O., Burrows, D. N., Garmire, G., Nousek, J. A., Olbert, C. M., & Keohane, J. W. 2001. *Astrophys. J.*, **559**, L153–L156
Hughes, J. P., Slane, P. O., Park, S., Roming, P. W. A., & Burrows, D. N. 2003. *Astrophys. J.*, **591**, L139–L142.
Hwang, U., Decourchelle, A., Holt, S. S., & Petre, R. 2002. *Astrophys. J.*, **581**, 1101–1115.
Khokhlov, A. M., Höflich, P., Oran, E. S., Wheeler, J. C., Wang, L., & Chtchelkanova, A. Yu. 1999. *Astrophys. J.*, **524**, L107–L110.
LeBlanc, J. M., & Wilson, J. R. 1970. *Astrophys. J.*, **161**, 541–551.
Marietta, E., Burrows, A., & Fryxell, B. 2000. *Astrophys. J. Suppl.*, **128**, 615–650.
MacFadyen, A. I., Woosley, S. E., & Heger A. 2001. *Astrophys. J.* **550**, 410–425.
Ostriker, J. P., & Gunn, J. E. 1971 *Astrophys. J.*, **164**, L95–L104
Paczyński, B. 1985. *Cataclysmic Variables and Low Mass X-Ray Binaries,* eds. D. Q. Lamb & J. Patterson Dordrecht, Reidel, 1.
Ruiz-Lapuente, P., Comeron, F., Smartt, S., Kurucz, R. Mendez, J. Canal, R., Filippenko, A., & Chornock, R. 2003. *From Twilight to Highlight: The Physics of Supernovae,* Proceedings of the ESO/MPA/MPE Workshop, Garching Germany, July 2002, p140.
Wang, L. *et al.* 2002. *Astrophys. J.*, **579**, 671–677.
Webbink, R. L. 1984. *Astrophys. J.*, **277**, 355–360.
Wheeler, J. C., Lecar, M., & McKee, C. F. 1975. *Astrophys. J.*, **200**, 145–157.
Wheeler, J. C., Yi, I., Höflich, P., & Wang, L. 2000. *Astrophys. J. Suppl.*, **537**, 810–823.
Wheeler, J. C., Meier, D., & Wilson, J. R. 2002. *Astrophys. J.*, **568**, 807–819.
Winkler, P. F., Gupta, G., & Long, K. 2003. *Astrophys. J.*, **585**, 324–335.

8

The evolution of supernovae in the winds of massive stars

V. Dwarkadas

ASCI FLASH Center, Dept. of Astronomy and Astrophysics, University of Chicago, 5640 S Ellis Ave, Chicago IL 60637

Abstract

We study the evolution of supernova remnants in the circumstellar medium formed by mass loss from the progenitor star. The properties of this interaction are investigated, and the specific case of a 35 $M_{\odot}$ star is studied in detail. The evolution of the SN shock wave in this case may have a bearing on other SNRs evolving in wind-blown bubbles, especially SN 1987A.

8.1 Introduction

Type II Supernovae are the remnants of massive stars (M $>$ 8 $M_{\odot}$). As these stars evolve along the main sequence, they lose a considerable amount of mass, mainly in the form of stellar winds. The properties of this mass loss may vary considerably among different evolutionary stages. The net result of the expelled mass is the formation of circumstellar wind-blown cavities, or bubbles, around the star, bordered by a dense shell. When the star ends its life as a supernova, the resulting shock wave will interact with this circumstellar bubble rather than with the interstellar medium. The evolution of the shock wave, and that of the resulting supernova remnant (SNR), will be different from that in a constant density ambient medium.

In this work we study the evolution of supernova remnants in circumstellar wind-blown bubbles. The evolution depends primarily on a single parameter, the ratio of the mass of the shell to that of the ejected material. Various values of this parameter are explored. We then focus on a specific simulation of the medium around a 35 $M_{\odot}$ star, and show how pressure variations within the bubble can cause the shock wave to be corrugated. Different parts of the shock wave collide with the dense shell at different times. Such a situation is reminiscent of the evolution of the shock wave around SN 1987A.

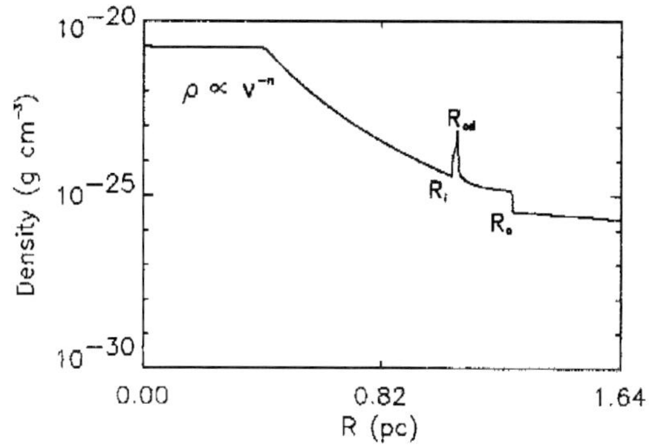

Fig. 8.1. The density profile of the ejected SN material in the initial stages, from a numerical calculation. Interaction of the ejecta with the ambient medium gives rise to an outer shock (R_o), inner shock (R_i) and contact discontinuity (R_{cd}).

8.2 The SN profile

The interaction of the SN shock wave with the surrounding medium in the early stages depends on the density profile of the SN and the surrounding medium. The density structure of the ejecta depends on the structure of the star and the shock acceleration in the outer layers (Chevalier & Fransson 1994). Although observational information on valid density profiles is scarce, numerical simulations, especially for SN 1987A, as well as semi-analytic calculations show that the ejecta density profile in the outer layers can be approximated by a power-law in radius (Fig. 8.1; see Chevalier & Fransson 2003 for further information). This approximation is used in the current work.

8.3 Structure of the Circumstellar Medium (CSM)

The general structure of a wind-blown nebula was first elucidated by Weaver *et al.* (1977). In the simplest, two-wind approximation, a fast wind from a star collides with slower material emitted during a previous epoch, driving a shock into the ambient medium. The pressure of the post-shock material causes the freely flowing fast wind to decelerate, driving a second shock that travels inwards towards the center. A double-shocked structure, separated by a contact discontinuity, is formed. Figure 8.2 shows the density and pressure profiles from a simulation of a wind-blown bubble. Proceeding in the direction of increasing radius from the central star we find the following regions delineated: freely flowing fast wind, inner or wind-termination shock (R_t), shocked fast wind, contact discontinuity (R_{cd}), shocked ambient medium, outer shock (R_o) and unshocked ambient medium.

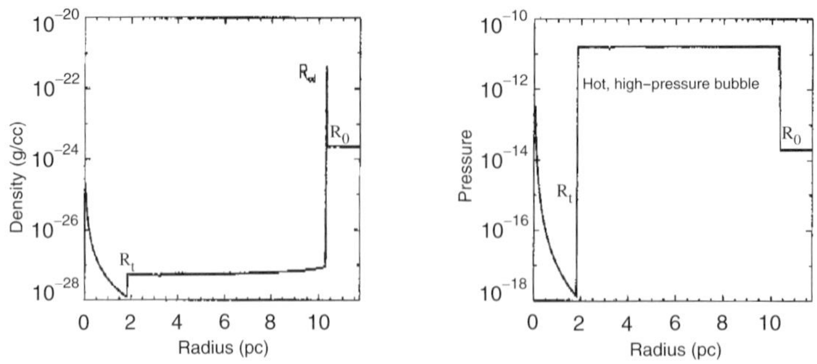

Fig. 8.2. a) Density and b) Pressure profiles from a numerical simulation of a wind-blown bubble around a massive star.

8.4 SNR-Circumstellar Medium interaction

The interaction of the SN ejecta with the freely expanding wind gives rise to a double-shocked structure, consisting of a forward shock driven into the wind and a reverse shock moving into the ejecta. Given the low density interior of the wind-blown bubble, the luminosity of the remnant is lower than if the explosion were to occur within the ISM. It is clear that in general most of the bubble mass is contained within the dense circumstellar (CS) shell. Thus the interaction of the ejecta with this shell is crucial to determining the evolution of the remnant. This interaction depends on a single parameter $\Lambda = \frac{M_{shell}}{M_{ejecta}}$, the ratio of the shell mass to the ejecta mass.

An exploration of the interaction of SN shock waves with CS bubbles described by the Weaver *et al.* (1977) model shows (eg. Dwarkadas 2002), that for small values of the parameter $\Lambda \leq 1$, the structure of the density profile is important (Figure 8.3a–d). Just after the shock-shell interaction has taken place, the density *decreases* outwards from the reflected shock to the contact discontinuity. However as the evolution proceeds, the supernova remnant begins to "forget" the existence of the shell, and loses memory of the interaction. The density structure changes to reflect this, and begins to *increase* from the reflected shock to the contact discontinuity. In this case it takes about 15 doubling times of the radius for the remnant to forget the interaction with the shell (Fig 8.3d). In another few doubling times, the remnant density profile will resemble that of a SNR evolving directly in the ambient medium. When computing the X-ray or optical emission from the remnant, which are functions of the density of the shocked material, it is imperative that this changing density structure be taken into account.

As the value of the parameter Λ increases, i.e. the mass of the wind-blown shell increases compared to the ejecta mass, the energy transmitted by the remnant to the shell also increases. Energy transfer to the shell becomes dynamically important,

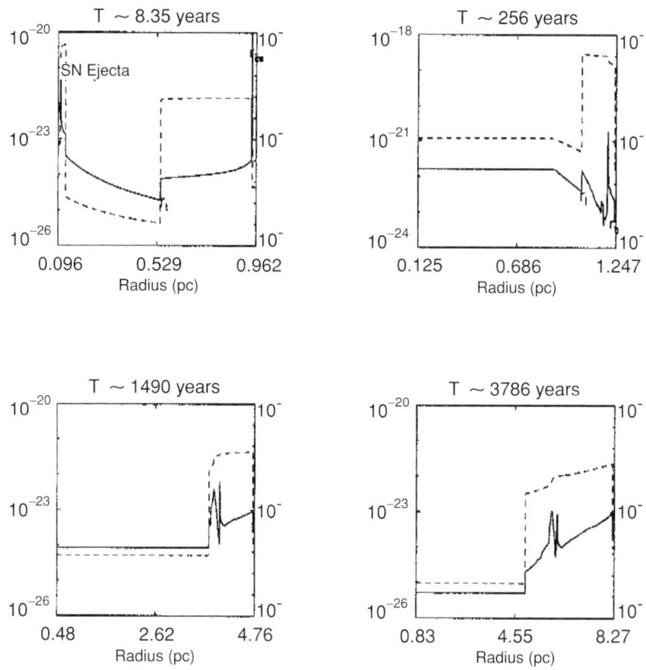

Fig. 8.3. Snapshots in time from a simulation of SN ejecta interacting with a CS bubble. The mass in the circumstellar shell is 14% of that in the ejecta. The solid lines display density, with the scale given on the LHS. The dashed lines displays the pressure, with the scale given on the RHS. All units are CGS. The time is given at the top of each figure in years. The labels (a) to (d) in the text go in order of increasing time, from top to bottom and left to right.

and the remnant evolution is speeded up. The reflected shock moves rapidly through the ejecta, and complete thermalization of the ejecta is achieved in a shorter time as compared to the SN reverse shock thermalizing the ejecta. If the value of Λ is large, the SN shock may become radiative, and the kinetic energy is converted to thermal energy. In extreme cases, the remnant may then go directly from the free-expansion stage to the radiative stage, by-passing the classical adiabatic or "Sedov" stage.

8.5 A 35 $M_{\odot}$ star

Using mass-loss data kindly provided to us by Norbert Langer, we have modeled the evolution of the medium around a 35 $M_{\odot}$. star, and the further interaction of the shock wave with this medium once the star explodes as a SN. The star goes through the sequence O-Star, Red-Supergiant Star (RSG) and Wolf-Rayet (WR) star. Below we describe, mainly through images of the fluid density, the subsequent evolution of the CSM around the star.

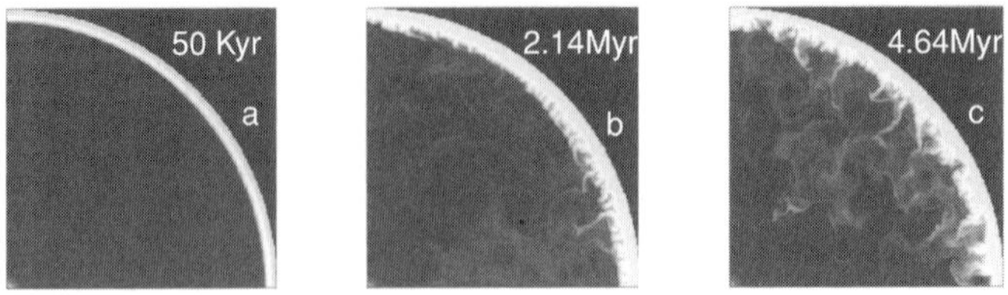

Fig. 8.4. Time-sequence of images of the formation of the CSM around a 35 $M_{\odot}$ O-Star during the main sequence.

MAIN SEQUENCE (MS) STAGE

The wind from the star, with velocity of a few (3–4) thousand km/s and mass loss rate on the order of 10^{-7} $M_{\odot}$/yr expands into a medium with density of about 1 particle/cc, giving rise to a bubble about 74 pc in radius. Fig 8.4 shows (from left to right) a time-sequence of density images of the formation of the MS bubble. Note that the main-sequence shell is unstable to a Vishniac-type thin-shell instability. The density inhomogeneities lead to pressure fluctuations which propagate within the interior, which soon develops into a turbulent state. The evolution of these perturbations distinguishes our results from those of Garcia-Segura *et al.* (1996), who considered the MS shell to be stable and therefore assumed spherical symmetry. However the 2D structure is quite different, and has significant implications for the succeeding evolution of the bubble.

RED-SUPERGIANT (RSG) STAGE

In the RSG stage the wind velocity falls to a low value of about 75 km/s, whereas the mass loss rate jumps up to a few times 10^{-5} $M_{\odot}$/yr. A new pressure equilibrium is established, and a RSG shell is formed in the interior, which is also unstable to thin-shell instabilities. Fig 8.5 (frames 1 and 2) shows images of the density during the RSG evolution.

WOLF-RAYET (WR) PHASE

The wind velocity in the WR phase climbs back up to almost 3000 km/s, whereas the mass loss rate drops by only a factor of a few from the RSG stage. The momentum of the WR wind is then about an order of magnitude larger than that of the RSG wind, and the wind pushes the RSG shell outwards, simultaneously causing it to fragment (Fig 8.5, frame 3). The RSG wind material is mixed in with the rest of the MS material (Fig 8.5, frame 4), a key result since the RSG wind velocity was so low that the material by itself could not have gone very far. Out of $\sim 26 M_{\odot}$ of material lost in the wind, about 19 $M_{\odot}$ is lost in the RSG stage, so much of the material within the nebula is composed of matter lost in the RSG phase.

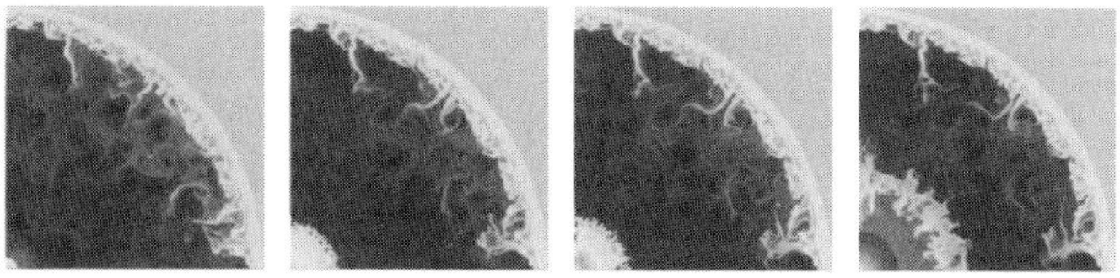

Fig. 8.5. The first two density images from left show the formation of the inner RSG shell, which is unstable to thin-shell perturbations. The next two display the onset of the WR wind and its collision with the RSG shell, causing it to fragment and the RSG material to be mixed in with the rest of the nebula.

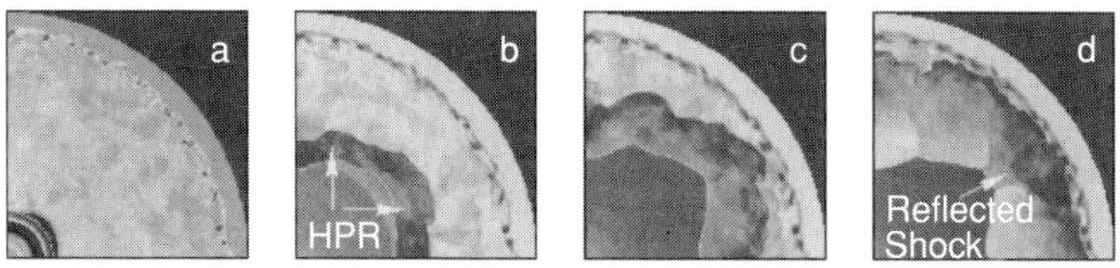

Fig. 8.6. Time-sequence of pressure images of the interaction of the SNR shock with the WR bubble. HPR represents the high-pressure region between the inner and outer SNR shocks. Note the rippled structure of the outer shock from frame 2 onwards, and its interaction at different times with various parts of the shell.

SN- CSM INTERACTION

At the end of the WR phase the stellar mass is about 9.1 $M_\odot$. We assume that about 1.4 $M_\odot$ remains as a neutron star, and the remaining mass is ejected in a supernova explosion, with a density profile that goes as a power-law in the outer parts, with a power-law index of 7. The interaction soon forms the usual double-shocked structure. In Figure 8.6 we show images of the fluid pressure. This variable is chosen to clearly illustrate the shocked region between the inner and outer shocks. The shock starts off as a spherical shock (Fig 8.6a), but the pressure within the turbulent interior soon causes it to become rippled (Fig 8.6b). The corrugated shock structure collides with the boundary of the bubble in a piecemeal fashion (Fig 8.6c), and as each small part collides with the outer boundary, a reflected shock arises in that region. There exist many pieces of reflected shock that arise from various interactions, have different velocities, and consequently reach the inner boundary at different times. The thermalization of the material then occurs in different stages, and X-ray images will reveal a very complicated structure which will differ considerably on scales of tens to hundreds of years.

HST images of SN 1987A have revealed the presence of various bright spots around the circumstellar ring, presumably due to the interaction of the SN shock front with the equatorial ring structure (eg. Sugerman *et al.* 2002). The collision of a highly wrinkled shock with various parts of the circumstellar shell, leading to the different parts brightening up at different times, is very similar to the current situation of the shock front in SN 1987A. The case of SN 1987A however is more complicated in that the region interior to the ring is presumed to be an ionized HII region (Chevalier & Dwarkadas 1995). It is possible though that an aspherical HII region would serve only to accentuate the asphericity in the shock front. The simulation described herein is for a 35 $M_{\odot}$ star, whereas in 87A the progenitor star was less massive, and possibly part of a binary system. Nevertheless the similarities are striking, and suggest the existence of such wrinkled shock fronts when SNe evolve in wind-blown bubbles.

Acknowledgements

This research is supported by Award # AST-0319261 from the National Science Foundation. Vikram Dwarkadas is also supported by the US. Department of Energy grant # B341495 to the ASCI Flash Center (U Chicago). I would like to thanks the organizers for inviting me to a most stimulating conference, and to wish Craig Wheeler a very happy 60th birthday.

References

Chevalier, R. A., & Fransson, C., 2003, *in Supernovae and Gamma-Ray Bursters, Lecture Notes in Physics 598*, ed. K. Weiler, (Springer-Verlag).
Chevalier, R. A., & Fransson, C., 1994, *ApJ*, **420**, 268.
Chevalier, R. A., & Dwarkadas, V. V., 1995, *ApJ*, **452**, L45.
Dwarkadas, V. V., 2002, in *Interacting Winds from Massive Stars, ASP Conference Proceedings, Vol. 260.* Edited by A. F. J. Moffat and N. St-Louis, (San Francisco: ASP), 141
Garcia-Segura, G., MacLow, M.-M., & Langer, N., 1996, *A&A*, **316**, 133.
Sugerman, B., *et al.*, 2002, *ApJ*, **572**, 209
Weaver, R., McCray, R., Castor, J., *et al.*, 1977, *ApJ*, **218**, 377.

9

Types for the galactic supernovae

B. E. Schaefer

Louisiana State University, Baton Rouge Louisiana 70803 USA

Abstract

The six galactic supernovae within the last millennium are critical to all work on the relationships between supernovae and their remnants. Yet this field has been dogged by controversy and discarded arguments. Even during the Wheeler Symposium, we had successive speakers give different type assignments to individual events. In an effort to at least define the confusion, I have polled a group of leading experts as to their current thinking on the types for each of the historical events. This complements a similar poll made a decade ago. We must realize that these results are not voting-on-the-truth, but is rather an expression of community opinion. The recent poll has the following results. SN1006 is universally agreed to be a Type Ia event. SN1054 (the Crab) is puzzling in many ways, but it must be from some sort of a core collapse event. SN1181 is thought to be a core collapse event primarily on the basis of its remnant being a plerion like the Crab. SN1572 (Tycho's) is agreed to be a Type Ia event. SN1604 (Kepler's) has no consensus, with *all* types being claimed and denied. Cas A is unanimously agreed to not be a Type Ia event, but after that all possibilities find their champions.

9.1 The polls

The six galactic supernovae within the last millennium (SN1006, SN1054, SN1181, SN1572, SN1604, and Cas A) all have very well observed remnants. A key question for understanding these remnants is the type of the original explosion. The 'type question' also feeds into the dynamics of the interstellar medium and the frequency of 'non-standard' events. Unfortunately, there has been substantial controversy and confusion on the question of the explosion type for each of the six events.

The basic trouble is that we do not have spectra of the supernovae near peak, and type assignments require such a spectrum. In its place, we necessarily must use measured quantities that are correlated (hopefully strongly) with the standard types (Ia, Ib, Ic, II, IIL, IIP, IIb, and IIn). No one quantity is conclusive, so generally multiple indicators are used although often with reliance on one primary correlation. Embarrassingly, even until the late 1980's, a frequent primary indicator in the literature was that 'the light curve is that of a Type Ia event'. (This ignores the fact that all supernovae, except Type IIP, have essentially the same light curves when viewed with the accuracy of the historical data, and this logic was even applied to SN1006 which does not have a light curve.) Further troubles began when the historical supernovae were used to calibrate the distance scales and the type assignments became involved in the Hubble constant debates. Currently, indicators include the remnants' iron mass, composition, distance from the galactic plane, and residual evidence for a dense circumstellar shell.

At the Wheeler Symposium, I heard various speakers giving contradictory claims on types. So starting at the banquet, I asked various experts for their best bets on the historical types. These experts were Craig Wheeler, Peter Hoflich, Chris Gerardy, Brian Schmidt, Jim Truran, Stirling Colgate, Rob Fesen, Bob Kirshner, Roger Chevalier, David Branch, Adam Burrows, and myself; while Virginia Trimble and Sidney van den Bergh were questioned by email. They were simply asked to assign a type to each of the six events, or at least to express their best knowledge.

The good news is that the community is virtually unanimous on the classification of two out of the six surveyed SNe. The bad news is that no consensus emerged for any of the other four events, other than that three of them are likely to be some version of a core collapse event. Here are my summaries of our community opinion:

SN1006 in Lupus. Everyone agrees confidently that this is a Type Ia event.

SN1054 (Crab). Everyone agrees that this is a core collapse of some type. But after that, people used words like 'weird', 'puzzling', and 'peculiar'. Few cared to name a particular class within the core-collapse possibility. Many pointed out that they thought that the Crab was not *any* of the usual SN types.

SN1181. Few knew enough about this event to hazard a guess. Those that did all pointed to its plerion to say that it must be like the Crab and hence a core collapse event.

SN1572 (Tycho's). Virtually all respondents claimed that this was a Type Ia event.

SN1604 (Kepler's). The responses were all over the map on this one. Here are all of my responses; not IIP, Ia, (Ib,Ic,IIb), Ia?, ?, don't know,?, ?, core collapse, not Ia, maybe Ia, core collapse, no response, and?. All types are claimed and denied. Thus I firmly conclude that there is widespread confusion within the community as to the type of Kepler's supernova.

Cas A. There was unanimous agreement that Cas A was not a Type Ia event, but after that there is no agreement. People claimed that it was IIb, IIn, Ib, Ic, or just simply 'II'.

This survey in 2003 is actually the second such survey that I have made. The first was in 1993, when I interviewed Stephen Reynolds and Sydney van den Bergh as well as used references Raymond (1984), Weiler & Sramek (1988), Hill (1993), Bandiera (1987), and van den Bergh (1988; 1994). The results are summarized in the Table. We see that the community opinion has not changed in the past decade.

9.2 Implications

I view the results of this community survey to be challenging. That is, only 2-out-of-6 times can we match a particular remnant with a particular type of explosion, and for three of the remaining four we are left with only saying that it is not Type Ia. Indeed, there is full confusion for one event (SN1604) while two are generally thought to *not* correspond to any known SN type. Our community has tried long and hard to connect up the observed remnants with a particular type of explosion, but this important program has apparently not gotten far.

SUPERNOVA	1993 POLL RESULTS	2003 POLL RESULTS
SN1006	Ia	Ia
SN1054 (Crab)	Core collapse (Peculiar)	Core collapse (not a 'standard' type)
SN1181	Core collapse	Core collapse (?)
SN1572 (Tycho's)	Ia	Ia
SN1604	Total confusion	Total confusion
Cas A	Core collapse	Core collapse

At a higher level, I wonder about what this census says about the fraction of SNe that are 'normal'. (Here, I'll take 'normal' to be any event that would be routinely classified as Ia [the Branch normals], Ib, Ic, IIP, IIL, IIn, or IIb.) The Crab and SN1181 do not fit into this SN classification scheme. Cas A was apparently a *very* subluminous event as it was not seen (apparently, even by Flamsteed, as we heard from Rob Fesen's talk) despite being circumpolar at a time when many eyes were charting the skies. SN1006 peaking at $-4.1 < V_{max} < -6.1$ (Schaefer 1996a), so it has to be greatly fainter than a Branch normal Type Ia event. Even SN1572 and SN1604 have Δm_{15} values of 1.73 ± 0.1 and 1.88 ± 0.3 mags (Schaefer 1996a), making them more like SN1991bg than Branch normals. It seems that *all* six out of six galactic SNe do not fit into the usual classification schemes. Perhaps this might not be surprising since the criteria are based only on peak spectra and not on light curve or remnant properties. Nevertheless, it appears that a volume limited sample of events is giving a greatly different mix of types than is present in the large scale magnitude limited surveys (c.f. Schaefer 1996b). Maybe we should ponder whether the classification scheme needs changes for application to volume limited samples, and maybe we should be seeking all the 'abnormal' events that must be popping off with high frequency in a volume limited sample.

References

Bandiera, R. 1987. *Astrophys. J.*, **319**, 885.
Hill, L. C. 1993. *QJRAS*, **34**, 73.
Raymond, J. C. 1984. *ARA&A*, **22**, 75.
Schaefer, B. E. 1996a. *Astrophys. J.*, **459**, 438.
Schaefer, B. E. 1996b. *Astrophys. J.*, **464**, 404.
van den Bergh, S. 1988. *Astrophys. J.*, **327**, 156.
van den Bergh, S. 1994. *Astrophys. J.*, **425**, 208.
Weiler, K. W. & Sramek, R. A. 1988. *ARA&A*, **26**, 247.

Part III

Theory of Thermonuclear Supernovae

10

Semi-Steady burning evolutionary sequences for CAL 83 and CAL 87: Super Soft X-ray binaries are supernova la Progenitors

S. Starrfield and S. Dwyer

Department of Physics and Astronomy, ASU, Tempe, AZ 85287–1504

F. X. Timmes

Center for Astrophysical Thermonuclear Flashes, University of Chicago, Chicago, IL 60637

W. R. Hix

Physics Division, ORNL, Oak Ridge, TN 37831

E. M. Sion

Department of Astronomy, Villanova University, Villanova, PA

W. M. Sparks

X-4, LANL, Los Alamos, NM, 87545

10.1 Introduction

The assumption commonly made is that Supernovae of Type Ia (SN Ia) are the result of thermonuclear runaways (TNR) in the cores of carbon-oxygen white dwarfs (WD) which are members of binary systems and have accreted material from a companion until their masses exceed the Chandrasekhar Limit (Liebundgut 2000, 2001). However, the binary star systems that end in this explosion are not yet known although there have been numerous proposals. Nevertheless, the importance of SN Ia, both to our understanding of the evolution of the Universe and to the formation of iron in the Galaxy, demands that we determine the progenitors of these explosions.

Originally proposed by Whelan and Iben (1973), virtually every type of close binary which contains a WD has been suggested at one time or another. However, based purely on observational concerns, most of the systems that have been proposed cannot be the progenitors (Starrfield 2003). For example, one of the first suggestions was a classical nova system (CN), but the amount of core material ejected during the outburst implies strongly that the WD is losing mass as a result of the outburst (Gehrz *et al.* 1998). Other suggestions such as Symbiotic Novae (T CrB or RS Oph, for example) can probably be ruled out.because there is too much hydrogen present in the system (the explosion takes place inside the outer layers of a red giant) and the defining characteristic of a SN Ia outburst is the absence of hydrogen in the spectrum (Filippenko 1997). (We note, however, that SN 2002ic, a SN Ia, did show a *shell* of hydrogen in its spectrum [Hamuy *et al.* 2003]).

One recent development has been the discovery of the Super Soft Sources (SSS: c.f., Branch *et al.* 1995; Kahabka & van den Heuvel 1997 [KVDH97]). While two of the SSS were discovered by EINSTEIN (CAL 83 and CAL 87) they were not recognized as a class until the ROSAT survey of the LMC. Since the SSS designation is heterogeneous, we refer here only to those members that are close binaries containing a WD. They are luminous, $L_* \sim 10^{37-38}$ erg s^{-1}, with surface temperatures ranging from 30 to 50 ev and, at low resolution [ROSAT-PSPC], the spectrum resembles a broad "emission" line with a peak around 0.4 keV. After their discovery as a class, it was proposed that nuclear burning was occurring on the surface of the WD at the same rate at which material was being accreted, the mass of the WD was increasing, and it could reach the Chandrasekhar Limit ("Steady Burning": van den Heuvel *et al.* 1992 [VDH92]; KVDH97). Thus, they were predicted to be one class of SN Ia progenitor (VDH92). Nevertheless, no self-consistent, stellar evolution calculations were done for massive WDs accreting at the high mass accretion rates required to test the Steady Burning hypothesis. Such calculations were done for lower mass WDs by Iben (1982) and Sion and Starrfield (1994) but the luminosities and effective temperature were too low to agree with the observations of CAL 83 and CAL 87.

It is the purpose of this paper to report on our new calculations of accretion onto massive WDs in which we have found conditions under which the WD accretes material and does not suffer TNRs that eject material in a nova outburst. An exciting implication of our study is that we strengthen the connection between a CN and a SN Ia explosion and we, therefore, propose that a CN and a SSS may be two different phases of the evolution of the same binary system. In the next section we briefly describe the characteristics of the two best studied SSS, CAL 83 and CAL 87. We follow that with a discussion of our evolutionary sequences and end with a Summary.

10.2 Two Super Soft X-ray source binaries: CAL 83 and CAL 87

Although these two systems were discovered by EINSTEIN in the early 1980's (Long *et al.* 1981), they were not established as binaries until optically identified by Cowley *et al.* (1990) and Smale *et al.* (1988). The ROSAT all sky survey (and later pointed observations) found additional SSS and established them as a class of X-ray emitting systems (Trümper *et al.* 1991). A recent catalog can be found in Greiner (2000) and a discussion of their binary properties can be found in Cowley *et al.* (1998). The two systems that we model have orbital periods of 1.04 d (CAL 83) and 10.6 h (CAL 87) which are longer than those of the typical CN (though consistent with a few outliers such as GK Per). A period this long implies that the secondary must be evolved to fill its Roche Lobe which is necessary for the large mass transfer rates assumed for Steady Burning (VDH92).

Although both systems are in the LMC where the extinction is low, a large range in their observed luminosities and effective temperatures can be found in the literature. Part of the difficulty is that for the systems studied by ROSAT, only a small region of the emitted spectrum is visible in the detectors and large corrections have to be applied to the observations (KVDH97). In addition, it was assumed until recently that these systems could be modeled by black-body fits which, however, give the wrong luminosity and effective temperature (Krautter *et al.* 1996). We also note that the XMM observations of CAL 83 by Paerels *et al.* (2001) show convincingly that the spectrum is not that of a black-body.

Given these difficulties, therefore, the luminosities and effective temperatures of these two sources are (1) CAL 83: $L = (0.7–2) \times 10^{37}$ erg s^{-1} (Gänskicke *et al.* 1998) and $T_{eff} = (4.5–7) \times 10^5$K (Greiner 2000). The analyses of XMM Reflection Grating Spectrometers spectra of CAL 83 imply a temperature $\sim 5 \times 10^5$K (Paerels *et al.* 2001) and, thereby, a radius of $\sim 6 \times 10^3$km. (2) CAL 87: $L = (3–5) \times 10^{36}$erg s^{-1} and $T_{eff} = (6–9) \times 10^5$K (Parmar *et al.* 1997).

While CAL 87 is apparently less luminous (although hotter) than CAL 83, it is also eclipsing with a decline in light in the optical of 2 mag which must hide some of the light from the central WD (Cowley *et al.* 1990). In addition, the large rates of mass transfer proposed for these systems imply that a reasonable amount of optical light is coming from their accretion disk. Finally, a number of SSS systems show spectral features indicating that material is being lost in jets (Cowley *et al.* 1998). Therefore, the mass accreted onto the surface of the WD is not the amount flowing through the accretion disk and measured by the optical emission. This complicates the comparison of our theoretical results with the observations.

10.3 Evolutionary models for CAL 83 and CAL 87

Before discussing the stellar evolution code and presenting our calculations, we must distinguish our results from the assumption of "Steady Burning." Steady burning (mass accretion at a few $\times 10^{-7} M_{\odot} yr^{-1}$ onto a *low* luminosity WD) was defined by Paczynski and Zytkow (1978), Iben (1982), and Fujimoto (1982a,b) and used by VDH92 to explain the surface properties of SSS. These authors *assume* both quasi-static evolution and, in addition, that the material burns at "exactly" the mass accretion rate. We show in our self-consistent calculations that the material can burn at rates that are lower than the assumed steady burning rate and still build up a layer of thermonuclear ashes below the surface without experiencing a violent TNR because the WD is sufficiently hot that hydrogen burns as soon as it is accreted. We refer to our results as Semi-Steady Burning (SSB) to prevent any confusion with the previous results.

We use the one-dimensional, fully implicit, Lagrangian, hydrodynamic computer code described in Starrfield *et al.* (2000) with one major change and a few minor changes. The nuclear reaction network used in that paper is very similar to that of Weiss and Truran (1990), only the nuclear reaction rates have been updated. We found, however, in our studies for this paper, in the deeper layers where hydrogen was absent and fusion was occurring through the 3α reaction sequence, that this network did not conserve nuclei. We then switched to the pp+cno+rp network of Timmes* which conserves nuclei under the same physical conditions. Another change from previous work is to use the equation of state of Timmes (Timmes and Arnett 1999; Timmes and Swesty 2000) which can also be obtained from his web pages. We also include the accretion luminosity as described in Shaviv and Starrfield (1988)

An extremely important difference between this work and our previous work is that we do not choose an initial model in equilibrium at low luminosities (Starrfield *et al.* 2000). For these studies, we either choose evolved WDs which have undergone at least three outbursts, or an initial model which has been evolved to the same surface luminosity.

We first evolve a CN through an outburst. Once *all* the ejected material is expanding faster than the escape speed, has reached radii exceeding 10^{13}cm, and is optically thin; we remove it from the calculations. We then take the remnant WD and either wait until it has cooled to minimum and restart accretion or begin accretion when the WD is still hot. We do this for three cycles and begin accretion when the WD luminosity is $\sim$30 $L_\odot$. We did this for 1.0 $M_\odot$, 1.25 $M_\odot$, and 1.35 $M_\odot$ WDs. We find regions of SSB at all three WD masses but, because of space considerations, only present the results for 1.35 $M_\odot$. We assume that the accreting composition is Solar although we realize that the metallicity of the LMC is about one-third that of the Solar neighborhood. We plan to study the effects of metallicity on the evolution in future papers since it can clearly affect the evolution of SN Ia (Timmes *et al.* 2003).

A summary of our results is given in Table 10.1. The rows are the luminosity, effective temperature and radius of the initial model which has a homogeneous composition of 50% carbon and 50% oxygen. These are followed by the mass accretion rate (both in gm s^{-1} and $M_\odot$ yr^{-1}), the length of time in years that we followed the evolution, the total amount of mass accreted ($M_\odot$), the SSB temperature and rate of energy generation in the sequence (maximum temperature and ϵ_{nuc} always occur at the surface mass zone), and the luminosity and effective temperature (both in K and ev) of the sequence at the end of the SSB evolution. The latter values did not

* http://flash.uchicago.edu/fxt/code-pages/net-pphotcno.shtml

change after the early phases of the evolution. The final rows give the composition at the interface between the accreted matter and the core matter (CI). Note that as the mass accretion rate increases, the abundance of ^{12}C decreases and that of ^{16}O increases. In no cases (except for sequence 5 which expanded to large radii after 30 years of evolution) was there any hydrogen left in the deep layers. Sequence 1 shows helium in the deepest layers but it experienced a helium shell flash after 2.4×10^4 yr of evolution. About 25% of the accreted material reached radii exceeding 10^{11}cm and would probably have been ejected.

10.4 Summary

In this paper we have investigated the effects of accretion at high rates onto massive WDs. We report only the results for 1.35 $M_\odot$ both because of space limitations and because the surface conditions of these sequences fit those observed for CAL 83 and CAL 87 better than the surface conditions of the 1.25 $M_\odot$ evolutionary sequences. Earlier work has investigated accretion onto WDs of lower mass (Iben 1982; Sion and Starrfield 1994). We find SSB at much broader values of the mass accretion rate than commonly accepted (Kahabka 2002) which makes it much more likely that systems such as CAL 87 and CAL 83 are the progenitors of SN Ia. In addition we find:

- If CAL 83 is a 1.35 $M_\odot$ WD, then it is accreting at a rate between 2×10^{-8} and 8×10^{-8} $M_\odot$ yr^{-1}. These values are less than the accepted Steady Burning mass accretion rate. We are investigating the evolutionary properties of lower mass WDs to see if they could better fit CAL 83.
- If CAL 87 is a 1.35 $M_\odot$ WD, then it is accreting at a rate between 8×10^{-8} and 2×10^{-7} $M_\odot$ yr^{-1}. Here we assume that we are observing only a fraction of the energy emitted by the WD since the system is eclipsing.
- In none of the cases that we have studied at a mass of 1.35 $M_\odot$ do we find that the layers accreted onto the WD end with a pure helium composition as is assumed. The final composition, as given in Table 10.1 ranges from 60% C, 20%N, and 20%O (by mass) to 31%C, 7%N, and 36%O. Sequences 1 and 4 produce significant ^{20}Ne. However, this is the most massive nucleus in this part of the network and, more than likely, if we were to extend the network, then the composition in the sequences would reach higher mass nuclei and produce more energy. However, the surface conditions of Sequence 4 exceed those observed for CAL 83 and CAL 87 so we did not extend the network for this initial survey.
- Our initial models were computed by accreting onto WDs that had just experienced a CN explosion but had not yet cooled to low luminosities. If this evolution is realized in Nature, then we predict a link between CN and SSS which results in SN Ia explosions.
- Sequences 2, 3, and 4 have accreted sufficient material and we feel confident that if we were to continue the evolution for the necessary time, then they would reach and exceed the Chandrasekhar Limit. *Thus, they satisfy the conditions necessary to be considered as strongly viable candidates for the progenitors of SN Ia explosions*

Table 10.1. *Results of the 1.35$M_\odot$ Hot White Dwarf Evolutionary Sequences*

Sequence	1[a]	2	3	4	5[b]
L(init)(erg s^{-1})	1.2×10^{35}	1.2×10^{35}	1.2×10^{35}	1.2×10^{35}	1.2×10^{35}
T_{eff}K (init)	2.3×10^{5}	2.3×10^{5}	2.3×10^{5}	2.3×10^{5}	2.3×10^{5}
R (km)	2391	2391	2391	2391	2391
M (gm s^{-1})	1.0×10^{18}	5.0×10^{18}	1.0×10^{19}	5.0×10^{19}	1.0×10^{20}
M ($M_\odot$ yr^{-1})	1.6×10^{-8}	8.0×10^{-8}	1.6×10^{-7}	8.0×10^{-7}	1.6×10^{-6}
T_{evol} (yr)	2.4×10^{4}	3.3×10^{4}	2.9×10^{4}	5.7×10^{4}	30
$\delta M_{acc}(M_\odot)$	3.9×10^{-4}	2.6×10^{-3}	1.4×10^{-3}	1.4×10^{-2}	
T_{SSB} (10^6K)	94	173	210	296	
ϵ_{nuc}(*SSB*: 10^8erg $gm^{-1}s^{-1}$)	1.4	7.3	15.0	76.0	
L_{SSB} (erg s^{-1})	5.1×10^{36}	2.6×10^{37}	5.9×10^{37}	3.0×10^{38}	
T_{eff} (SSB:K)	4.6×10^{5}	$8. \times 10^{5}$	1.0×10^{6}	1.5×10^{6}	
T_{eff} (SSB:ev)	40	69	86	129	
^{1}H(CI)[c]	0.0	0.0	0.0	0.0	
^{4}He(CI)[c]	0.27	0.0	0.0	0.0	
^{12}C(CI)[c]	0.44	0.55	0.44	0.17	
^{13}C (CI)[c]	<0.01	0.06	0.12	0.14	
^{14}N (CI)[c]	0.01	0.19	0.16	0.07	
^{16}O (CI)[c]	0.10	0.19	0.26	0.36	
^{20}Ne (CI)[c]	0.17	<0.01	<0.01	0.17	

[a] Sequence 1 experienced a helium flash at 2.4×10^4 s but ejected less than 25% of the accreted material.
[b] This sequence begins expanding to large radii after 30 years of evolution.
[c] CI = Composition Interface: All abundances are mass fractions.

Acknowledgements

We are grateful to J. Truran for valuable discussions which influenced us to continue these calculations. S. Starrfield acknowledges partial support from NSF and NASA grants to ASU. FXT is supported by the Department of Energy under Grant No. B341495 to the Center for Astrophysical Thermonuclear Flashes at the University of Chicago.

References

Branch, D., Livio, M., Yungelson, L. R., Boffi, F. R., Baron, E. 1995, PASP, 107, 1019
Cowley, A., Schmidtke, P., Crampton, D., Hutchings, J. 1990, ApJ, 350, 288
Cowley, A., Schmidtke, P., Crampton, D., Hutchings, J. 1998, ApJ, 504, 854
Filippenko, A. V. 1997, ARAA, 35, 309
Fujimoto, M. Y. 1982a, ApJ, 257, 752
Fujimoto, M. Y. 1982b, ApJ, 257, 767
Gehrz, R. D., Truran, J. W., Williams, R. E., & Starrfield, S. 1998, PASP, 110, 3.

Gänsicke, B. T., van Teeseling, A., Beuermann, K, & de Martine, D. 1998, A&A, 333, 163
Greiner, J. 2000, New Astronomy, 5, 137
Hamuy, M., *et al.* 2003, Nature, 424, 651
Iben, I. 1982, ApJ, 259, 244
Kahabka, P. 2002, in Compact Stellar X-ray Sources, ed. W. Lewin & M. van der Klis, Cambridge University Press (astro-ph:0212037)
Kahabka, P., van den Heuvel, E. P. J. 1997, ARAA, 35, 69
Krautter, J., *et al.* 1996, ApJ, 456, 788
Leibundgut, B. 2000, A&A Reviews, 10, 179
Leibundgut, B. 2001, ARAA, 39, 67
Long, K., S., Helfand, D. J., Grabelsky, D. A. 1981, 248, 925
Paczynski, B., & Zytkow, A. N. 1978, ApJ, 222, 604
Paerels, F., *et al.* 2001, A&A, 365, L308
Parmer, A. N. *et al.* 1997, A&A, 323, L33
Shaviv, G., & Starrfield, S. 1988, ApJ, 335, 383
Sion, E. M., & Starrfield, S. 1994, ApJ, 421, 261
Smale, A. P. Corbet, R. H. D., Charles, P. 1988, MNRAS, 233, 51
Starrfield, S. 2003, in From Twilight to Highlight: The Physics of Supernovae, ed. W. Hillebrandt & B. Leibundgut, Springer, Heidelberg, p. 128
Starrfield, S., Sparks, W. M., Truran, J. W., Wiescher, M. C. 2000, ApJS, 127, 485
Timmes, F. X., & Arnett, D. A. 1999, ApJS, 125, 277
Timmes, F. X., Brown, E. F., Truran, J. W. 2003, ApJ, 590, L83
Timmes, F. X., & Swesty, D. 2000, ApJS, 126, 501
Trümper, J., Hasinger, G., Aschenbach, B., Bräuninger, H., Briel, E. G. *et al.* 1991, NATURE, 349, 579
van den Heuvel, E. P. J., Bhattacharya, D., Nomoto, K., Rappaport, S. A. 1992, A&A, 262, 97
Weiss, A., & Truran, J. W. 1990, A&A, 238, 178
Whelan, J., Iben, I. 1973, ApJ, 186, 1007

11

Type Ia progenitors: effects of spin-up of white dwarfs

S. C. Yoon and Norbert Langer

Astronomical Institute, Utrecht University,
Princetonplein 5, NL-3584 CC, Utrecht, The Netherlands

Abstract

The effects of rotation in progenitor models for Type Ia supernovae are addressed. After discussing processes of angular momentum transport in carbon+oxygen white dwarfs, we investigate pre-explosion conditions of accreting white dwarfs. It is shown that differential rotation will persist throughout the mass accretion phase, with a shear strength near the threshold value for the dynamical shear instability. It is also found that rotational effects stabilise the helium shell source and reduce the carbon abundance in the accreted envelope.

11.1 Introduction

Unlike core collapse supernovae, Type Ia supernovae (SNe Ia) occur exclusively in binary systems (e.g. Livio 2000). Although it is still unclear which kinds of binary systems lead to SNe Ia, non-degenerate stars such as main sequence stars, red giants or helium stars are often assumed as the white dwarf companion (e.g. Hachisu *et al.* 1999, Langer *et al.* 2000, Han & Podsiadlowski 2003, Yoon & Langer 2003). This leads us to consider the spin-up of the white dwarf, since the transfered matter from those companions should form a Keplerian disk that carries a large amount of angular momentum. The observation that white dwarfs in cataclysmic variables rotate much faster than isolated ones (Sion 1999) provides evidence that accreting white dwarfs are indeed spun up. A rapidly rotating progenitor may also explain the asphericity implied by the polarizations observed in SNe Ia explosions (Wang, this volume). Here we discuss implications of the spin-up of accreting white dwarfs for the progenitors of SNe Ia.

11.2 On the transport of angular momentum in a white dwarf

In a white dwarf, angular momentum can be transported by the Eddington Sweet circulation and by turbulent diffusion induced by hydrodynamic instabilities such as the shear instability.

11.2.1 Eddington-Sweet circulation

The time scale for the Eddington circulation is roughly given by $t_{\mathrm{ES}} \simeq t_{\mathrm{KH}}/\chi^2$ (Maeder & Meynet 2000), where t_{KH} is the Kelvin-Helmoltz time scale and χ is the angular velocity normalized to the Keplerian value, i.e., $\chi = \omega/\omega_{\mathrm{Kep}}$. In an accreting white dwarf with accretion rates of $\dot{M} > 10^{-7}\ \mathrm{M}_{\odot}/\mathrm{yr}$, this time scale is much shorter in the non-degenerate envelope than the accretion time scale ($t_{\mathrm{ES}} \ll t_{\mathrm{acc}}$), where the thermal time scale is rather short due to accretion heating and nuclear burning, and χ is close to 1. In the degenerate core, however, we have $t_{\mathrm{ES}} > t_{\mathrm{acc}}$ since the Kelvin-Helmoltz time scale is typically larger than 10^7 yr. Consequently, the role of the Eddington-Sweet circulation in the angular momentum redistribution is restricted to the non-degenerate outer envelope.

11.2.2 Dynamical and secular shear instabilities

The condition for the dynamical shear instability (DSI) is given by $N^2/\sigma^2 < R_{\mathrm{i,c}} \approx 1/4$ (e.g. Meader & Meynet 2000), where N^2 is the Brunt-Väisälä frequency and σ is the shear factor: $\sigma := \partial\omega/\partial \ln r$. Fig. 11.1 shows the critical value of σ for the dynamical shear instability (i.e, $|\sigma_{\mathrm{dyn,crit}}| = \sqrt{N^2/R_{\mathrm{i,c}}}$) as function of density, with the physical conditions as indicated in the figure caption. The stability criterion can be relaxed if thermal diffusion reduces the buoyancy force such that $\mathcal{P}_{\mathrm{r}} R_{\mathrm{e,c}} N^2/\sigma^2 < R_{\mathrm{i,c}}$, where $\mathcal{P}_{\mathrm{r}}$ and $R_{\mathrm{e,c}}$ are the Prandtl number and the critical Reynolds number, respectively (Heger *et al.* 2002). This process is often called 'secular shear instability (SSI)'. The threshold value of the shear factor for this instability (i.e. $|\sigma_{\mathrm{sec,crit}}| := \sqrt{\mathcal{P}_{\mathrm{r}} R_{\mathrm{e,c}} N^2/R_{\mathrm{i,c}}}$) is also plotted in Fig. 11.1. As shown in the Figure, the effect of relaxation of the buoyancy force due to thermal diffusion holds no more for $\rho \gtrsim 5 \times 10^6\ \mathrm{g/cm^3}$, because the thermal diffusion time scale becomes larger than the turbulent viscous time scale at such high densities. The diffusion coefficient for the SSI is $D_{\mathrm{ssi}} = \frac{1}{3} K \sigma^2 R_{\mathrm{i,c}}/N^2$ where $K[= (4acT^3)/(3C_{\mathrm{P}}\kappa\rho^2)]$ is the thermal diffusivity (Zahn 1992). The diffusion time scale of the SSI when $\sigma = \sigma_{\mathrm{dyn,crit}}$ thus becomes simply the thermal diffusion time scale: τ_{ssi} (at $\sigma = \sigma_{\mathrm{dyn,crit}}$) $\approx 3R^2/K$. In the highly degenerate core of a CO white dwarf, we have $\tau_{\mathrm{ssi}} \approx 9.5 \times 10^6\ (R/10^8\ \mathrm{cm})^2/(K/100\ \mathrm{cm^2 s^{-1}})$ yr.

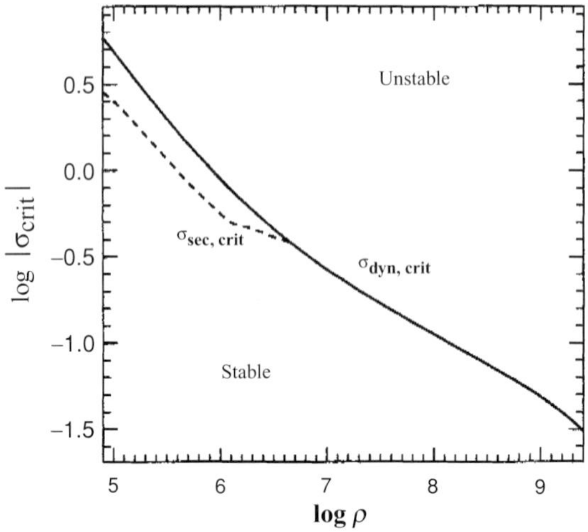

Fig. 11.1. Threshold values of the shear factor $\sigma = (\partial \omega)/(\partial \ln r)$ for the dynamical (solid line) and the secular (dashed line) instability. Constant gravity and temperature, i.e., $g = 10^9$ cm/s^2, $T = 5 \times 10^7$ K, are assumed in the calculation. The chemical composition is also assumed to be constant, with $X_{\rm C} = 0.43$ and $X_{\rm o} = 0.54$. For the critical Richardson number, $R_{\rm i,c} = 1/4$ is employed. $R_{\rm e,C} = 2500$ is used for the critical Reynolds number.

Implications of this consideration are straightforward. Any shear motion with $\sigma > \sigma_{\rm dyn,crit}$ will decay such that σ approaches $\sigma_{\rm dyn,crit}$ in a dynamical time scale. Further angular momentum transport will operate on the thermal diffusion time scale until σ reaches $\sigma_{\rm sec,crit}$ for $\rho \lesssim 5 \times 10^6$ g/cm^3. For $\sigma < \sigma_{\rm sec,crit}$ and $\sigma < \sigma_{\rm dyn,crit}$, angular momentum will be transported only via the electron viscosity and the Eddington Sweet circulation, unless other kinds of instabilities are invoked. Therefore, if we consider a fast accretion as required by the single degenerate SN Ia progenitor scenario ($\dot{M} > 10^{-7}$ M$_\odot$/yr), the degree of shear may not be far from the thresh-old point for the dynamical shear instability throughout the degenerate white dwarf interior. This conjecture is confirmed by the numerical results presented in Sect. 1.3.

11.2.3 GSF instability and magnetic instabilities

The Goldreich, Schubert and Fricke instability (GSF instability) can be induced if a star is in a baroclinic condition (Goldreich & Schubert 1967, Fricke 1968). In an accreting white dwarf, this instability may be important in the non-degenerate envelope, but is likely suppressed in the degenerate core where the baropic condition will be retained through a dynamical meridional circulation (Kippenhahn &

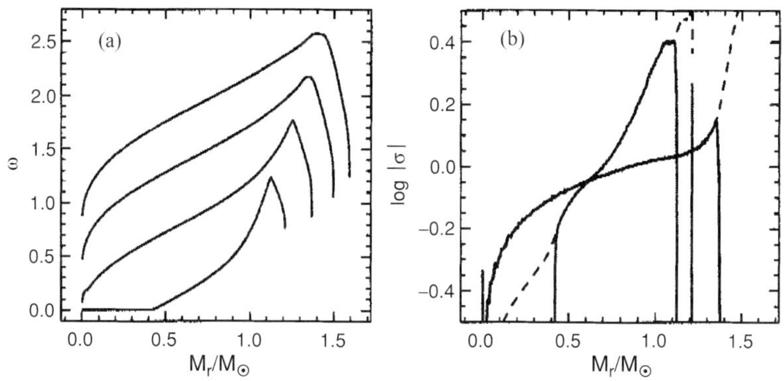

Fig. 11.2. (a) Angular velocity in accreting white dwarf models at 4 different masses (1.21, 1.37, 1.50 and 1.59 $M_\odot$). The initial mass is 1.0 $M_\odot$ and the accretion rate is 10^{-6} $M_\odot$/yr with carbon and oxygen rich matter. (b) The shear factor ($\sigma = \partial\omega/\partial \ln r$, solid line) in the accreting white dwarf models when $M_{\rm WD} = 1.21$ and 1.59 $M_\odot$. The dashed lines denote the threshold value for the dynamical shear instability ($\sigma_{\rm dyn,crit}$).

Möllenhoff 1974). Magnetic instabilities such as the Taylor instability (Spruit 2002) may be potentially important and their role will be investigated in the near future.

11.3 Differential rotation and critical mass for the explosion

In an accreting white dwarf, rotation can change the white dwarf structure significantly. Fig. 11.2 shows the angular velocity and the corresponding shear factor in our accreting white dwarf models, at four different masses: 1.21, 1.37, 1.50 and 1.59 $M_\odot$. The accretion of CO rich matter started at 1.0 $M_\odot$ with a constant accretion rate of $\dot{M} = 10^{-6}$ $M_\odot$/yr. As expected from the previous discussions, the white dwarf does not rotate rigidly as often assumed by other authors (e.g. Piersanti *et al.* 2003, Uenishi *et al.* 2003), but differential rotation persists at the threshold value of the dynamical shear instability. As a result, the central density remains well below the carbon ignition density even when the white dwarf mass is 1.59 $M_\odot$ ($\rho_c = 2.4 \times 10^8$ g/cm^3). This might imply that a super-Chandrashekhar mass is required for the explosion. However, a differentially rotating white dwarf becomes unstable if the ratio of the rotational energy to the gravitational potential energy (T/W) becomes significant (e.g. Ostriker & Tassoul 1969). I.e., if T/W $\gtrsim 0.14$, the white dwarf becomes secularly unstable and loses angular momentum via gravitational radiation. Our white dwarf model reaches this point when $M_{\rm WD} \simeq 1.58$ $M_\odot$. The critical value of T/W is, however, uncertain, and Yoshida & Eriguchi (1995) suggest smaller values. It is under investigation how this critical mass varies according to

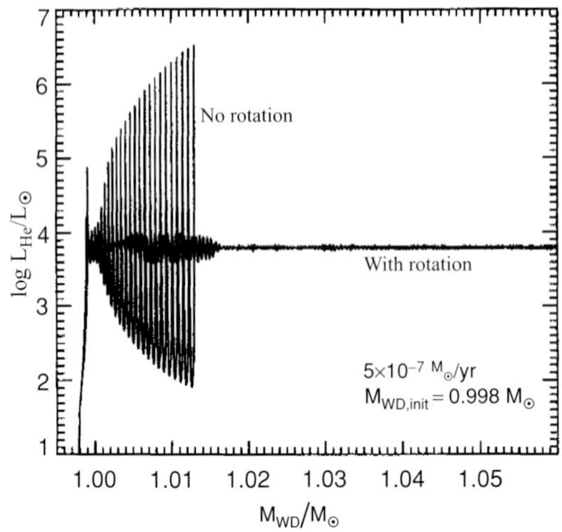

Fig. 11.3. Luminosity due to the helium shell burning as function of the total mass of white dwarf. The initial mass is 0.998 $M_\odot$ and the accretion rate of helium rich matter is set to 5×10^{-7} $M_\odot$/yr.

different initial masses and accretion rates. The physics of explosion with rotation is also of great interest (e.g. Woosley *et al.* 2003).

11.4 Stabilisation of the helium shell source

Rotation results in another interesting consequence, affecting the behavior of the helium shell source. Fig. 11.3 shows the luminosity due to helium shell burning in a helium accreting white dwarf with an accretion rate of 5×10^{-7} $M_\odot$/yr. In the non-rotating model, the helium shell burning is not stable but undergoes thermal pulses, while it is stablised in the rotating model. The thermal instability is mainly induced by the thinness of the shell source (Schwarzshild & Härm 1965) but also determined by the degree of degeneracy and the temperature of the shell source. I.e., a shell source is more susceptible to the instability if it is thinner, more degenerate and colder (Yoon *et al.* 2003a). In the rotating model, as the accreted helium is mixed into the CO core due to the rotationally induced chemical mixing, the shell source becomes wider compared to the non-rotating case, which is the main reason for the stabilisation of the shell source. The $^{12}C(\alpha,\gamma)^{16}O$ reaction becomes more active as a result, rendering a lower carbon abundance in the accreted layer. Given that the realization of the single degenerate Chandrasekhar mass scenario has been questioned due to the unstable helium shell burning and the consequent mass loss (e.g. Cassisi *et al.* 1999, Hachisu & Kato 2001), the fact that rotation favors the stable shell source seems to give a plausible solution to this question. Readers are referred to Yoon *et al.*(2003b) for more detailed discussions.

References

Cassisi, S., Iben, I., Tornambé, A., 1998, *Astrophys. J*, **496**, 376
Fricke, K., 1968 *Zeitschrift für Astrophysik*, **150**, 571
Goldreich, P., Schubert, G., 1967 *Astrophys. J*, **150,** 571
Hachisu, I., kato, M., Nomoto, K., 1999, *Astrophys. J*, **522**, 487
Han, Z., Podsiadlowski, Ph., 2003 *Mon. Not. R. Astron. Soc.*, submitted.
Heger, A., Langer, N., Woosley, S. E., 2002, *Astrophys. J*, **528**, 368
Kippenhahn, R., Möllenhoff, C., 1974, *Astrophys. Space Science,* **528**, 368
Langer, N., Deutschmann, A., Wellstein, S., Höflich P., 2000 *Astron. Astrophys.*, **362**, 1046
Livio, M., 2000, In: *Type Ia Supernovae: Theory and Cosmology*, ed. by Niemeyer, J. C. and Truran, J. W., Cambridge Univ. Press
Maeder, A., Meynet, G., 2000, *Annu. Rev. Astron. Astrophys.*, **38**, 143
Ostriker, J. P., Tassoul, J. L., 1969, ApJ *Astrophys. J*, **155**, 987
Piersanti, L., Gagliardi, S., Iben, I., Tornambé, A., 2003, *Astrophys. J*, **583**, 885
Sion, E. M., 1999 *Pub. Astron. Soc. Pac.,* **111**, 532
Spruit, H. C., 2002, *Astron. Astrophys.*, **381**, 923
Schwarzshild, M., Härm, R., 1965, *Astrophys. J.*, **142**, 855
Uenishi, T., Nomoto, K., Hachisu, I., 2003, *Astrophys. J.*, **595**, 1094
Woosley, S. E., Wunch, S., Kuhlen, M., 2003, *Astrophys. J.*, submitted
Yoon, S.-C., Langer, N., 2003, *Astron. Astrophys.*, submitted.
Yoon, S.-C., Langer, N., van der Sluys, M., 2003a., *Astron. Astrophys.,* to be submitted.
Yoon, S.-C., Langer, N., Scheithauer, S., 2003b, *Astron. Astrophys.,* to be submitted.
Yoshida, S., Eriguchi, Y., 1995, *Astrophys. J.*, **438**, 830
Zahn, J.-P. *Astron. Astrophys.*, **265**, 115

12

Terrestrial combustion: feedback to the stars

E. S. Oran

Laboratory for Computational Physics and Fluid Dynamics
U.S. Naval Research Laboratory
Washington, DC 20375 USA

Abstract

This paper describes how we have used numerical simulations and laboratory combustion experiments to learn about Type Ia thermonuclear supernova explosions. We discuss detonations, deflagrations, and the transition from deflagrations to detonations, and how these relate to exploding white dwarf stars.

12.1 Introduction

This paper is for Craig Wheeler (a*k*a Professor J. Craig Wheeler, Captain, *ISS Bunbry,* often stationed in the Virgo Cluster), who has been a good friend and fellow traveler for many years. Craig is wonderfully enthusiastic, persistently curious, and always asking those painfully "simple" questions for which we have no answers. He has motivated and driven research programs that have brought combustion science to astrophysics.

A cursory study of the multivolume *Proceedings of the Combustion Institute* shows that combustion can now be loosely defined as the result of *fluid dynamics combined with exothermic reactions, and everything this implies.* The definition has expanded with the understanding of the controlling phenomena and the range of applications. In the early 1900's, there was *combustion* and *detonation*, and the concepts seemed separated. Combustion was defined as oxidation with energy release, with an emphasis on specific chemical reactions. Detonation studies emphasized the fluid dynamics with shocks and explosions. Now these fields have merged and expanded. We now consider exothermic reactions, including the physics, chemistry, structure and dynamics of flames and detonations, including the production products such as pollutants, soot, diamonds, fullerenes, microparticles, and nanoparticles.

The purpose of this paper is to introduce some aspects of combustion and the combustion community to astrophysicists. These two communities have many overlapping interests in physics, chemistry, and even the techniques used to investigate the problems. This paper summarizes some of the results and the philosophy of our joint work on Type Ia supernovae (SNIa), as we try to answer the question: How does an SNIa explode? Other overlapping areas of burning interest, such as those related to astrobiology and the formation and stability of large, carbon-based particles, are alluded to at the end.

12.2 Reaction waves in homogeneous media

There are three basic types of reaction waves that occur in a homogeneous, exothermic material: laminar flames, turbulent flames, and detonations. *Laminar flames* propagate at velocities very much less than the Mach number of the exothermic material. They are controlled by energy release, expansion, and diffusive transport processes such as thermal conduction, molecular diffusion, and radiation transport. *Turbulent flames* or *deflagrations* travel at subsonic but high speeds. Fluid-chemical instabilities, obstacles, and boundaries can cause flame wrinkling, increase the surface area and therefore increase the energy-release rate. *Detonations* are very powerful supersonic reaction waves, traveling at speeds substantially over the speed of sound. The most important physical processes are compressibility leading to acoustic waves and shocks, energy release, and complex shock structures. All of these types of reaction waves can occur in the thermonuclear combustion in a white dwarf (WD) before it explodes.

The possibility of detonations occurring in SNIa strongly motivated our recent work in astrophysical combustion. The basic physics of detonation propagation is straightforward. A shock moves through an energetic materials, compresses it and raises its temperature. This accelerates the chemical reactions behind the shock, substantially shortening the time for exothermic reactions. The exothermic reactions actually begin at some induction distance behind the shock. These reactions generate pressure waves that move in both directions. Those that move downstream interact with the leading shock and accelerate it. Thus the high speed of a detonation is maintained by a feedback process between the leading shock and the pressure waves that increase its velocity. If the exothermicity of the material varies, which could happen if the background material changes, the induction distance changes to reflect this change in composition and material properties. If the exothermicity decreases so that the induction zone becomes very large, the detonation structure decouples into a shock followed by a flame. If the exothermicity increases, the size of the induction zone decreases and the detonation velocity increases

Now there are limitations and problems with this somewhat idealized picture. It does not take into account multidimensional effects, the effects of boundaries and other heat or mass losses, obstacles internal to the flow, condensed or multiple

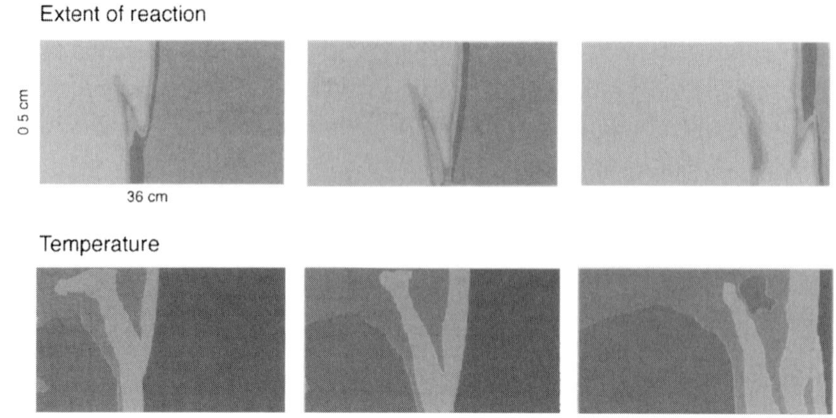

Fig. 12.1. First computations of a detonation in a hydrogen-oxygen gas forming unreacted pockets behind the leading front.

phases, or turbulence. Such a model cannot describe the complex shock structure at the front, or the transient properties of detonations, such as ignition, diffraction, and quenching.

Figure 12.1 is taken from one of our first attempts to compute the multidimensional structure of a detonation propagating in an energetic gaseous mixture of hydrogen and oxygen (Oran *et al.* 1982). (This is also the figure that started our interactions with Craig.) The simulation is of a detonation propagating in a relatively narrow (0.5 cm), long (36 cm) channel containing an argon-diluted, low-pressure mixture of hydrogen and oxygen. The top row of figures show the extent of reaction. The detonation propagates to the right into the unreacted gas mixture (aqua) and fully reacted gas (yellow) is on the lefthand side. The colors in between represent the progress through the reaction from fuel to product. In the third frame, we see a detached pocket of partially reacted gas trapped behind the detonation front. The bottom row of figures on the bottom show the temperature, where dark blue is the coldest, red the hottest, and pink is intermediate. All together, these show that there is a relatively cold, incompletely reacted pocket of gas cut off by interacting shock waves.

The appearance of these unburned pockets in the calculations was completely unexpected. It was also a potentially important phenomenon that had been seen in experiments, but not reported until they were seen in numerical simulations (D. H. Edwards, private communication). From the calculations, we learned that the formation of these pockets can be traced to the curvature of the transverse shock waves. When two transverse waves collide, or one hits a reflecting wall, the interaction can cut off a portion of unreacted, cold material. If the material in the pockets burns slowly enough, the process effectively draws energy out of the detonation. This could provide a mechanism for detonation extinction. If the pockets burn rapidly, they could generate new pressure pulses that perturb the system and affect the structure of the leading shock front.

These findings are curious because they show how an initially homogeneous material can develop an extremely inhomogeneous structure as shocks move through it. These findings are important because the existence of unreacted pockets and their properties can determine whether a detonation lives or dies. The existence of unburned pockets provides a multidimensional fluid-dynamic mechanism by which a detonation can die in a situation where, in one dimension, it might be expected to propagate.

In 1987, when Craig first saw Fig.12.1 (Oran & Kailasanath 1987), he asked whether this same type of effect could occur in thermonuclear detonations propagating in SNIae. Boisseau *et al.* (1996) then showed that steadily propagating detonations on the carbon (smallest) scale in thermonuclear carbon-oxygen matter do produce unreacted pockets. Then Gamezo *et al.* (1999) showed that pockets are produced at all scales, even at the largest silicon scale.

12.3 Deflagration of a Type Ia supernova

There are several possibilities for SNIae: (1) a detonation; (2) a turbulent flame or deflagration; (3) a turbulent flame that undergoes a transition to a detonation (DDT); (4) some other combination of a turbulent flame and a detonation. Whatever the answer, the model must be consistent with the spectra and observed variations of these spectra with time. The pure detonation model (1) has been ruled out, even if it does produce pockets, because it does not produce the a final composition of elements needed to produce observed spectra. As described now, recent results from purely deflagration models do not predict enough energy release (Gamezo *et al.* 2003), which rules out (2).

A detailed description of the numerical model we are using is given by Gamezo *et al.* (2003) and by other articles in this volume (Gamezo, Khokhlov, & Oran 2003; Khokhlov 2003). We assume that the WD has suddenly accreted enough mass so that the temperature and density at its core become high enough for the center of the star to ignite spontaneously. The result is a thermonuclear flame. Because of the very large gravitational field, the flame becomes unstable and turbulent, in large part due to Rayleigh-Taylor instability. As the flame propagates outward, the star expands and material races outward at very high velocities. The computational model solves the unsteady, compressible fluid dynamics equations for the flow and reactions in an octant of the star, assuming symmetry at the octant boundaries, and with an equation of state for a degenerate electron gas, and uses a flame-tracking algorithm to follow the flame front.

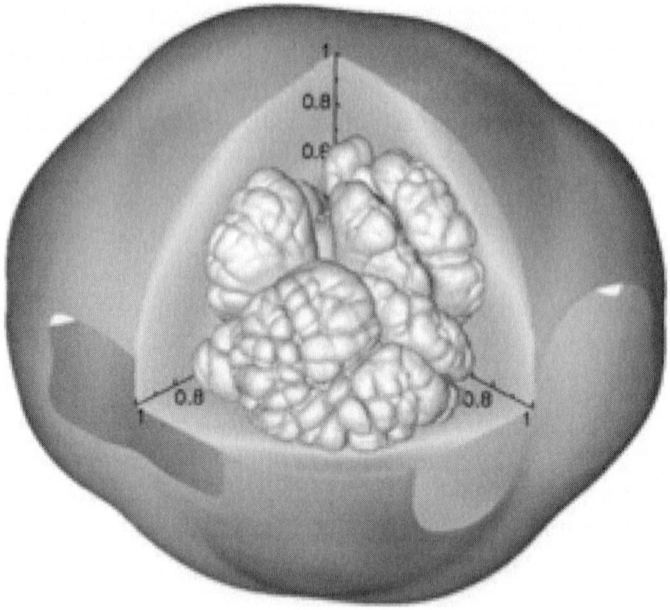

Fig. 12.2. Late time ($\simeq$2 s) in the computation of the deflagration stage of an SNIa.

Figure 12.2 is taken from a time near the end of the simulation of the development of the deflagration stage of a WD explosion (Gamezo *et al.* 2003). The initial condition was a spherical laminar flame at $\simeq 10^6$ cm from the center. As the flame propagates outward, it becomes more and more turbulent. Gravitational forces pull cold, high-density unburned material down into the funnels between the lobes, towards the center of the star. When the flame lobes become large enough, secondary instabilities develop on their surfaces. When these secondary structures become large enough, more instabilities develop on their surfaces, and so forth. As the flame develops, the energy released causes the star to expand.

After many similar calculations, in which the numerical resolution and initial composition and flame were varied, we conclude that none of these thermonuclear flame computations show enough energy release or produce the composition of elements that could account for observations. Something is missing. What could solve the problem is the appearance of a detonation during the period of flame expansion.

12.4 Transition from a deflagration to a detonation

The next step is to determine if and how a detonation could arise in this system. But the question of how a detonation develops from a deflagration is not new, nor is it one that only applies to the problem of exploding WDs. It is a question that has been considered for many years, both because it is a very fundamental question and because of its practical importance. It is an important question for explosion hazards in mines, ammunition stores, and transportion of chemicals. It arises for propulsion devices, such as pulsed detonation engines, ram accelerators, ramjets, and scramjets. It affects the operation of shaped charges, igniters, and nuclear devices. The difficulty for experiments is that to observe DDT requires diagnostics of very high-speed, reactive flows, with excellent time resolution. The difficulty for numerical simulation is again the disparity of time scales. It is necessary to resolve the flame front structure and turbulent flame, in a large enough system to see the interaction and watch the evolution.

There has been a considerable amount of work on DDT that has both shed light on the problem and confused the issues, sometimes at the same time. The most important concept was developed by Zeldovich *et al.* (1970) and Lee, Knystautas, & Yoshikawa (1978), who showed that a gradient in induction time (the time to energy release) is an essential part of the process of DDT because it provides the environment in which the reaction wave and shock wave couple to form a detonation. Early, important experimental work was done by Oppenheim and colleagues (see, for example, (Zajac & Oppenheim 1971; Meyer & Oppenheim 1971), who observed that hot spots arise because of fluctuations in a system.

Our work has focused primarily on experiments, as summarized by Thomas, Bambrey, & Brown (2001). Fill a shock tube with an energetic mixture that supports both flames and detonations. Ignite the mixture with a spark or a set of sparks that create a laminar flame. Then release a shock from the far end of the tube. The shock interacts with the flame, distorts it, and through a series of shock reflections, quickly produces a turbulent flame. The intensity of the turbulence and the reactivity of the background energetic mixture is controlled by the strength of the shock: A stronger shock will create more flame with more intense turbulence and, by raising the temperature and pressure more, a faster-reacting background mixture.

As an example, consider Fig.12.3, which shows the initial conditions from a typical set of simulations (Khokhlov, Oran, & Thomas, 1999). A shock of Mach number $M_s = 1.5$ is released into a background mixture of low-pressure acetylene-air. Before the shock was released, and at the other end of the tube, a series of sparks generated six small flames that eventually grew and merged. These simulations were done to simulate specific experiments designed to study DDT. The basic idea is that if we can reproduce DDT, we can use the simulations to study the physical mechanisms that control the process. Then, by analogy, this might shed some light on DDT in SNIa. As shown in the figure, the simulation describes a small part of the shock-flame interaction in the center of the channel.

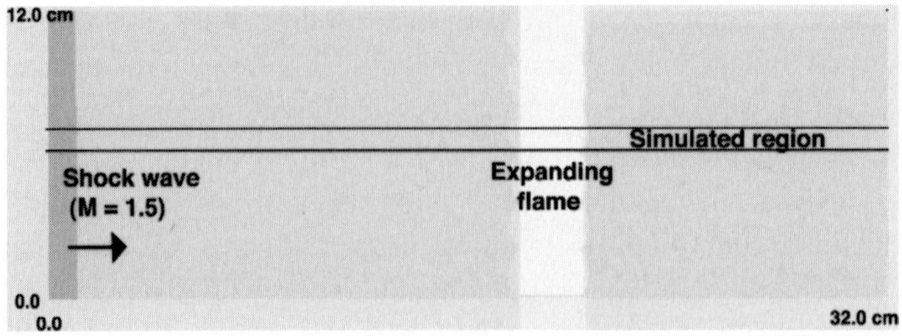

Fig. 12.3. Initial conditions for the simulation of the shock-flame interaction and DDT in an acetylene-air mxiture.

Figure 12.4 is a selection of frames from this simulation. The initial shock-flame interaction results in a funnel of unburned material penetrating a region of burned material (a Richtmyer-Meshkov (RM) interaction), then a shock is transmitted through the flame, and a rarefaction that moves backwards into the shocked material. When the shock exits the second flame surface, there is another RM interaction, this time generating a transmitted and and a reflected shock. Eventually, a series of shocks have passed through the flame, reflected from the back wall, moved back through the flame, and finally all re-emerged from the flame and become one reflected shock. This has left the flame in a highly perturbed state, and it has left the unreacted material in a turbulent, noisy state. Pressure waves are continually generated from energy release in the expanding turbulent flame.

In this case, DDT occurs in an unreacted, shocked, noisy region of the flow. Once it occurs, it spreads quickly and consumes all of the unreacted material, and further shocks the already burned material. A detailed investigation of how this occurs was presented by Khokhlov, Oran, & Thomas (1999) and Khokhlov & Oran (1999).

There we showed that the detonation occurs in a relatively small region containing a gradient of reactivity, consistent with the Zeldovich gradient mechanism of ignition. In fact, the fluctuations in the unreacted material are consistent with the size of the critical gradient in this material.

There are several results from other computations that should be mentioned here. First, as the Mach number of the incident shock increases, the background fluctuations created by the turbulent flame are more intense, hot spots occur closer to the flame surface, and DDT occurs sooner. When the Mach number was 1.4, we did not see DDT through the course of the calculation. When the mach number was 1.6, however, DDT occurs in the narrow funnel of unreacted gas created by the initial shock-flame interaction. The unreacted material in which DDT occurs is highly perturbed, repeatedly shocked, and subject to a series of weaker pressure fluctuations from the surrounding flame.

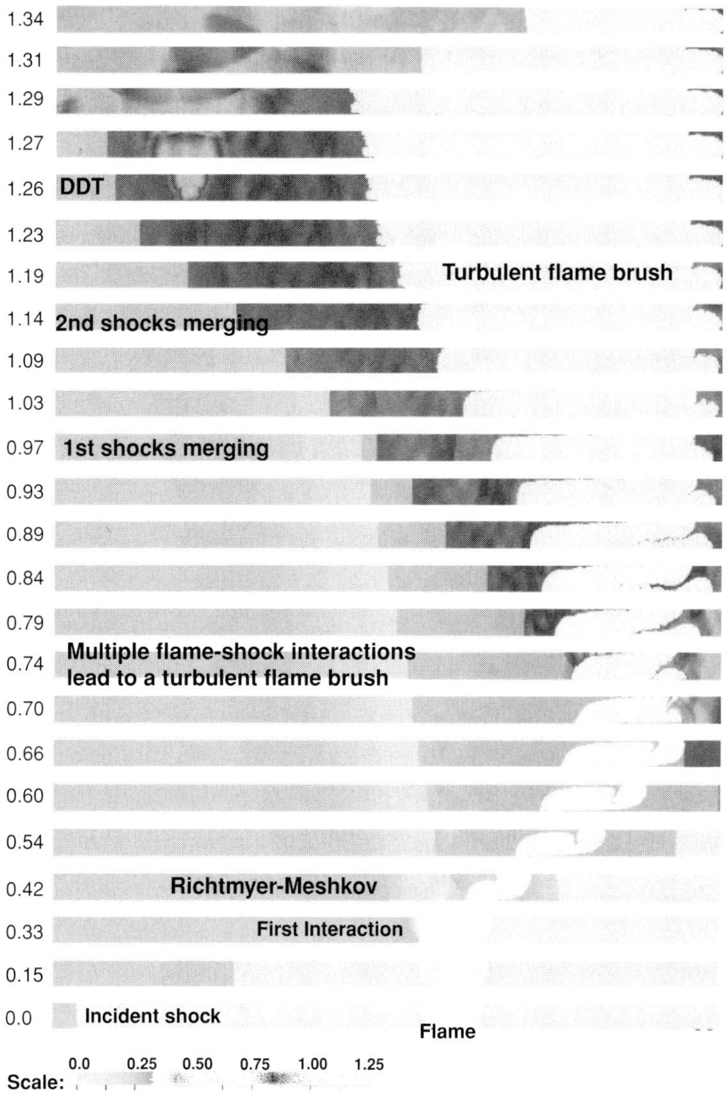

Fig. 12.4. Density at a sequence of timesteps showing the overall flow development for a Mach 1.5 incident shock. The number on the left of each frame is the physical time in microseconds. The computational domain for each frame is 32 cm by 1 cm. Units on the scale are 10^{-3} g/cm^3.

12.5 Where do we stand? (Far away, I hope.)

The question now is whether DDT can occur in the funnels of the SNIa shown in Fig.12.2. This particular computation could not predict DDT even if it did occur, given the level of model input and the numerical resolution. The expanding flame is not computed from a consistent model as we did for laboratory DDT computations.

An analysis of the unreacted gas in the funnels in Fig. 12.2 shows that the temperatures and densities are not high enough to allow detonation ignition. Most important, the SNIa model is not resolving fluctuations and events on the scale of critical ignition sizes for hot spots. And, there is no way that we will be able to do this calculation completely for another twenty years!!

There is a road map we can follow from here, one that involves several steps that should take us part of the way. We can ignite a detonation in the turbulent thermonuclear flame artificially and evaluate the energy released and the resulting materials formed. This will tell us whether the occurrence of a detonation solves the problem. Then, if these studies show that DDT provides a solution, we must proceed with detailed studies of the physics of turbulent thermonuclear flames and the conditions necessary to create hot spots that would detonate. If none of this works, we will have to look again and try to evaluate what is missing.

12.6 Astrobiology and science fiction

There are other areas of overlapping interests, some of which take us into the realms of astrobiology and science fiction. For example, carbon-based nanoparticles, fullerenes, and polycylic aromatic hydrocarbons (PAHs), are large molecules that can be quite stable, especially the large fullerenes (C_n, with $n > 60$). These have been detected spectroscopically in the interstellar medium and in circumstellar objects. They form on earth, under extreme conditions: in flames and detonations, or by shocking, irradiating and subjecting carbon to intense processes such as laser ablation, ion bombardment, and lightning. The are formed in carbon-rich as well as carbon-lean processes. Are these related to, or are they essentials or precursors to life?

Finally, there are science-fiction aspects of combustion right here on earth. Combustion is used by truly alien life forms for defense and propulsion. Consider the African bombardier beetle, an insect that ejects hot spray from the tip of its abdomen, especially when it is attacked (Eisner & Aneshansley 1999). It can spray a truly noxious material in any direction with extreme precision. The source of this hot, high-speed gas is combustion of material stored in pair adjacent sacs, one with fuel and the other with oxidizer. The beetle mixes the material from the two sacs, and this reacts explosively as it is ejected. We ask: Can studies of these very odd insects prepare us for encounters with extraterrestrial life? Maybe Craig knows. . . .

Acknowledgements

This work done in collaboration Alexei M. Khokhlov, Vadim N. Gamezo, J. Craig Wheeler, Geraint Thomas, and Peter Höflich. The work was sponsored in part by the National Aeronautics and Space Agency in the Astrophysical Theory Program, and in part by the Office of Naval Research throught he Naval Research Laboratory.

References

Boisseau, J. R., Wheeler, J. C., Oran, E. S., & Khokhlov, A. M., 1996. *Astrophy. J.*, **471**, L99–L102.

Eisner, T., & Aneshansley, D. J., 1999, *Proc. Natl. Acad. Sci.*, **96**, 9705–9707.

Gamezo, V. N., Wheeler, J. C., Khokhlov, A. M., & Oran, E. S., 1999. *Astrophys.* J., **512**, 827–842.

Gamezo, V. N., Khokhlov, A. M., and Oran, E. S., 2003. – this volume

Gamezo, V. N., Khokhlov, A. M., Oran, E. S., Chtchelkanova, A. Y., & Rosenberg, R. O., 2003. *Science*, **299** (January 3), 77–81.

Khokhlov, A. M., 2003–this volume.

Khokhlov, A. M., & Oran, E. S., 1999, *Combust. Flame*, **119**, 400–416.

Khokhlov, A. M., Oran, E. S., & Thomas, G. O., 1999. *Combust. Flame*, **117**, 323–339.

Lee, J. H. S., Knystautas, R., & Yoshikawa, N., 1978. *Acta Astro.*, **5**, 971–982.

Meyer, J. W, & Oppenheim, A. K., 1971. *Combust. Flame*, **17**, 65–68.

Oran, E. S., & Kailasanath, K, 1987. *Physics News*, AIP, New York.

Oran, E. S., Young, T. R., Boris, J. P., Picone, J. M., & Edwards, D. H., 1982. *Proc. Combust. Inst.*, **19**, 573.

Thomas, G., Bambrey, R., & Brown, C., 2001. *Combust. Theory Mod.*, **5**, 573–594.

Zajac, L. J., & Oppenheim, A. K., 1971. *AIAA J.*, **9**, 545–553.

Zeldovich, Ya. B., Librovich, V. B., Makhviladze, G. M., & Sivashinsky, G. I, 1970. *Astro. Acta*, **15**, 313–321.

13

Non-spherical delayed detonations

E. Livne

Racah Institute of Physics, The Hebrew University, Jerusalem 91904, ISRAEL

Abstract

Delayed detonations in exploding carbon-oxygen (C-O) white dwarfs, are bound to ignite and propagate in an expanding Rayleigh-Taylor (R-T) unstable region. Therefore, non-spherical detonations are expected to evolve due to a possible off-center ignition and due to the inhomogeneous composition ahead of the detonation front. We examine some of the possible consequences of such non-spherical explosions, using two-dimensional axisymmetric simulations.

We find that the explosion products, namely the amount of energy released and the composition of the burnt material, are rather sensitive to the asphericity. This sensitivity follows from the fact that the expansion speed is not negligible with respect to the detonation speed. With lower transition density we get less Fe group elements, smaller explosion energy and higher asphericity in the distribution of elements. We also show that the delayed detonation cannot directly induce a second detonation in a nearby isolated bubbles or channels of cold fuel. Therefore, pockets of unburnt C-O mixture may survive deep inside the ejecta.

13.1 Introduction

The delayed detonation model for Type Ia supernovae assumes that transition from deflagration to detonation occurs during the combustion of a carbon oxygen (C-O) Chandrasekhar mass white dwarf. In order to fit observations, the transition should occur after a significant expansion that reduces the density of the fuel ahead of the front. Traditionally, the transition point is parametrized by a transition density ρ_{tr}, which is the density ahead of the deflagration front at the transition moment. Observations impose a narrow range of a few times 10^7 g/cm^3 for that (parametric) transition density. The following references – Dominguez & Hoflich (2000), Hoflich *et al.* (1995), Hoflich *et al.* (1996) –, represent a large volume of publications which provide the main features of 1D delayed detonation models. Since the

Table 13.1. *C-J values for $Q = 5 \times 10^{17}$ erg/g*

ρ_0	T_0	P_0	ρ_{cj}	P_{cj}	C_{cj}	V_{cj}	D_{cj}
1.E07	7.E08	9.34E23	1.60E07	5.18E24	6.66E08	3.99E08	1.06E09
2.E07	7.E08	2.44E24	3.14E07	1.12E25	6.98E08	3.99E08	1.10E09
3.E07	7.E08	4.26E24	4.67E07	1.77E25	7.20E08	3.99E08	1.12E09
4.E07	7.E08	6.33E24	6.17E07	2.45E25	7.36E08	4.00E08	1.14E09
5.E07	7.E08	8.59E24	7.67E07	3.16E25	7.50E08	4.00E08	1.15E09

Table 13.2. *C-J values for $Q = 7 \times 10^{17}$ erg/g*

ρ_0	T_0	P_0	ρ_{cj}	P_{cj}	C_{cj}	V_{cj}	D_{cj}
1.E07	7.E08	9.34E23	1.63E07	6.54E24	7.40E08	4.65E08	1.20E09
2.E07	7.E08	2.44E24	3.21E07	1.40E25	7.70E08	4.66E08	1.24E09
3.E07	7.E08	4.26E24	4.77E07	2.19E25	7.90E08	4.67E08	1.26E09
4.E07	7.E08	6.33E24	6.32E07	3.01E25	8.06E08	4.68E08	1. 27E09
5.E07	7.E08	8.59E24	7.86E07	3.87E25	8.19E08	4.68E08	1.29E09

deflagration leaves a R-T unstable region, the detonation wave is born and propagates in a disturbed, inhomogeneous region. Therefore, it is most likely that the transition occurs at a point rather than in a spherical shell, as discussed in Livne 1999. Here we examine the consequences of such multidimensional processes in more detail and give additional information about the fate of non-spherical delayed detonations.

The most important physical parameter in the combustion of degenerate fuel is the burning density. Due to low heat capacity, the density determines the burning temperature and thus determines also the final composition of the ashes. There are several other interesting characteristics of detonations in degenerate C-O fuel that result from the very different rates of different reactions. Depending on the density, detonations show multi-scale front structure (Khokhlov 1989), become perpendicularly unstable (Khokhlov 1993) and form cellular patterns (Gamezo *et al.* 1999, Fryxell *et al.* 2000). For the large scale effects considered here, the important character of a detonation in a pre-expanded C-O mixture is its *weakness* compared to terrestrial processes. In Tables 13.1–13.2 we present Chapman-Jouguet (C-J) detonations for several background densities and two typical Q-values. As shown in the tables, the detonation speed in all cases is roughly 50% faster than the speed of sound behind the shock, while the pressure jumps only by a factor of 3–5. Such a detonation is weakly supersonic and when running in an expanding media would be susceptible to expansion effects.

Given an expansion speed u and a detonation speed D, the expansion effect can be estimated as follows. During a time interval dt the density drops according to

$$\frac{\rho(r,t+dt)}{\rho(r,t)} = \left[\frac{r}{r+u \times dt}\right]^3 \tag{13.1}$$

The relevant time interval is $dt = r/D$, where r is some average radius of the (slow moving) deflagration front. This gives

$$\frac{\rho(r,t+dt)}{\rho(r,t)} = [1+u/D]^{-3} \approx 1 - 3u/D \tag{13.2}$$

The expansion speed during the deflagration phase depends strongly on the amount of burnt fuel and weakly on the history of the deflagration speed. For the relevant transition densities u is $1 - 2 \times 10^8$ *cm/s*. Thus, according to eq. [13.2] and the detonation speeds presented in tables [13.1–13.2], the expansion effect is very significant.

13.2 Numerical techniques and method

We carry out 1D and 2D numerical simulations using the appropriate versions of the code VULCAN (Livne 1993). The code incorporates a realistic equation of state for degenerate matter and an *alpha* network of reaction rates. For simplicity we simulate the deflagration phase under spherical symmetry where the deflagration speed v_{def} is taken from Dominguez *et al.* (2001). Those 1D simulations terminate when the density ahead of the front drops below a given transition density ρ_{tr}. We consider three cases with different transition densities –$2 \times 10^7\ g/cm^3$, $3 \times 10^7 g/cm^3$ and $4 \times 10^7 g/cm^3$. For each case we continue the deflagration phase with three kinds of simulations with artificial transition to detonation – a) spherical 1D simulation, b) 2D spherical simulation which serves as an accuracy test, c) 2D simulation with off-center ignition at the intersection of the deflagration front with the symmetry axis.

13.3 Results

Figures 13.1–13.3 show temperature maps of the off-center case with $\rho_{tr} = 2 \times 10^7$ *g/cm*^{3} at three epochs. One can see the almost stalled deflagration front and the overlapping detonation that propagates from right to left. The expansion can be easily identified by the different contours of the outer boundary and the (spherical) deflagration position. When the detonation reaches the left side of the domain the density has already decreased by a considerable fraction. In the final stage of the detonation, the shock converges to the symmetry axis and as a result an axial jet is formed. Exploring those jets is beyond the scope of this study.

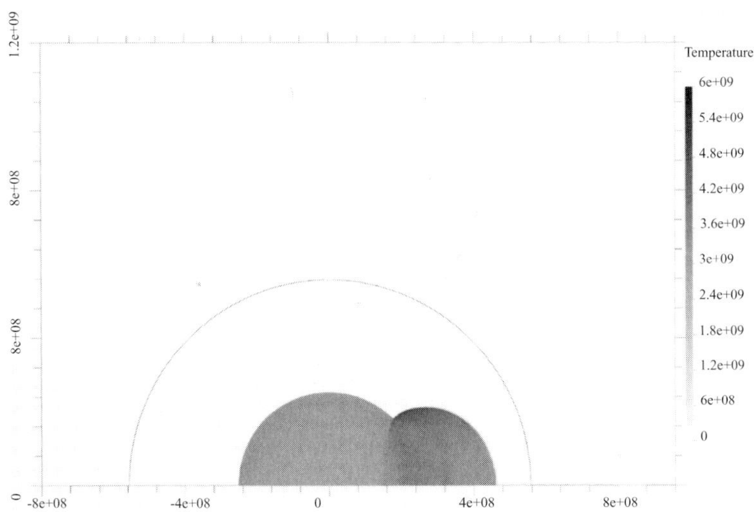

Fig. 13.1. Off Center Detonation, $\rho_{tr} = 2 \times 10^7\ g/cm^3$, t = 0.2 s

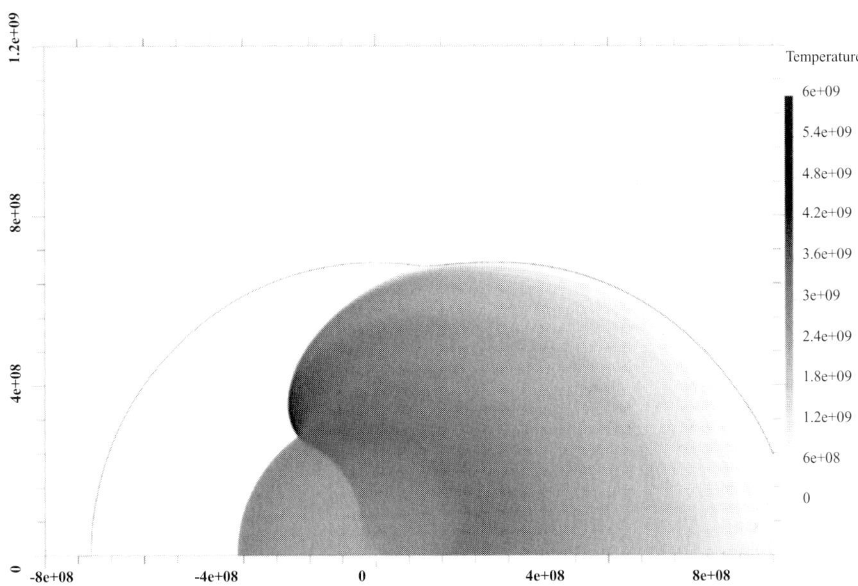

Fig. 13.2. Off Center Detonation, $\rho_{tr} = 2 \times 10^7\ g/cm^3$, t = 0.6 s

In Table 13.3 we summarize the total abundances for the three different transition densities. Note that the spherical (Eulerian) 2D simulations agree well with the (Lagrangian) 1D simulations, at least for the most abundant species, despite the many differences between 1D and 2D simulations. The results of the off-center simulations, however, are significantly different, with lower Fe-group products and

Table 13.3. *Nucleosynthesis Composition (mass in solar units)*

	1D		2D – Spherical		2D Non-spherical	
Element	mass	fraction	mass	fraction	mass	fraction
	Nucleosynthesis for $\rho_{tr} = 2 \times 10^7\ g/cm^3$					
He	1.80E-05	1.29E-05	2.38E-05	1.70E-05	3.68E-06	2.74E-06
C	1.41E-04	1.01E-04	3.95E-03	2.82E-03	1.95E-02	1.46E-02
O	1.06E-01	7.60E-02	1.05E-01	7.53E-02	2.70E-01	2.01E-01
Ne	2.01E-05	1.43E-05	1.84E-03	1.31E-03	1.24E-02	9.21E-03
Mg	4.12E-02	2.94E-02	1.67E-02	1.20E-02	6.98E-02	5.20E-02
Si	3.09E-01	2.21E-01	3.19E-01	2.28E-01	3.87E-01	2.89E-01
S	2.19E-01	1.56E-01	2.60E-01	1.86E-01	2.38E-01	1.78E-01
Ar	5.72E-02	4.08E-02	6.96E-02	4.97E-02	4.91E-02	3.66E-02
Ca	7.35E-02	5.25E-02	7.85E-02	5.61B-02	4.11E-02	3.06E-02
Ti	6.18E-05	4,41E-05	6.20E-05	4.43B-05	1.60E-05	1.19E-05
Cr	2.17E-03	1.55E-03	2.23E-03	1.59B-03	7.57E-04	5.64E-04
Fe	2.61E-02	1.86E-02	2.57E-02	1.83E-02	7.77E-03	5.79E-03
Ni	5.66E-01	4.04E-01	5.17B-01	3.69E-01	2.46E-01	1.83E-01
	Nucleosynthesis for $\rho_{tr} = 3 \times 10^7\ g/cm^3$					
He	3.69E-03	2.63E-03	1.10E-03	7.84E-04	9.15E-06	6.78E-06
C	1.58E-05	1.13E-05	2.36E-03	1.69E-03	1.55B-03	1.15E-03
O	5.62E-02	4.02E-02	5.87E-02	4.19E-02	1.50E-01	1.11E-01
Ne	7.19E-07	5.14E-07	2.90E-04	2.07E-04	1.58E-03	1.17E-03
Mg	1.84E-02	1.31E-02	6.77E-03	4.83E-03	3.38E-02	2.50E-02
Si	2.17E-01	1.55E-01	2.08E-01	1.48E-01	3.68E-01	2.72E-01
S	1.49E-01	1.06E-01	1.68E-01	1.20E-01	2.59E-01	1.92E-01
Ar	3.86E-02	2.76E-02	4.60E-02	3.28E-02	6.11E-02	4.53E-02
Ca	4.96E-02	3.54E-02	5.38E02	3.83E-02	6.39E-02	4.74E-02
Ti	1.86E-04	1.33E-04	9.97E-05	7.11E-05	3.35E-05	2.48E-05
Cr	1.67E-03	1.20E-03	1.94E-03	1.38E-03	1.57E-03	1.16E-03
Fe	2.11E-02	1.51E-02	2.68E-02	1.91E-02	1.74E-02	1.29E-02
Ni	8.46E-01	6.04E-01	8.29E-01	5.91E-01	3.91E-01	2.90E-01
	Nucleosynthesis for $\rho_{tr} = 4 \times 10^7\ g/cm^3$					
He	6.6835E-03	4.7722E-03	5.8707E-03	4.1892E-03	6.0413E-04	4.4533E-04
C	1.3022E-05	9.2978E-06	1.3297E-03	9.4884E-04	4.9175E-04	3.6249E-04
O	3.3115E-02	2.3645E-02	3.5878E-02	2.5602E-02	9.0458E-02	6.6682E-02
Ne	2.8464E-07	2.0324E-07	1.6248E-04	1.1595E-04	3.4673E-04	2.5559E-04
Mg	8.9766E-03	6.4094E-03	2.5410E-03	1.8132E-03	1.5626E-02	1.1519E-02
Si	1.6534E-01	1.1805E-01	1.5429E-01	1.1010B-01	2.9130E-01	2.1473E-01
S	1.1305E-01	8.0720E-02	1.2441E-01	8.8775E-02	2.1620E-01	1.5937E-01
Ar	2.9605E-02	2.1139E-02	3.3580E-02	2.3962E-02	5.4628E-02	4.0269E-02
Ca	3.8257E-02	2.7316E-02	3.9367E-02	2.8092E-02	6.2767E-02	4.6269E-02
Ti	2.5067E-04	1.7898E-04	1.2781E-04	9.1201E-05	2.3282B-04	1.7163B-04
Cr	1.4079E-03	1.0053E-03	1.5819E-03	1.1288E-03	1.9325E-03	1.4246E-03
Fe	1.7500E-02	1.2495E-02	2.1211E-02	1.5135E-02	2.3618E-02	1.7410E-02
Ni	9.8633E-01	7.0426E-01	9.8104E-01	7.0005E-01	5.9836E-01	4.4108E-01

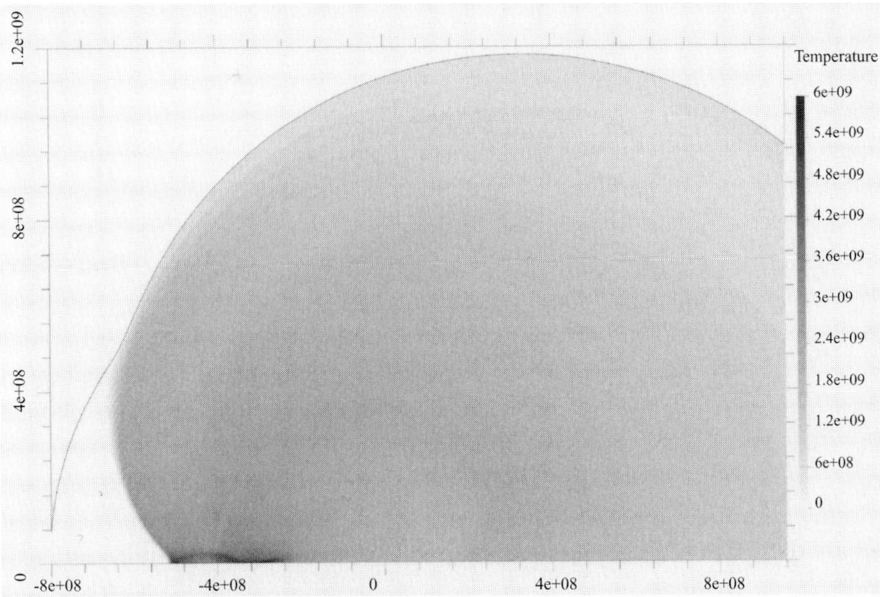

Fig. 13.3. Off Center Detonation, $\rho_{tr} = 2 \times 10^7\ g/cm^3$, t = 1.0 s

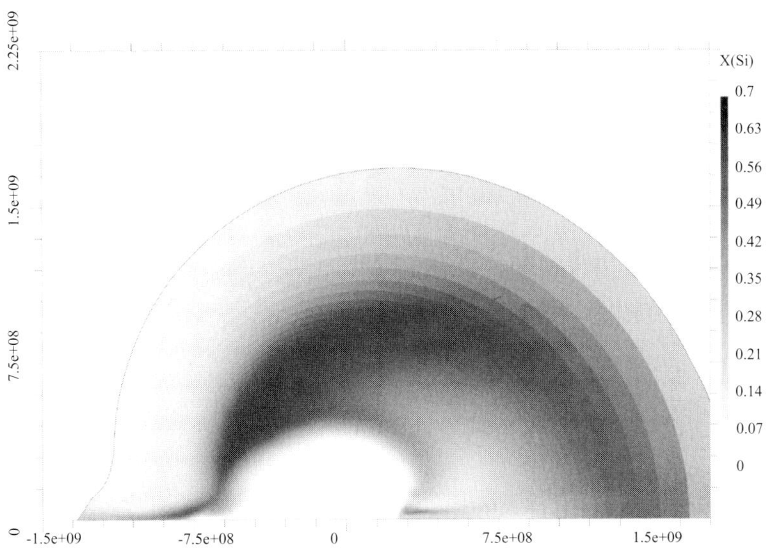

Fig. 13.4. Si mass fraction at end of detonation phase: $\rho_{tr} = 2 \times 10^7\ g/cm^3$, t = 1.5 s

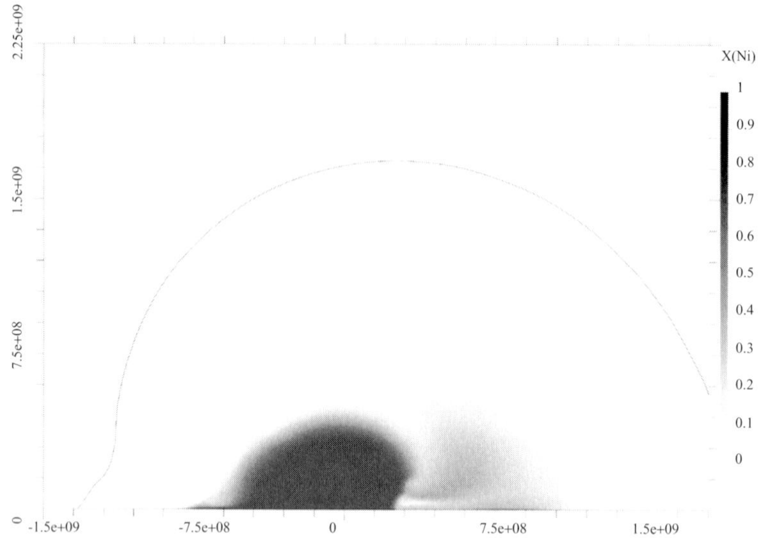

Fig. 13.5. Ni mass fraction at end of detonation phase: $\rho_{tr} = 2 \times 10^7\ g/cm^3$, t = 1.5 s

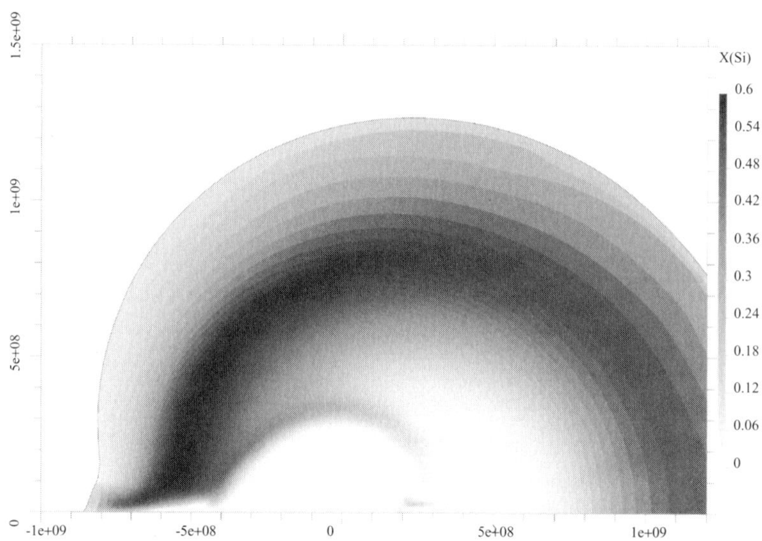

Fig. 13.6. Si mass fraction at end of detonation phase: $\rho_{tr} = 3 \times 10^7\ g/cm^3$, t = 1.0 s

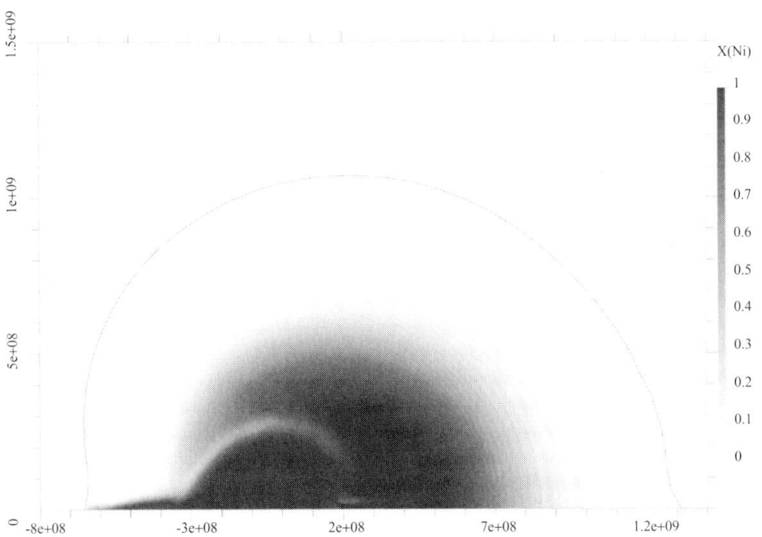

Fig. 13.7. Ni mass fraction at end of detonation phase: $\rho_{tr} = 3 \times 10^7 \ g/cm^3$, t = 1.0 s

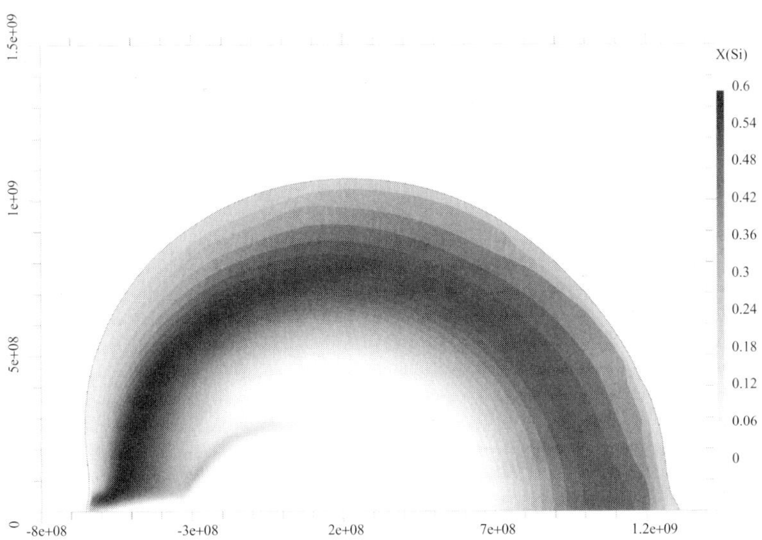

Fig. 13.8. Si mass fraction at end of detonation phase: $\rho_{tr} = 4 \times 10^7 \ g/cm^3$, t = 0.8 s

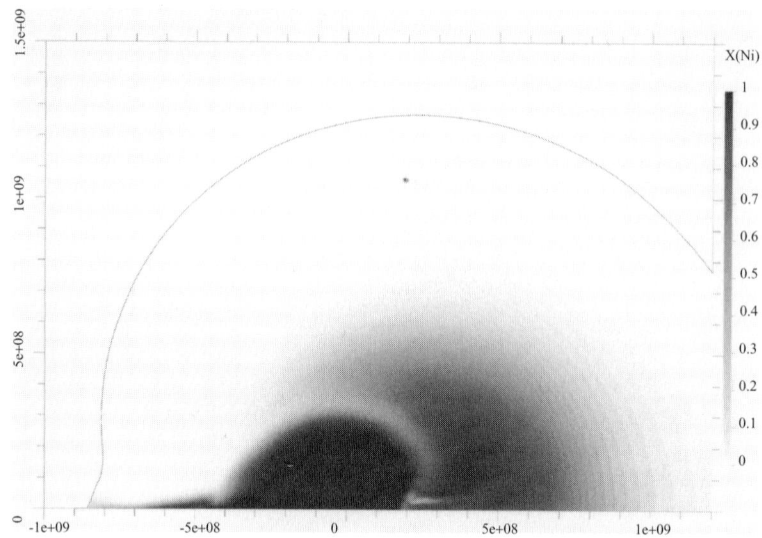

Fig. 13.9. Ni mass fraction at end of detonation phase: $\rho_{tr} = 4 \times 10^7\ g/cm^3$, t = 0.8 s

smaller burning energy. The differences are higher for smaller transition densities because in those cases the expansion effects are higher. In figures 13.4–13.9 we present the distribution of Si mass fraction and Ni mass fraction for the three cases. Again, non-spherical distributions are seen in all cases, where deviations from sphericity are larger for smaller transition density.

13.4 Propagation of detonations in a R-T disturbed region

Transition to detonation presumably occurs in a R-T unstable region that had enough time to form a distribution of bubbles of hot ashes and spikes of cold fuel. An important question is – can the detonation incinerate the fuel entirely or can some unburnt pockets survive the detonation? The latter could happen if some pockets of fuel are disconnected from the fuel channels in which the detonation wave propagates. To answer this question we performed two small scale tests, where regions (channel/bubble) of fuel are separated by non reactive barriers. The dimensions of the barriers and the reactive regions are similar in those tests. Due to space limits we do not show graphs of these numerical experiments. In both cases, channels or bubbles, the simulations show that the emerging shock waves diverge in the non-reactive barriers and eventually become too weak to ignite a successive detonation in the nearby reactive region. These results are consistent with previous work (Livne & Glasner 1990) on the interaction between helium detonations and the boundary of a C-O core. Similar conclusions, based on 3-D SPH simulations, can be found in Bravo & Garcia-Senz (2002).

Table 13.4. *Energy Productions*

ρ_{tr}	1D Eburn	2D – Spherical Eb(def)+Eb(det)	2D Non-spherical Eb(def)+Eb(det)
2.E7	1.818E51	3.858E50 + 1.414E51	3.858E50 + 1.042E51
3.E7	1.951E51	3.687E50 + 1.585E51	3.687E50 + 1.288E51
4.E7	2.013E51	3.512E50 + 1.651E51	3.512E50 + 1.446E51

The structure of the inner region at the end of the deflagration phase is yet poorly resolved by 3-D simulations (Gamezo *et al.* 2002, Reinecke *et al.* 2002). In those few simulations all fuel channels were connected and therefore the above problem does not exist. However, it is well known that the structure of the spatial distribution of materials in R-T unstable simulations depends strongly on the spatial resolution. With higher resolution smaller scale structures appear, and more complicated distribution evolve. It is also still unclear how much mass of C-O should be present in the inner region for being detectable by observations.

Conclusions

Non-spherical delayed detonations, which ignite off-center and propagate through inhomogeneous disturbed matter, produce significantly different products from spherical ones. The main effect comes from the expansion effect that reduces the density of the fuel before being consumed by the detonation front. As a result, both the energy released and the amount of synthesized Fe-group elements are lower in the non-spherical cases. Moreover, off-center detonation produces non-spherical distribution of elements at the end of the combustion phase. These effects grow with lower transition density.

The structure of the R-T disturbed region could determine the post-burning composition and distribution of species in the inner region. If this region contains isolated bubbles and channels of fuel that are not connected, than some of them will not be swept by she detonation directly. We claim that the shock wave induced by the detonation wave on the surrounding is too weak for igniting subsequent detonations in those isolated pockets. Unless those pockets are mixed later with hot ashes from their vicinity, some unburnt fuel may remain in the inner region. At this point, it is impossible to estimate the mass of such unburnt pockets of fuel, as the exact structure of the disturbed region is not yet resolved.

Acknowledgements

The author is grateful to P. Hoflich for many fruitful discussions.

References

Bravo, E., Gracia-Senz, D., 2002. *IAUS*, **187**, 220.
Dominguez, I., Hoflich, P., 2000. *Astrophys. J.*, **528**, 854.
Dominguez, I., Hoflich, P., Straniero, Q., 2001. *Astrophys. J.*, **557**, 279.
Fryxell, B., Timmes, F. X., Zingale, M., Dursi, L. J., Ricker, P., Olson, K., Calder, A. C., Tufo, H., MacNeice, P., Truran, J. W., Rosner, R., 2000. *Astrophys. J.*, **543**, 938.
Gamezo, V. N., Wheeler, J. C., Khokhlov, A. M., Oran, E. S., 1999. *Astrophys. J.*, **512**, 827.
Gamezo, V. N., Khokhlov, A. M., Oran, E. S., 2002. *AAS*, **200**, 1401.
Hoflich, P., Khokhlov, A. M., Wheeler, J. C, 1995. *Astrophys. J.*, **444**, 831.
Hoflich, P., Khokhlov, A. M., Wheeler, J. C, Nomoto, K., Thielemann, F. K., 1996. *AAS*, **28**, 1332.
Khokhlov, A. M., 1989. *Soviet Science Rev.*, **8**
Khokhlov, A. M., 1993. *Astrophys. J.*, **419**, 200.
Livne, E., Glasner, A., 1990. *Astrophys. J.*, **361**, 244.
Livne, E., 1993. *Astrophys. J.*, **412**, 634.
Livne, E., 1999. *Astrophys. J.*, **527**, L97.
Reinecke, M., Hillebrandt, W., Neimeyer, J. C., 2002. *Astron. Astrophys.* **391**, 1167.

14

Numerical simulations of Type Ia supernovae: deflagrations and detonations

V. N. Gamezo and E. S. Oran

Laboratory for Computational Physics and Fluid Dynamics, Naval Research Laboratory, Washington, D. C. 20375, USA

A. M. Khokhlov

Department of Astronomy and Astrophysics, University of Chicago, Chicago, IL 60637

Abstract

We study a thermonuclear explosion of a carbon-oxygen white dwarf (WD) using a three-dimensional hydrodynamic model with a simplified mechanism for nuclear reactions and energy release. The explosion begins as a deflagration with the flame front highly distorted by the Rayleigh-Taylor instability. Turbulent combustion and convective flows produce an inhomogeneous mixture of burned and unburned materials that extends from the center to about 0.8 of the radius of the expanding WD. At this stage, a detonation is ignited and propagates through the layers of unburned material with the velocity about 12,000 km/s, which is comparable to the expansion velocities induced in outer layers of the WD by the subsonic burning. During the period of detonation propagation, the density of the expanding unreacted material ahead of the shock can decrease by an order of magnitude compared to its value before the detonation started. Because the detonation burns material to different products at different densities, it can create a large-scale asymmetry in composition if it starts far from the WD center. In contrast to the 3-D deflagration model, the 3-D delayed-detonation model of SN Ia explosions does not leave carbon, oxygen, and intermediate-mass elements in central parts of a WD. This removes the key disagreement between simulations and observations, and confirms that the delayed detonation is currently the most promising mechanism for SN Ia explosions.

14.1 Introduction

Type Ia supernovae (SNe Ia) [1–10] result from the most powerful thermonuclear explosions in the Universe. The explosion is now believed to occur in a carbon-oxygen white dwarf (WD) in a binary system, so that the WD can increase its own mass by attracting the material from outer layers of the companion star. When the WD mass approaches the Chandrasekhar limit, 1.4 solar masses, any small mass increase results in a substantial contraction of the star. The compression increases the temperature, accelerates thermonuclear reactions, and eventually ignites a thermonuclear burning near the WD center. This starts a SN Ia explosion that lasts only a few seconds, but releases 10^{51} ergs, about as much energy as the Sun would radiate during 8 billion years. The energy is produced by a network of thermonuclear reactions that begins from original ^{12}C and ^{16}O nuclei and ends in ^{56}Ni and other iron-group elements. Considerable amounts of intermediate-mass elements, such as Ne, Mg, Si, S, Ca, are created as well. The main energy-producing reactions occur in a thin layer, called a thermonuclear flame. At the beginning of the explosion, the flame is laminar and its propagation velocity is defined by nuclear reaction rates and the electron heat conduction of the degenerate matter. As the flame moves away from the center, it becomes turbulent and accelerates. At the same time, the WD expands due to the energy release. Eventually, the deflagration can undergo a transition to a detonation.

General ideas about possible explosion mechanisms have been extensively tested using one-dimensional (1D) numerical models [11–14,4,6]. Delayed-detonation models [15–22], that postulate a deflagration-to-detonation transition (DDT) at some stage of the thermonuclear explosion are the most successful in reproducing observed characteristics of SNe Ia. Many important details, however, including the mechanism of DDT, are still unknown because SN Ia explosions are intrinsically three-dimensional (3-D) phenomena. Only a full-scale 3-D numerical model can reproduce all key features of the explosion that involves propagation of a turbulent thermonuclear flame in gravitational field of a WD. Building such a model is a complicated interdisciplinary problem on the leading edge of astrophysics, nuclear physics, combustion physics, and computational physics.

Full-scale 3-D numerical simulations of thermonuclear supernova explosions have become a reality during the last few years [23–26], in great part owing to the progress in computational technology. Here, we describe a 3-D supernova model, present numerical results for deflagration and detonation stages of the explosion, and compare the results with observations of SN Ia.

14.2 Reactive fluid dynamic model

The numerical model is based on reactive Euler equations

$$\frac{\partial \rho}{\partial t} = -\nabla \cdot (\rho \mathbf{U}),$$

$$\frac{\partial \rho \mathbf{U}}{\partial t} = -\nabla \cdot (\rho \mathbf{U}\mathbf{U}) - \nabla P + \rho \mathbf{g},$$

$$\frac{\partial E}{\partial t} = -\nabla \cdot (\mathbf{U}(E + P)) + \rho \mathbf{U} \cdot \mathbf{g} + \rho \dot{q},$$

where ρ, $E = E_i + \rho U^2/2$, E_i, $\mathbf{U}$, g and $\dot{q}$ are the mass density, energy density, internal energy density, flow velocity, gravitational acceleration, and nuclear energy release rate per unit mass, respectively. These equations describe mass, momentum, and energy conservation laws for an inviscid fluid.

The thermodynamic properties of the fluid are defined by the equation of state of degenerate matter, which is well described by basic theory and includes contributions from ideal Fermi-Dirac electrons and positrons, equilibrium Planck radiation, and ideal ions. Pressure $P = P(\rho, E_i, Y_e, Y_i)$ and temperature $T = T(\rho, E_i, Y_e, Y_i)$ are determined by the equation of state as functions of ρ, E_i, the electron mole fraction Y_e, and the mean mole fraction of ions Y_i. The relation between P and ρ is close to that in a polytropic gas with γ varying from 4/3 to 5/3. This equation of state is valid for the thermodynamic parameters and compositions expected in the computations, ranging from relatively cold highly degenerate carbon-oxygen matter to partially degenerate hot products of thermonuclear reactions.

Fusion reactions involved in the thermonuclear burning of the carbon-oxygen mixture [27–29] can be separated into three consecutive stages. First, the $^{12}C + {}^{12}C$ reaction leads to the consumption of C and formation of mostly Ne, Mg, protons, and α-particles. Then begins the nuclear statistical quasi-equilibrium (NSQE) stage, during which O burns out and Si-group (intermediate mass) elements are formed. Finally, Si-group elements are converted into the Fe-group elements and the nuclear statistical equilibrium (NSE) sets in. The reaction time scales associated with these stages strongly depend on temperature and density and may differ from one another by several orders of magnitude [30–34]. The full nuclear reaction network includes hundreds of species that participate in thousands of reactions. Integration of this full network is too time-consuming for it to be used in multidimensional numerical models. Therefore, we used a simplified four-equation kinetic scheme [15,23] that describes all major stages of carbon burning.

The fluid dynamic equations, coupled to the nuclear reaction mechanism, are integrated using an explicit, second-order, Godunov-type, adaptive-mesh-refinement code [35,36]. A Riemann solver is used to evaluate fluxes at cell interfaces. The computational mesh is comprised of cubic cells of various sizes that are organized in a fully threaded tree (FTT) [36]. The FTT-based parallel adaptive mesh refinement algorithm dynamically adjusts cell sizes in accordance with changing physical conditions in the vicinity of each cell. Here, the mesh was refined around shock waves, flame fronts, and in regions of steep gradients of density, pressure, composition, and tangential velocity. The cell size dx varies within predefined limits dx_{min} and dx_{max} in such a way that neighboring cell sizes can be the same or differ by a factor of two. The code has been used extensively to solve terrestrial combustion ([37,38], and references therein) and in astrophysical [39,23,26] problems.

14.3 Reaction front propagation

During the explosion of a WD, the thermonuclear burning occurs mostly inside thin reaction fronts that separate burned and unburned materials, and can propagate in the form of deflagrations or detonations. The physical thickness of these fronts differs from the WD radius by up to 12 orders of magnitude, and is not resolved in the large-scale simulations described here. The unresolved reaction zone has no significant effect on the detonation propagation, but special treatment is required for deflagrations.

A detonation front is always supersonic relative to the unburned material and involves a strong shock that triggers energy-producing reactions. The steady-state detonation velocity is a thermodynamic parameter that depends on the released energy and the equation of state, but not on the particular kinetics of energy release or transport properties of the material. In numerical simulations, these basic properties of detonation waves make the detonation velocity and the equilibrium composition of detonation products practically independent of the numerical resolution. The reactive fluid dynamic model described above produces correct detonation parameters even though the physical thickness of the thermonuclear reaction zone of a detonation wave is not resolved. Transient detonation phenomena, such as detonation initiation and extinction, cannot be correctly reproduced without resolving the detonation wave thickness, and are not included in the simulations.

A deflagration, or a flame, always propagates subsonically with a speed that depends on small-scale physics. In our simulations, the flame speed is given by an additional model that takes into account physical processes at scales smaller than the computational cell size. The flame-capturing algorithm [35,23] ensures the flame propagation with the prescribed speed S:

$$S = \max(S_l, S_t), \tag{14.1}$$

where S_l and S_t are the laminar and the turbulent flame speed, respectively.

The speed of a laminar thermonuclear flame in a WD is defined by reaction rates and transport properties of the material and governed by the same laws that describe laminar flame structure in terrestrial chemical systems [40–42]. The only substantial difference is that transport properties of degenerate matter are dominated by the electron heat conduction at high densities, and by both electron and photon heat conduction at low densities. The steady-state laminar flame speed S_l in carbon-oxygen degenerate matter is a known function of temperature, density, and composition [43,44], but the flame can be laminar only very near the center of a WD. Away from the center, the flame is turbulent and propagates with a higher effective speed [13,35,45,9].

We assume that turbulent burning on small unresolved scales is driven by the gravity-induced Rayleigh-Taylor (RT) instability. This instability distorts the flame surface at multiple scales and generates turbulent motions in the surrounding fluid. The turbulent energy propagates from large to small scales, thus further disturbing the flame surface on small scales. A developed turbulent flame is statistically steady-state and forms a dynamic hierarchical self-similar 3-D structure where the flame surface is distorted at multiple scales. The flame distortions increase the flame surface and, therefore, the burning rate. The larger the scale, the higher the burning rate at this scale due to the increase of the flame surface resulting from distortions at smaller scales. The burning rate defines the turbulent flame speed S_t for any given length scale. This turbulent flame structure was analyzed in 3-D numerical simulations [35,46] of the thermonuclear burning of carbon-oxygen degenerate matter in a uniform gravitational field. It was found that a turbulent flame in a vertical column of width L becomes quasi-steady-state and propagates with the speed

$$S_t \simeq 0.5\sqrt{AgL} \tag{14.2}$$

independent of the laminar speed S_l, where $A = (\rho_0 - \rho_1)/(\rho_0 + \rho_1)$ is the Atwood number, and ρ_0 and ρ_1 are the densities ahead and behind the flame front, respectively. We used this result in the subgrid model with $L = 2dx$, assuming that at $L \ll R_{WD}$ burning can be considered as locally steady state [47,16,35].

The subgrid model is based on the two main properties of a turbulent flame [35,46]: self-similarity of the flame structure and self-regulation of the flame speed. Self-similarity means that the 3-D distortions of the flame surface at different scales are similar. Self-regulation means that changing the flame speed at small scales does not affect the flame speed at larger scales. This occurs because a higher flame speed at small scales causes small flame wrinkles to burn out, thus decreasing the flame surface. The resulting burning rate, defined as a product of the flame speed at small scales and the flame surface, does not change. This subgrid model makes it possible to reproduce the correct flame propagation in numerical simulations while explicitly resolving only the large-scale flame structure. If the resolved flame structure is self-similar and self-regulating, and behaves according to the Eq.(14.2), the subgrid model just extends this behavior to unresolved small scales. Shifting the boundary between resolved and unresolved scales by changing the numerical resolution should not affect the turbulent flame propagation. The solution obtained should then be independent of numerical resolution and on the exact value of S_t. Resolution tests have shown [26] that this is the case for $dx_{min} = 2.6 \times 10^5$ cm.

14.4 Simulations and results

The initial conditions for the simulations were set up for $1.4M_{\odot}$ WD in hydrostatic equilibrium with initial radius $R_{WD} = 2 \times 10^8$ cm, initial central density $\rho_c = 2 \times 10^9$ g/cm^3, uniform initial temperature $T = 10^5$ K, and uniform initial composition with equal mass fractions of ^{12}C and ^{16}O nuclei. Starting from the central pressure $P(\rho_c)$, the equations of hydrostatic equilibrium, $dP/dr = -GM\rho/r^2$ and $dM/dr = 4\pi \rho r^2$, were integrated outward until $P = 0$ was reached (here G is the gravitational constant, and M is the mass of the material inside a sphere of radius r). The resulting WD configuration was interpolated onto a 3-D mesh extended from the WD center $x = y = z = 0$ to $x = y = z = 2.6R_{WD}$. Thus, we model one octant of the WD assuming mirror symmetry along the $x = 0$, $y = 0$ and $z = 0$ planes. The burning was initiated at the center of WD by filling a small spherical region at $r < 0.015R_{WD}$ with hot reaction products without disturbing the hydrostatic equilibrium.

Because a Chandrasekhar-mass WD is close to collapsing, its gravitational equilibrium is sensitive to the discretization errors that appear when the spherical body is mapped into a Cartesian mesh. To minimize these errors, the mesh is initially refined to the finest level near the WD center ($r < 0.4R_{WD}$). During the simulations, we keep the mesh unchanged until the flame reaches the boundary of the fine grid. Then the adaptive mesh refinement algorithm is turned on.

14.4.1 Deflagration

The development of the thermonuclear flame is shown in Fig. 14.1 by a series of 3-D snapshots of the flame surface. At the beginning, the initially spherical flame ignited at the center of the WD propagates outwards with the laminar flame speed S_l. As it moves away from the center, the gravitational acceleration g and the Atwood number A increase. This increases the amplitude and rate of development of the RT instability controlled by g and A. Due to the RT instability, small perturbations of the flame surface grow and form a few plumes that have characteristic mushroom shapes. The turbulent flame speed S_t also increases with g and A according to Eq.(14.2), and eventually dominates S_l in Eq.(14.1). The flame plumes continue to grow, partially due to the flame propagation and partially to gravitational forces that cause the hot, burned, low-density material inside the plumes to rise towards the WD surface. The same gravitational forces also pull the cold, high-density unburned material between the plumes down towards the center. The resulting shear flows along the flame surface are unstable (Kelvin-Helmholtz (KH) instability) and quickly develop vortices. These vortices further distort the flame surface, and also contribute their energy into the turbulent cascade that creates turbulent motions at smaller scales, down to a few dx.

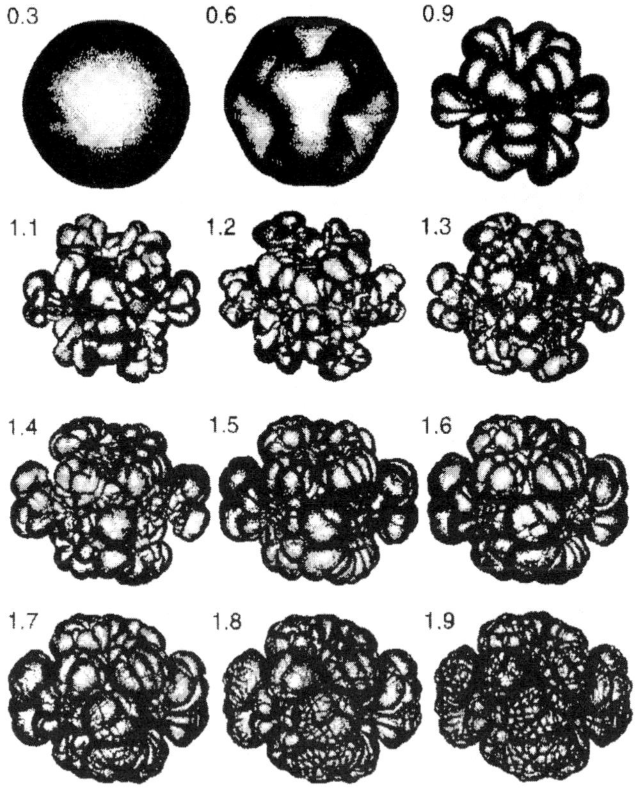

Fig. 14.1. Development of turbulent thermonuclear flame in carbon-oxygen white dwarf. Numbers show time in seconds after ignition. All flames are plotted at different scales adjusted to the flame radius R_f^{max} that grows with time t as $R_f^{max} = exp(2.95t - 5.5)$ (time is in seconds, radius is scaled by 5.35×10^8 cm). $L = 2dx$, $dx_{min} = 2.6 \times 10^5$ cm.

When the original flame plumes grow large enough, secondary RT instabilities develop on their surface, thus producing the next level of "mushrooms" that also grow and may become subject to the RT instability at a smaller scale, etc. These smaller gravity-induced mushrooms interact with the turbulence created by the previous generation of larger flame plumes, and also produce some turbulence themselves through the KH instability. The resulting complicated turbulent flame surface is shown in Fig. 14.1.

As the turbulent flame develops, the energy released by the thermonuclear burning causes the WD to expand. The expansion accelerates and becomes nonuniform as the rising plumes approach the surface of the star. We continued the simulations until the surface reached the boundary of the computational domain. By that time, about half of the material was burned, the radius of the expanding star increased

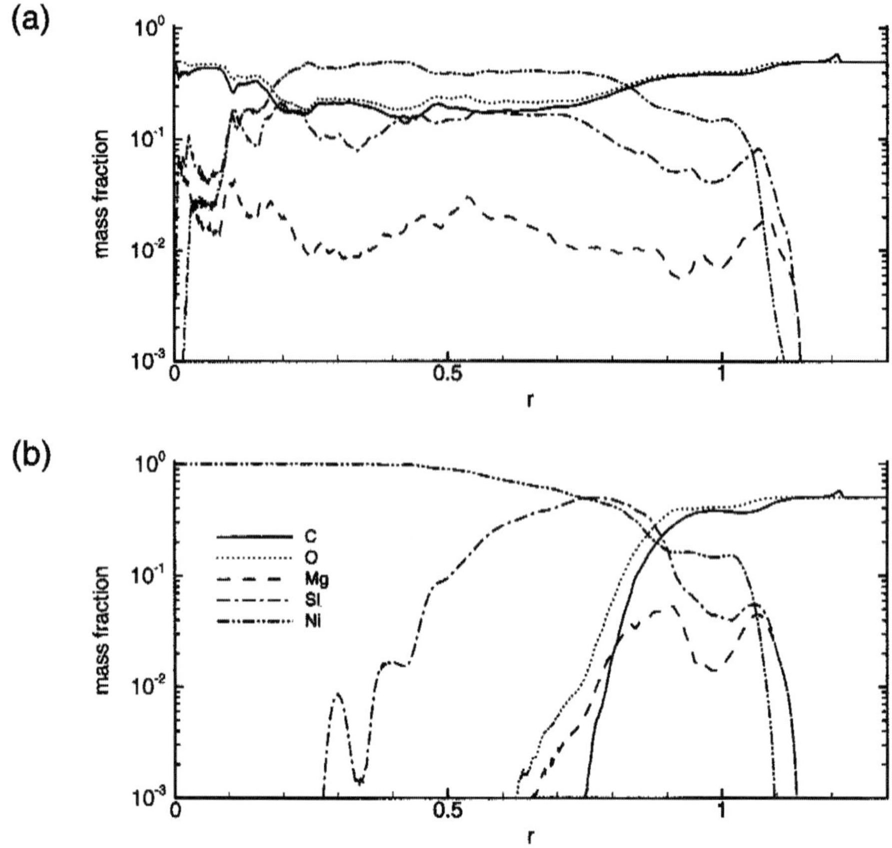

Fig. 14.2. Angle-averaged mass fractions of main elements in the exploding WD as functions of distance from the WD center at 1.9 s after the beginning of the explosion. Mass fractions are estimated for the deflagration model (a) and the delayed-detonation model with central ignition (b) using the simplified four-equation kinetics. The distance r is scaled by the computational domain size 5.35×10^8 cm.

by about a factor of 2.6, the outer layers accelerated to about 12,000 km/s, and the density of unburned material near the star center decreased to about 5×10^7 g/cm^3. The area around the center still contains a significant amount of unburned material that sinks at 1000 km/s towards the center between large flame plumes. This material continues to burn, but it will not burn out completely as long as convective flows supply fresh unburned material from outer layers. As the WD continues to expand, the density decreases. The deflagration begins to produce intermediate-mass elements when the density of unburned material becomes lower than $\simeq 5 \times 10^7$ g/cm^3, and the burning stops when the density drops below $\simeq 10^6$ g/cm^3. This means that the final ejecta produced by the 3-D deflagration model will contain unburned material and intermediate-mass elements at any distance from the center.

The unburned carbon and oxygen that remain between the flame plumes and intermediate-mass elements that form at low densities at different radii should produce spectral signatures in a wide range of expansion velocities, including very low velocities that correspond to central parts of WD. For intermediate-mass elements, minimum observed velocities [2,48] are large enough ($\sim$10,000 km/s for Si) to rule out the presence of these elements near the WD center. For carbon and oxygen, analyses of SN Ia spectra usually imply high velocities [49–52], as would be produced by the acceleration of expanding outer layers. There is some evidence [48] that SN Ia ejecta can contain carbon at lower velocities, down to 11,000 km/s. Still, these velocities are too high for the material produced in central parts of WD.

14.4.2 Delayed detonation

The disagreement between deflagration-based simulation results and observations strongly suggests that the turbulent flame may trigger a detonation. The process of deflagragration-to-detonation transition involves small scales comparable to the detonation wave thickness, and, therefore, cannot be directly modeled in large-scale simulations. To study the detonation stage of the explosion, we use the deflagration results as initial conditions and create an artificial hot spot to ignite the detonation at 1.573 s near the WD center or 10^8 cm off center. At this time, 1/3 of the WD mass is burned, the density of unburned material near the WD center is approximately 2×10^8 g/cm^3, and the turbulent flame surface extends from the center to about 0.8 of the radius of the expanding WD that reaches 3×10^8 cm.

The detonation propagates through the layers of unburned material with the velocity about 12,000 km/s, which is comparable to the expansion velocities induced in outer layers of the WD by the subsonic burning. Simulations with the off-center detonation initiation indicate that, during the period of detonation propagation, the density of the expanding unreacted material ahead of the shock can decrease by an order of magnitude compared to its value before the detonation started. Because the detonation burns material to different products at different densities, it can create a large-scale asymmetry in composition if it starts far from the WD center. A similar result was obtained in 2D simulations [53].

The detonation releases an additional energy, transforms all carbon and oxygen in the central parts of the WD into iron-group elements, and produces intermediate-mass elements in outer layers, where the density of the unburned material drops below 5×10^7 g/cm^3. Angle-averaged mass fractions of the main elements in a WD produced by the deflagration and the delayed-detonation models are compared in Fig. 14.2. These results show that, in contrast to the 3-D deflagration model, the 3-D delayed-detonation model of SN Ia explosion does not leave carbon, oxygen, and intermediate-mass elements in central parts of a WD. This removes the key disagreement between simulations and observations, and confirms that the delayed detonation is a promising mechanism for SN Ia explosion. Further analysis of 3-D delayed detonations requires varying the time and location of the detonation initiation, 3-D radiation transport simulations to produce spectra, and a detailed comparison between the calculated and observed spectra of SN Ia.

Acknowledgements

This work was supported in part by the NASA ATP program (NRA-02-OSS-01-ATP) and by the Naval Research Laboratory (NRL) through the Office of Naval Research. Computing facilities were provided by the DOD HPCMP program.

References

[1] J. C. Wheeler, R. P. Harkness, *Rep. Prog. Phys.* **53**, 1467 (1990)
[2] A. V. Filippenko, *Annu. Rev. Astron. Astrophys.* **35**, 309 (1997)
[3] J. C. Wheeler, *Am. J. Phys.*, **71**, 11 (2003)
[4] S. E. Woosley, T. A. Weaver, *Annu. Rev. Astron. Astrophys.* **24**, 205 (1986)
[5] D. Branch, A. M. Khokhlov, *Phys. Rep.* **256**, 53 (1995)
[6] J. C. Wheeler, R. P. Harkness, A. M. Khokhlov, P. A. Höflich, *Phys. Rep.* **256**, 211 (1995)
[7] K. Nomoto, K. Iwamoto, N. Kishimoto, *Science* **276**, 1378 (1997)
[8] D. Branch, *Annu. Rev. Astron. Astrophys.* **36**, 17 (1998)
[9] W. Hillebrandt, J. C. Niemeyer, *Annu. Rev. Astron. Astrophys.* **38**, 191 (2000)
[10] A. Burrows, *Nature* **403**, 727 (2000)
[11] W. D. Arnett, *Astrophys. Space Sci.* **5**, 180 (1969)
[12] C. J. Hansen, J. C. Wheeler, *Astrophys. Space Sci.* **3**, 464 (1969)
[13] K. Nomoto, D. Sugimoto, S. Neo, *Astrophs. Space Sci.* **39**, L37 (1976)
[14] K. Nomoto, F.-K. Thielemann, K. Yokoi, *Astrophys. J.* **286**, 644 (1984)
[15] A. M. Khokhlov, *Astron. Astrophys.* **245**, 114 (1991)
[16] D. Arnett, E. Livne, *Astrophys. J.* **427**, 315 (1994)
[17] D. Arnett, E. Livne, *Astrophys. J.* **427**, 330 (1994)
[18] H. Yamaoka, K. Nomoto, T. Shigeyama, F.-K. Thielemann, *Astrophys. J.* **393**, L55 (1992)
[19] A. M. Khokhlov, E. Müller, P. A. Höflich, *Astron. Astrophys.* **270**, 223 (1993)
[20] P. A. Höflich, *Astrophys. J.* **443**, 89 (1995)
[21] P. A. Höflich, A. M. Khokhlov, J. C. Wheeler, *Astrophys. J.* **444**, 831 (1995)
[22] P. A. Höflich, A. M. Khokhlov, *Astrophys. J.* **457**, 500 (1996)
[23] A. M. Khokhlov, http://www.arxiv.org/astro-ph/0008463 (2000)
[24] M. Reinecke, W. Hillebrandt, J. C. Niemeyer, *Astron. Astrophys.* **386**, 936 (2002)
[25] M. Reinecke, W. Hillebrandt, J. C. Niemeyer, *Astron. Astrophys.* **391**, 1167 (2002)
[26] V. N. Gamezo, A. M. Khokhlov, E. S. Oran, A. Y. Chtchelkanova, and R. O. Rosenberg, *Science* 2003, **299**, 77 (2003)
[27] W. A. Fowler, G. R. Caughlan, B. A. Zimmerman, *Annu. Rev. Astron. Astrophys.* **13**, 69 (1975)
[28] S. E. Woosley, W. A. Fowler, J. A. Holmes, B. A. Zimmerman, *Atomic Data and Nuclear Data Tables*, **22**, 371 (1978)

[29] F.-K. Thielemann, M. Arnould, J. W. Truran, in *Advances in Nuclear Astrophysics*, E. Vangioni-Flam, Ed. (Editions frontières, Gif-sur-Yvette, 1987), p.525
[30] J. W. Truran, A. G. W. Cameron, A. Gilbert, *Canadian J. of Phys.* **44**, 563 (1966)
[31] D. Bodansky, D. D. Clayton, W. A. Fowler, *Astrophys. J. Suppl. Ser.* **16**, 299 (1968)
[32] S. E. Woosley, W. D. Arnett, D. D. Clayton, *Astrophys. J. Suppl. Ser.* **26**, 231 (1973)
[33] A. M. Khokhlov, *Mon. Not. R. Astron. Soc.* **239**, 785 (1989)
[34] V. N. Gamezo, J. C. Wheeler, A. M. Khokhlov, E. S. Oran, *Astrophys. J.* **512**, 827 (1999)
[35] A. M. Khokhlov, *Astrophys. J.* **449**, 695 (1995)
[36] A. M. Khokhlov, *J. Comput. Phys.* **143**, 519 (1998)
[37] A. M. Khokhlov, E. S. Oran, *Combust. Flame* **119**, 400 (1999)
[38] V. N. Gamezo, A. M. Khokhlov, E. S. Oran, *Combust. Flame* **126**, 1810 (2001)
[39] A. M. Khokhlov *et al.*, *Astrophys. J.* **524**, L107 (1999)
[40] D. A. Frank-Kamenetskii, *Diffusion and Heat Transfer in Chemical Kinetics* (Plenum, New York, 1969), chap. 6.
[41] Ya. B. Zeldovich, G. I. Barenblatt, V. B. Librovich, G. M. Makhviladze, *The Mathematical Theory of Combustion and Explosions* (Consultants Bureau, New York and London, 1985), chap. 4.
[42] F. A. Williams, *Combustion Theory* (Benjamin-Cummings, Menlo Park, ed. 2, 1985), chap. 5.
[43] F. X. Timmes, S. E. Woosley, *Astrophys. J.* **396**, 649 (1992)
[44] A. M. Khokhlov, E. S. Oran, J. C. Wheeler, *Astrophys. J.* **478**, 678 (1997)
[45] J. C. Niemeyer, S. E. Woosley, *Astrophys. J.* **475**, 740 (1997)
[46] A. M. Khokhlov, E. S. Oran, J. C. Wheeler, *Combust. Flame* **105**, 28 (1996)
[47] E. Livne, *Astrophys. J.* **406**, L17 (1993)
[48] D. Branch *et al.*, *Astron. J.* **126**, 1489 (2003)
[49] D. J. Jeffery *et al.*, *Astrophys. J.* **397**, 304 (1992)
[50] A. Fisher, D. Branch, P. Nugent, E. Baron, *Astrophys. J.* **481**, L89 (1997)
[51] P. A. Mazzali, *Mon. Not. R. Astron. Soc.* **321**, 341 (2001)
[52] R. P. Kirshner *et al.*, *Astrophys. J.* **415**, 589 (1993)
[53] E. Livne, *Astrophys. J.* **527**, L97 (1999)

15

Type Ia supernovae: spectroscopic surprises

D. Branch

Department of Physics and Astronomy, University of Oklahoma, Norman, OK 73019, USA

Abstract

Recent observations have extended the range of diversity among spectra of Type Ia supernovae. I briefly discuss SN Ia explosion models in the spectroscopic context, the observed diversity, and some recent results obtained with the **Synow** code for one normal and two peculiar SNe Ia. Relating the observational manifestations of diversity to their physical causes is looking like an ever more challenging problem.

15.1 Introduction

"Surprises" refers not only to some recent developments in Type Ia supernova (SN Ia) spectroscopy that will be discussed below, but also to additional recent discoveries that I will be able only to mention, such as the polarization signal in SN 2001el (Wang *et al.* 2003; see also the chapter by Wang); the unusual properties of SN 2001ay (see the chapter by Howell); and the circumstellar Hα emission of SN 2002ic (Hamuy *et al.* 2003; see also the chapter by Hamuy). The scope of this chapter is restricted to photospheric–phase optical spectra. For recent results on infrared spectra see, e.g., Marion *et al.* (2003).

Some background, including mention of the various kinds of SN Ia explosion models in the spectroscopic context, is in §15.2. An overview and update of the SN Ia spectroscopic diversity is in §15.3. Some recent results from direct analysis of the spectra of three events (the normal SN 1998aq and the peculiar SNe 2000cx and 2002cx), obtained with the parameterized, resonance scattering code **Synow**, are discussed in §15.4. The final section (§15.5) contains more questions than conclusions.

15.2 Background

Around the time of maximum light the optical spectrum of a normal SN Ia consists of a thermal continuum with superimposed features due to lines of ions such as Si II, S II, Ca II, and O I. Ubiquitous blends of Fe II develop shortly after maximum. The features are formed by resonance scattering of the photospheric continuum and have P Cygni–type profiles characteristic of expanding atmospheres. Emission components peak at the rest wavelength and absorption components are blueshifted according to the velocity of the matter at the photosphere, ordinarily $\sim$10,000 km s^{-1}.

SNe Ia are thought to be thermonuclear disruptions of accreting or merging carbon–oxygen white dwarfs. The classic SN Ia explosion model is model W7 of Nomoto, Thielemann, & Yokoi (1984), a 1D model that was constructed by parameterizing the speed of the nuclear burning front. Because the speed remained subsonic, W7 is known as a deflagration model (but see below). The composition structure of W7 is radially stratified, with a low-velocity ($<$10,000 km s^{-1}) $\sim$1 $M_\odot$ core of iron-peak elements (initially mostly radioactive ^{56}Ni), surrounded by $\sim$0.3 $M_\odot$ of intermediate-mass elements such as silicon, sulfur, and oxygen expanding at $\sim$10,000 to $\sim$15,000 km s^{-1}, capped by $\sim$0.1 $M_\odot$ of unburned carbon and oxygen moving at $\sim$15,000 to $\sim$22,000 km s^{-1}. Synthetic-spectrum calculations of various levels of complexity (Branch *et al.* 1985; Wheeler & Harkness 1990; Nugent *et al.* 1995; Salvo *et al.* 2001; Lentz *et al.* 2001a) have shown that the spectra of W7, with or without mixing of the layers above $\sim$8000 km s^{-1}, generally resemble the observed spectra of normal SNe Ia. Recent detailed calculations with the **Phoenix** code indicate that spectra of model W7 begin to differ from observed spectra after maximum light (Lentz *et al.* 2001a), and that they may fail to quantitatively reproduce the dependence of the *R(Si II)* parameter on temperature (S. Bongard *et al.*, in preparation). Nevertheless, there seems to be something right about the basic composition structure of W7: an iron–peak core surrounded by lighter elements.

In 1D delayed-detonation models (Khokhlov 1991) the subsonic deflagration makes a transition to a supersonic detonation when the burning front reaches a transition density that is treated as a free parameter. Composition structures of 1D delayed-detonations are not extremely different from that of model W7. The main differences are in the outer, high-velocity layers, because in delayed-detonations the burning front extends farther into the outer layers of the white dwarf and leaves less unburned carbon than W7. [A nice comparison of the composition structures of W7 and a 1D delayed-detonation can be found in Sorokina *et al.* (2000).] Höflich and colleagues (e.g., Höflich 1995; Höflich, Wheeler, & Thielemann 1998; Wheeler *et al.* 1998; see also the chapter by Höflich) have maintained that spectra of delayed-detonation models agree well with the observed spectra of normal SNe Ia.

Only recently have 3-D deflagrations been calculated (Khokhlov 2000; Gamezo *et al.* 2003; see also the chapter by Gamezo, Khokhlov, & Oran and references therein). The composition structure of the 3-D deflagration presented by Khokhlov and Gamezo *et al.* is quite unlike that of a 1D deflagration. The 3-D model contains clumps of iron-peak elements surrounded by shells of intermediate-mass elements, all embedded in a substrate of unburned carbon and oxygen. The angle-averaged composition is *not* strongly radially stratified; in fact, unburned carbon is present all the way to the center. Although the model has not yet been evolved to homologous expansion, it is unlikely that it will be compatible with the observed spectra of normal SNe Ia. The presence of unburned carbon at low velocity may not be a problem during the first months after explosion (Baron *et al.* 2003; see also the chapter by Lentz), but a strong argument can be made for a smooth distribution of silicon in normal SNe Ia in order to consistently produce the deep 6100 Å absorption (Thomas *et al.* 2002; see also the chapter by Thomas). The 3-D model also would produce [C II] and [O I] lines in the nebular phase (C. Kozma *et al.*, in preparation), which are not observed. However, some future 3-D deflagration models may be different: much may depend, for example, on whether the deflagration originates at a single point or practically simultaneously at several or even a multitude of points.

In the chapter by Gamezo *et al.* a 3-D *delayed-detonation* model is discussed, in which the composition structure is radially stratified, rather like 1D explosion models. If the composition structures of future 3-D deflagrations continue to be very different from that of model W7, perhaps W7 should be referred to as a parameterized 1D model, rather than as a deflagration, because its composition structure may be more like 3-D delayed-detonations.

The effects on explosion models of rapid rotation of the white-dwarf progenitors (Uenishi *et al.* 2003; see also the chapter by Yoon and Langer) has not yet begun to be taken into account.

15.3 Spectroscopic diversity

Nugent *et al.* (1995) showed that retaining the composition structure of model W7 and varying the temperature produced a sequence of synthetic spectra that resembled a sequence of observed SN Ia spectra ranging from the peculiar powerful SN 1991T, with its high-excitation Fe III features, through the normals, to the peculiar weak SN 1991bg, with its low-excitation Ti II features. Quantitatively, the sequence was arranged according to increasing values of the parameter *R(Si II)*, the ratio of the depth of an absorption feature near 5700 Å to the depth of the deeper absorption near 6100 Å that is primarily due to Si II λ6355. [The major contributors to the 5700 Å aborption remain uncertain (Garnavich *et al.* 2001, S. Bongard *et al.*, in preparation)]. As shown by Nugent *et al.* (1995), and Benetti *et al.* (2003), the

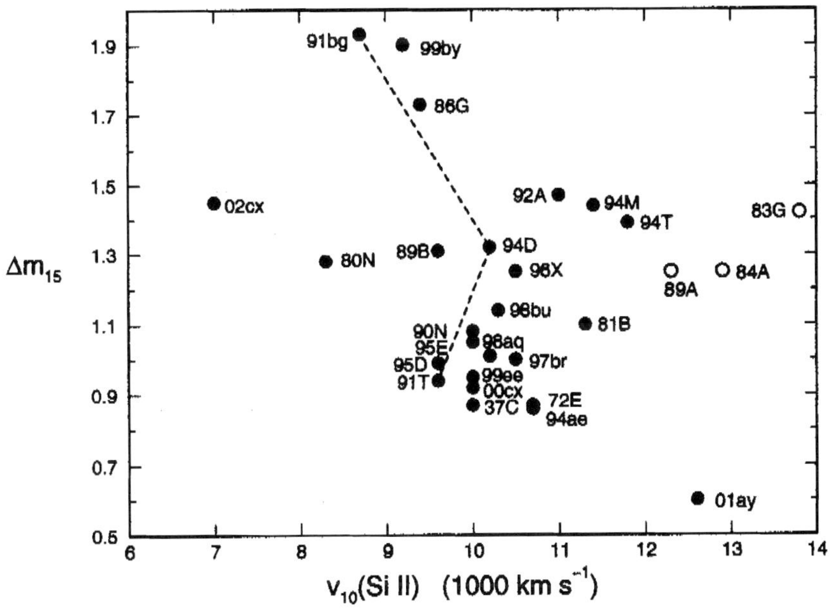

Fig. 15.1. The photometric Δm_{15} parameter is plotted against the spectroscopic V_{10}(*Si II*) parameter. Open circles mean that the value of Δm_{15} has been estimated from the observed value of the *R(Si II*) parameter. The dashed line is intended to represent the temperature sequence of Nugent *et al.* (1995).

R(*Si II*) parameter correlates with the light-curve decline-rate parameter Δm_{15} of Phillips (1993).

But it has become well known that not all SNe Ia can be forced into a single spectroscopic sequence. Events such as SN 1984A have the same spectral features as normal SNe Ia but higher expansion velocities at a given epoch. I will refer to the most extreme of these as HPV (high photospheric velocity) SNe Ia. Lentz *et al.* (2001b) found that certain delayed-detonation models, with their high densities in the high-velocity ($\sim$20,000 km s^{-1}) layers, were able to account well for the spectra of SN 1984A. The HPV SNe Ia have high values of the V_{10}(*Si II*) parameter – the velocity corresponding to the blueshift of the Si II λ6355 absorption at 10 days after maximum light (Branch & van den Bergh 1993). Hatano *et al.* (2000) presented a plot of *R*(*Si II*) against V_{10}(*Si II*) to illustrate the extent to which some events having similar values of *R*(*Si II*) have different values of V_{10}(*Si II*). Hatano *et al.* suggested that it may be useful to think of *R*(*Si II*) (like Δm_{15}) as a measure of the ejected nickel mass and V_{10}(*Si II*) as a measure of the density in the high-velocity layers – whether the high density actually is produced by a delayed detonation or not. Similarly, one can plot Δm_{15} against V_{10}(*Si II*), as in Fig. 15.1. Plotting Δm_{15} allows some interesting events for which *R*(*Si II*) is unavailable to be included. The dashed line serves to represent the Nugent "temperature" sequence connecting the peculiar SN 1991bg to the peculiar SN 1991T via the normal SN 1994D (a different normal event, e.g., SNe 1989B, 1996X, or SN 1998bu, could just as well have been chosen). SN 2001ay, an HPV event with a very slow postmaximum light-curve decline (see the chapter by Howell) and SN 2002cx (Li *et al.* 2003) have extended the observed range of Δm_{15} and $V_{10}(Si\ II)$, respectively, and the value of $V_{10}(Si\ II)$ plotted for SN 2002cx is an upper limit because the Si II absorption was seen only near maximum light.

15.4 Recent results with Synow

The spectroscopically normal SN 1998aq was thoroughly observed by the CfA group. In Fig.15.1 it falls in a region heavily populated by normals. One result of a recent direct analysis of the photospheric-phase spectra with the **Synow** code (Branch *et al.* 2003) was good evidence for the presence of C II lines indicating the presence of carbon at velocities at least as low as 11,000 km s^{-1}. Mazzali (2000) identified the same lines, but at higher velocity, $\sim$16,000 km s^{-1}, in the normal SN 1990N. Carbon as slow as 11,000 km s^{-1} would be inconsistent with published 1D delayed-detonation models. Most of the other spectral features in the early spectra were securely identified, making SN 1998aq a useful benchmark for comparison with other SNe Ia.

SN 2000cx was extensively observed by Li *et al.* (2002), who referred to it as uniquely peculiar. In Figure 15.1 it falls among the normals, but the unusual properties discovered by Li *et al.* included a lopsided *B*-band light curve – quick to rise but slow to fall – and a strange $B - V$ color evolution – from redder than normal before maximum to bluer than normal after maximum. The many spectral peculiarities included Ca II infrared-triplet absorptions forming not only at the photospheric velocity of $\sim$10,000 km s^{-1} but also at much higher velocity, greater than 20,000 km s^{-1}. Thomas *et al.* (2003) investigated the high-velocity Ca II features with a parameterized 3-D spectrum-synthesis code and concluded that they probably formed in a nonspherical high-velocity structure that partially covered the photosphere. Kasen *et al.* (2003) already had reached a similar conclusion for the high-velocity Ca II features discovered by Wang *et al.* (2003) in SN 2001el, for which polarization measurements clearly indicated a strong departure from spherical symmetry in the high-velocity matter. I will refer to features like these as DHVFs (detached high velocity features). Direct analysis of the photospheric-phase spectra of SN 2000cx with **Synow** (D. Branch *et al.*, in preparation) reveals that in the blue, the spectra were composite, containing some features forming just above the photospheric velocity, but also containing additional DHVFs forming at the same high velocity as the Ca II DHVFs. In particular, blends of Ti II HDVFs

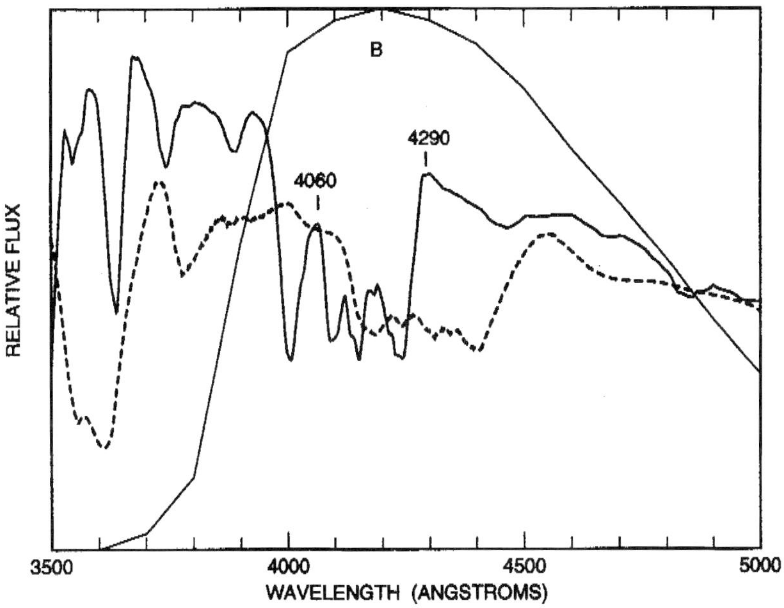

Fig. 15.2. Two synthetic spectra having $v_{phot} = 11{,}000$ km s^{-1} and containing only Ti II lines are compared. One spectrum (dashed line) contains only undetached lines and has features like those seen in the peculiar SN 1991bg. The other spectrum (thick solid line) contains only lines detached at 23,000 km s^{-1}, as required to fit the spectrum of the peculiar SN 2000cx. The *B*-band filter function also is shown (thin solid line).

suppressed the spectrum in the blue (see Figure 15.2). The flux blocking by Ti II HDVFs decreased with time, which is in the right sense to cause the *B*-band light curve to rise more quickly and fall more slowly than it otherwise would have done, and to cause the peculiar $B - V$ evolution. Quantitative evaluation of the effects of line blocking by DHVFs forming in an asymmetric matter distribution will require 3-D spectrum calculations such as those of Thomas *et al.* (2003). In any case, as an extreme case of DHVFs SN 2000cx shows the importance of learning how much of the SN Ia spectroscopic and photometric diversity is caused by DHVFs.

SN 2000cx spectra contained a distinct, persistent absorption feature near 4530 Å for which the most plausible identification that we (Thomas *et al.* 2003 and Branch *et al.*, in preparation) can suggest is an Hβ (λ4861) DHVF, blueshifted by the same high velocity as the Ca II and Ti II DHVFs. At this blueshift the Hα absorption happens to fall within the red Si II absorption. As discussed by Thomas *et al.* and Branch *et al.*, if Hα and Pα are confined to a clump in front of the photosphere and have source functions that are elevated relative to the resonance-scattering source function (as they generally do in SNe II), then they could be difficult to see. In principle their spectroscopic signatures even could vanish. The Hβ identification remains tentative, however, because there is no independent evidence for the presence of hydrogen.

SN 2002cx was just as surprising as SN 2000cx. Li *et al.* (2003) observed SN 2002cx and referred to it as the "most peculiar" SN Ia. Near maximum its spectra first contained features of Fe III; Fe II features developed later. Despite the fact that Si II and S II features apparently were not present, Li *et al.* argued that SN 2002cx should be regarded as a Type Ia – the first one observed to be blue, yet subluminous. At maximum light the velocity at the photosphere was only 7000 km s^{-1}, making it the first good example of an LPV (*low* photospheric velocity) SN Ia. Li *et al.* called attention to mysterious emission lines in the red part of spectra obtained three weeks postmaximum, and to what appeared to be an unusually early appearance of nebular emission lines two months postmaximum. The early development of the nebular phase was taken as an indication that the mass ejection may have been sub-Chandraskhar. Direct analysis with **Synow** (D. Branch *et al.*, in preparation) clarifies some of these issues. At maximum light the spectrum contained not only Fe III features but also weak features of Si III, Si II, and S II – confirming the Type Ia classification. The spectrum also contained Fe II and Ca II DHVFs forming at about 14,000 km s^{-1}. The **Synow** analysis reveals that in the spectrum several weeks postmaximum the mysterious emission lines in the red actually are blends of P Cygni features of permitted Fe II and Co II lines (see Figure 15.3), and that the spectrum two months postmaximum had not gone nebular – it was much like the two-week postmaximum spectrum, but with stronger Fe II and weaker Co II. SN 2002cx raises several interesting questions. For example, if the ejected composition was dominated by iron-peak elements, the kinetic energy per unit mass should have been high, so why were the velocities low? And if the mass ejection was Chandraskhar, why was the luminosity low? One possibility (see the chapter by Kasen) is that SN 2002cx was a low-luminosity SN 1991bg-like event, viewed right down the hole in the ejecta caused by the presence of the donor star (Marietta, Burrows, & Fryxell 2000).

15.5 Questions

Thinking of *R*(*Si II*) as a measure of the nickel mass and of V_{10}(*Si II*) as a measure of the density at high velocity may be a useful way to think about the diversity, but clearly it is not enough. Some of the diversity is produced by DHVFs, which we now recognize to be not uncommon in SNe Ia. Weak DHVFs such as those of SN 1994D (Hatano *et al.* 1999) may be caused by radial ionization variations in smooth ejecta density distributions (Höflich *et al.* 1998; Lentz *et al.* 2001a), but what is the origin of the stronger DHVFs – those that appear to require density enhancements

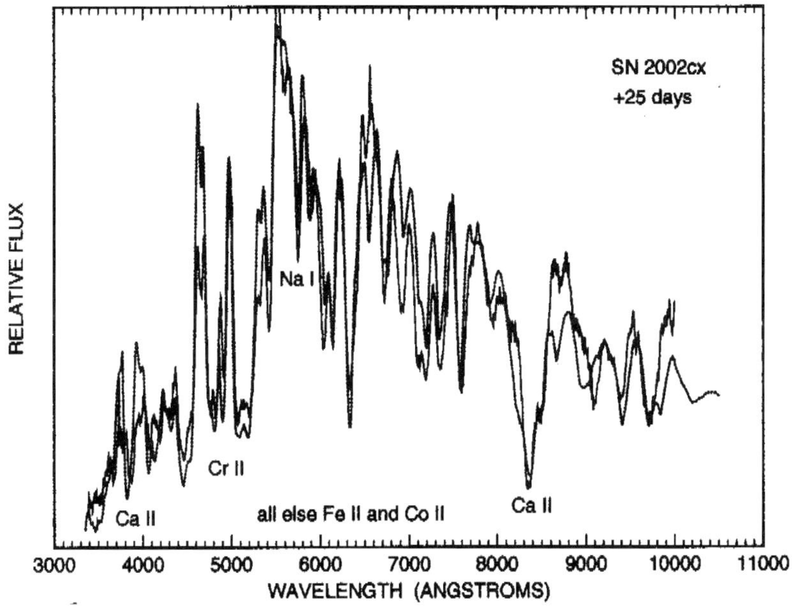

Fig. 15.3. A synthetic spectrum of having $v_{phot} = 7000$ km s^{-1} and dominated by strong permitted resonance scattering features of Fe II and Co II (dotted line) is compared with a spectrum of the peculiar SN 2002cx obtained by Li *et al.* (2003) 25 days after maximum light (solid line).

at high velocity? The most extreme case so far is SN 2000cx with its strong Ti II DHVFs. The DHVFs of SNe 2001el and 2000cx appear to be in asymmetric matter distributions. Are they a consequence of asymmetric matter ejection?

Has an Hβ DHVF been detected in SN 2000cx? If so, at least some of the matter producing the DHVFs of SN 2000cx must have come from a nondegenerate companion star. Do any other SNe Ia also show Hβ DHVFs, at less noticeable levels? Gerardy *et al.* (2003) suggest that strong DHVFs are produced by high-velocity density enhancements owing to interaction with solar-abundance circumstellar matter. If this is correct, then it may be DHVFs, rather than narrow Hα emission or absorption, that are the first to reveal the presence of the long-sought circumstellar matter associated with SN Ia single-degenerate progenitor systems.

Events such as SNe 2001ay, 2002cx, and 2002ic seem, so far, to be peculiar each in their own way. As mentioned above, the effects of the Marietta *et al.* hole in the ejecta may be responsible for some of the peculiarities of SN 2002cx, as well as some of the other SN Ia diversity (see the chapter by Kasen). SN 2002ic (Hamuy *et al.* 2003; see also the chapter by Hamuy) evidently is a very special case, since it shows a clear signature of circumstellar hydrogen to a degree that would not have been missed in spectra of previously observed SNe Ia. Note that the characteristic velocity of the Hα emission of SN 2002ic, $\sim$1800 km s^{-1}, is much lower than the velocity of the Hβ DHVF tentatively identified in SN 2000cx ($\sim$23,000 km s^{-1}).

Learning how to relate the various manifestations of observational diversity – the range of *R*(*Si II*) and V_{10}(*Si II*) values, the DHVFs, and the individually peculiar events – to their physical causes – e.g., nickel mass, flame propagation, asymmetric ejection, asymmetric circumstellar interaction, looking down the hole – is looking like an increasingly challenging problem.

Acknowledgements

I have had the benefit of many discussions of these matters with Eddie Baron, Peter Höflich, Dan Kasen, Peter Nugent, and Rollin Thomas. It is a pleasure to take this opportunity to thank Craig Wheeler for all he has done to make supernova research so interesting and enjoyable.

References

Baron, E., Lentz, E. J., & Hauschildt, P. H. 2003, ApJ, 588, L29
Benetti, S., *et al.* 2003, MNRAS, in press
Branch, D., Doggett, J. B., Nomoto, K., & Thielemann, F.-K. 1985, ApJ, 294, 619
Branch, D., & van den Bergh, S. 1993, AJ, 105, 2231
Branch, D., *et al.* 2003, AJ, 126, 1489
Gamezo, V. N., Khokhlov, A. M., Oran, E. S., Chtchelkanova, A. Y., & Rosenberg, R. O. 2003, Science, 299, 77
Garnavich, P., *et al.* 2001, astro-ph/0105490
Gerardy, C. L., *et al.* 2003, ApJ, submitted
Hamuy, M., *et al.* 2003, Nature, 424, 651
Hatano, K., Branch, D., Lentz, E. J., Baron, E., Filippenko, A. V., & Garnavich, P. M. 2000, ApJ, 543, L49
Hatano, K., Branch, D., Fisher, A., Baron, E., & Filippenko, A. V. 1999, ApJ, 525, 881
Höflich, P. 1995, ApJ, 443, 831
Höflich, P., Wheeler, J. C., & Thielemann, F.-K. 1998, ApJ, 495, 617
Kasen, D., *et al.* 2003, ApJ, 593, 788
Khokhlov, A. M. 1991, A&A, 245, 114
Khokhlov, A. M. 2000, astro-ph/0008463
Lentz, E. J., Baron, E., Branch, D., & Hauschildt, P. H. 2001b, ApJ, 547, 402
Lentz, E. J., Baron, E., Branch, D., & Hauschildt, P. H. 2001a, ApJ, 557, 266
Li, W., *et al.* 2003, PASP, 115, 453
Li, W., *et al.* 2002, PASP, 113, 1178
Marion, G. H., Höflich, P., Vacca, W. D., & Wheeler, J. C. 2003 ApJ, 591, 316
Marietta, E., Burrows, A., & Fryxell, B. 2000, ApJS, 128, 615
Mazzali, P. A., 2001, MNRAS, 321, 341
Nomoto, K., Thielemann, F.-K., & Yokoi, K., 1984, ApJ, 286, 644
Nugent, P., Phillips, M. M., Baron, E., Branch, D., & Hauschildt, P. 1995, ApJ, 455, L147
Phillips, M. M. 1993, ApJ, 413, L105

Salvo, M. E., Capppellaro, E., Mazzali, P. A., Benetti, S., Danziger, I. J., Patat, F. & Turatto, M. 2001, MNRAS, 321, 254
Sorokina, E. I., Blinnikov, S. I., & Bartunov, O. S. 2000, Astr. Let. 26, 67
Thomas, R. C., Branch, D., Baron, E., Nomoto, K., Li, W., & Filippenko, A. V. 2003, ApJ, in press
Thomas, R. C., Kasen, D., Branch, D., & Baron, E. 2002 ApJ, 567, 1037
Uenishi, T., Nomoto, K., & Hachisu, I. 2003, ApJ, 595, 1094
Wang, L., *et al.* 2003, ApJ, 591, 1110
Wheeler, J. C., Höflich, P., Harkness, R. P., & Spyromilio, J. 1998, ApJ, 496, 908
Wheeler, J. C., & Harkness, R. P. 1990, Rep. Prog. Phys. 53, 1467

16

Asphericity effects in supernovae

P. Höflich, C. Gerardy, R. Quimby

University of Texas, Austin, TX 78712, USA

Abstract

We present a brief summary of asphericity effects in thermonuclear and core collapse supernovae (SN), and how to distinguish the underlying physics by their observable signatures. Electron scattering is the dominant process to produce polarization which is one of the main diagnostical tools. Asphericities result in a directional dependence of the luminosity which has direct implications for the use of SNe in cosmology. For core collapse SNe, the current observations and their interpretations suggest that the explosion mechanism itself is highly aspherical with a well defined axis and, typically, axis ratios of 2 to 3. Asymmetric density/chemical distributions and off-center energy depositions have been identified as crucial for the interpretation of the polarization *P*. For thermonuclear SNe, polarization turned out to be an order of magnitude smaller strongly supporting rather spherical, radially stratified envelopes. Nevertheless, asymmetries have been recognized as important signatures to probe A) for the signatures of the progenitor system, B) the global asymmetry with well defined axis, likely to be caused by rotation of an accreting white dwarf or merging WDs, and C) possible remains of the deflagration pattern.

16.1 Introduction

During the last decade, advances in observational, theoretical and computational astronomy have provided new insights into the nature and physics of SNe and gamma-ray bursts. Due to the extreme brightness of these events, they are expected to continue to play important role in cosmology. SNe Ia allowed good measurements of the Hubble constant both by statistical methods and theoretical models. SNe Ia have provided one of the strong evidences for a non-zero cosmological constant, and will be used to probe the nature of the dark energy. In the future, core collapse SNe and the related GRBs may become the tool of choice to probe the very first generation of stars. The multidimensional nature of these objects has been realized and it has become obvious that measurements of asymmetries and their understanding is a key for understanding of both SNe and GRBs. In this contribution, we want to give a brief overview on the observable consequences of asphericity including a directional dependence of the luminosity with a special emphasis on polarization in Thomson scattering dominated envelopes. In the first part, we want to give a brief introduction to the configurations which produce polarization. Subsequently, we present examples for various mechanisms which produce polarization in core collapse and thermonuclear SNe. It is beyond the scope to present a review of the current literature. A more general overview about the physics core collapse SNe with references to the general literature can be found in Höflich *et al.* (2002). For thermonuclear SNe on scenarios and details of the nuclear burning front, we want to refer Höflich *et al.* (2003), and Hillebrandt & Niemeyer (2000), Khokhlov (2001) and Gamezo & Khokhlov (this volume), respectively. We will focus on the theoretical aspects. For a complementary discussion of the observations, see Wang (this volume).

16.2 General

Asymmetry can be probed by direct imaging of ejecta of the remnants, e.g. in SN1987A (Wang *et al.* 2002, Höflich *et al.* 2001b), or Cas A (Fesen & Gunderson 1997), proper motions of neutron stars (Strom *et al.* 1995), or, more generally by polarization measurements during the early phase of the expansion. In SN, polarization is mainly produced by Thomson scattering of photons in an aspherical configuration. It can be caused by asymmetries in the density, abundances or excitation structure of an envelope. In general, the ejecta cannot be spatially resolved. Although the light from different parts of a spherical disk is polarized, the resulting polarization $\bar{P}$ is zero for the integrated light (Fig.16.1). To produce $\bar{P}$, three basic configurations may be considered, in which I) the photosphere is aspherical, II) parts of the disk are shaded, and III) the envelope may be illuminated by an off-center light source. In case II, the shading may be either by a broad-band absorber such as dust or a specific line opacity. In the latter case, this would produce a change of $\bar{P}$ in a narrow line range (Fig. 16.2). In reality, a combination of all cases may be realized. Note that quantitative analyses of SNe need to take into account that the continua and lines are not formed in the same layers. Polarization carries the information about the apparent, global asymmetries. E.g., it increases with increasing axis ratios in ellipsoidal geometries or off-center energy sources but decreases due to multiple scattering (the optical depth) or steep density profiles. The observed size of P depends also on the position of the observer relative to the object. For a detailed discussion see Höflich (1991, 1995). As a consequence, the interpretation of polarization data are not unique. Another problem is due to

Electromagnetic wave : $\psi(z, t) = Ee^{i(kz-wt)}$

$\underline{E} = (E_x, E_y)$

Intensity is defined as the time average over many waves

$I = I_0 + I_{00} = \overline{E_x\ E_x^* + E_y\ E_y^*}\ =\ \overline{E_x^2 + E_y^2}$

Degree of polarization *P*

$P = (I_0 - I_{00})/(I_0 + I_{00})$

with position angle χ

Stokes Parameter (equivalent)

$Q - I_0 - I_{90}$

$U = I_{45} - I_{-45}$

$V = 0$ for linear polarization

Rem.: $tan\ 2\chi - U/Q$ and $P - \sqrt{Q^2 + U^2}$

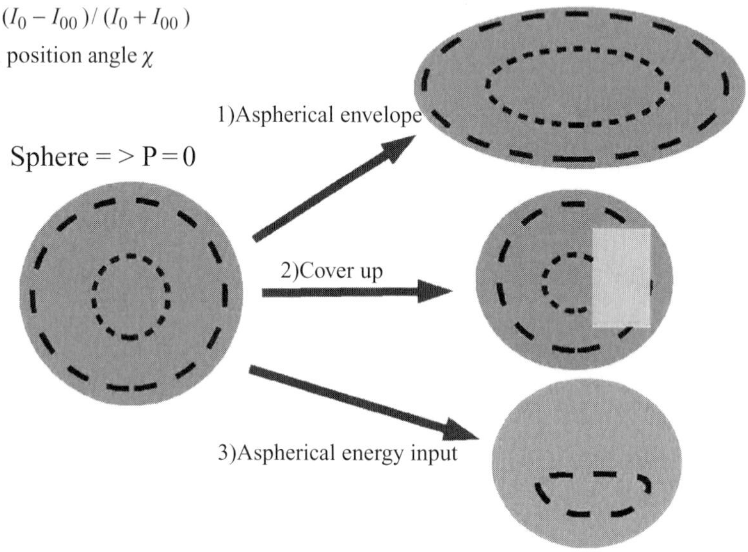

Fig. 16.1. Definition of the polarization and schematic diagram for its production. The dotted lines give the main orientation of the electrical vectors. For an unresolved sphere, the components cancel out (from Höflich 1995).

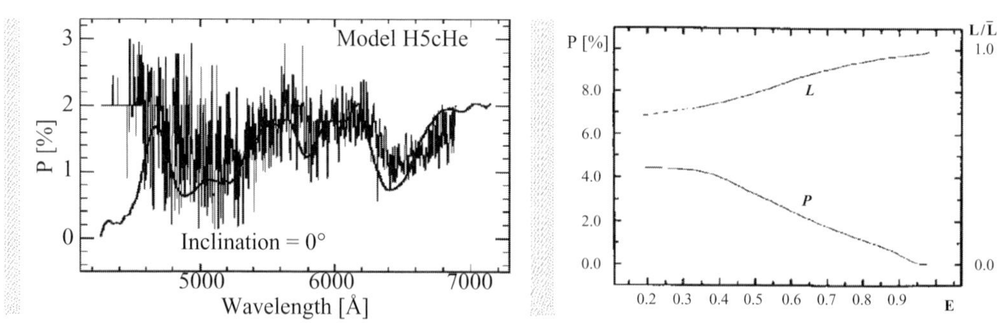

Fig. 16.2. Polarization spectrum for SN1993J for an axis ratio of 1/2 for an oblate ellipsoid in comparison with observations by Trammell *et al.* (1993) (left plot). On the right, the dependence of the continuum polarization (right) and directional dependence of the luminosity is shown as a function axis ratios for oblate ellipsoids seen from the equator (from Höflich, 1991 & Höflich *et al.* 1995).

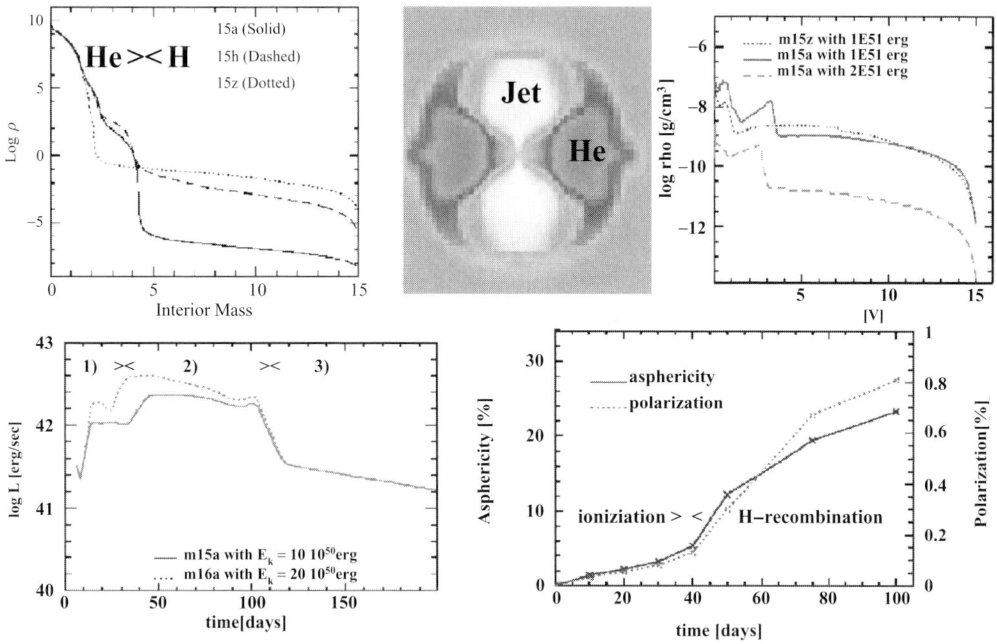

Fig. 16.3. Polarization produced by an aspherical, chemical distribution for an extreme SN IIp model such as SN1999em (see text).

polarization by the interstellar medium. In parts, these limitations can be overcome by spectropolarimetry and time series of observations which provides additional information due to spectral features (see Fig. 16.2) and their evolution. Still, the use of consistent physical models is mandatory to further constrain the variety of interpretation.

16.3 Core collapse supernovae

In recent years, there has been a mounting evidence that the explosions of massive stars (core collapse SNe) are highly aspherical. The spectra (e.g., SN87A, SN93J, SN94I, SN99em, SN02ap) are significantly polarized at a level of 0.5 to 3% (Méndez *et al.* 1988, Cropper *et al.* 1988, Höflich 1991, Jeffrey 1991) indicating aspherical envelopes by factors of up to 2 (see Fig. 16.2). The degree of polarization tends to vary inversely with the mass of the hydrogen envelope, being maximum for Type Ib/c events with no hydrogen (Wang *et al.* 2001). For SNeII, Leonard *et al.* (2000) and Wang *et al.* (2001) showed that the polarization and, thus, the asphericity increase with time. The orientation of the polarization vector tends to stay constant both in time and with wavelength. This implies that there is a global symmetry axis in the ejecta. Both trends suggest a connection of the asymmetries with the central engine which may be understood in terms of jet-induced explosions (Khokhlov *et al.* 1999, Höflich *et al.* 1998, 2002), or pulsational modes in neutrino driven explosions (Scheck *et al.* 2003).

However, even strongly asymmetric explosions do not produce asymmetries in the massive hydrogen-rich envelopes of SNeII which are sufficiently large to explain the polarization observed in SN1987A or SN1999em. Aspherical excitation by hard radiation is found to be crucial. As example, the extreme SNIIp 1999em is shown in Fig.16.3 and, for details, see Höflich *et al.* (2002). Our calculations of the initial stage of the explosion employ 3-D hydrodynamics. The explosion of a star with 15 solar masses is triggered by a low velocity, high density jet/bipolar outflow which delivers a explosion energies of of 1 and $2 \times 10^{51} erg$. The jet from the central engine stalls after about 250 seconds, and the abundance distribution freezes out in the expanding envelope. The resulting distribution of the the He-rich layers is given in Fig.16.3. The colors white, yellow, green, blue and red correspond to He mass fractions of 0., 0.18, 0.36, 0.72, and 1., respectively. The composition of the jet-region consists of a mixture of heavy elements with about 0.07 $M_\odot$ of radioactive ^{56}Ni. After about 100 seconds, the expansion of the envelope becomes spherical. Thus, for times larger than 250 seconds, the explosion has been followed in 1-D up to the

Fig. 16.4. Analysis of the subluminous SN1999 by using a combination of flux and polarization data. **Upper panel:** Comparison of the NIR spectrum on May 16 (left) with a spherical, subluminous delayed detonation model. For this object, the spectra are formed in layers of explosive carbon and incomplete silicon burning up to about 2 weeks after maximum light. This is in strict contrast to normal bright SNe Ia where the photosphere enters layers of complete Si burning already at about maximum light. On the right, we show a comparison of the observed and theoretical spectrum if we impose mixing of the inner 0.7 $M_\odot$ as can be expected based on detailed 3-D deflagration models (Khokhlov, 2001). Strong homogeneous mixing of the inner layers can be ruled out because the excess excitation of intermediate mass elements and the absorption by iron-group elements (from Höflich *et al.* 2002). **Lower, left panel:** Energy deposition by γ-rays at day 1 (left) and 23 (right) based on our full 3-D MC gamma ray transport based on a ^{56}Ni distribution of a typical deflagration models. The diameter of the WD is normalized to 100. At about day 23, the energy deposition is not confined to the radioactive ^{56}Ni ruling out clumpiness as a solution to the problem mentioned above (from Höflich 2002). **Lower, right panel:** Optical flux and polarization spectra at day 15 after the explosion for the subluminous 3-D delayed-detonation model in comparison with the SN1999 by at about maximum light. The interstellar component of P has been determined to $P = 0.25\%$ with a polarization angle of 140°. The observed flux and the smoothed polarization spectra are the solid black lines. The light grey line is the original data for P at a resolution of 12.5 Å. In the observations, the polarization angle is constant indicating rotational symmetry of the envelope. The structure of the spherical model has been mapped into oblate ellipsoids with axis [t]tios A/B of 1.17 (from Howell, Höflich, Wang & Wheeler 2001).

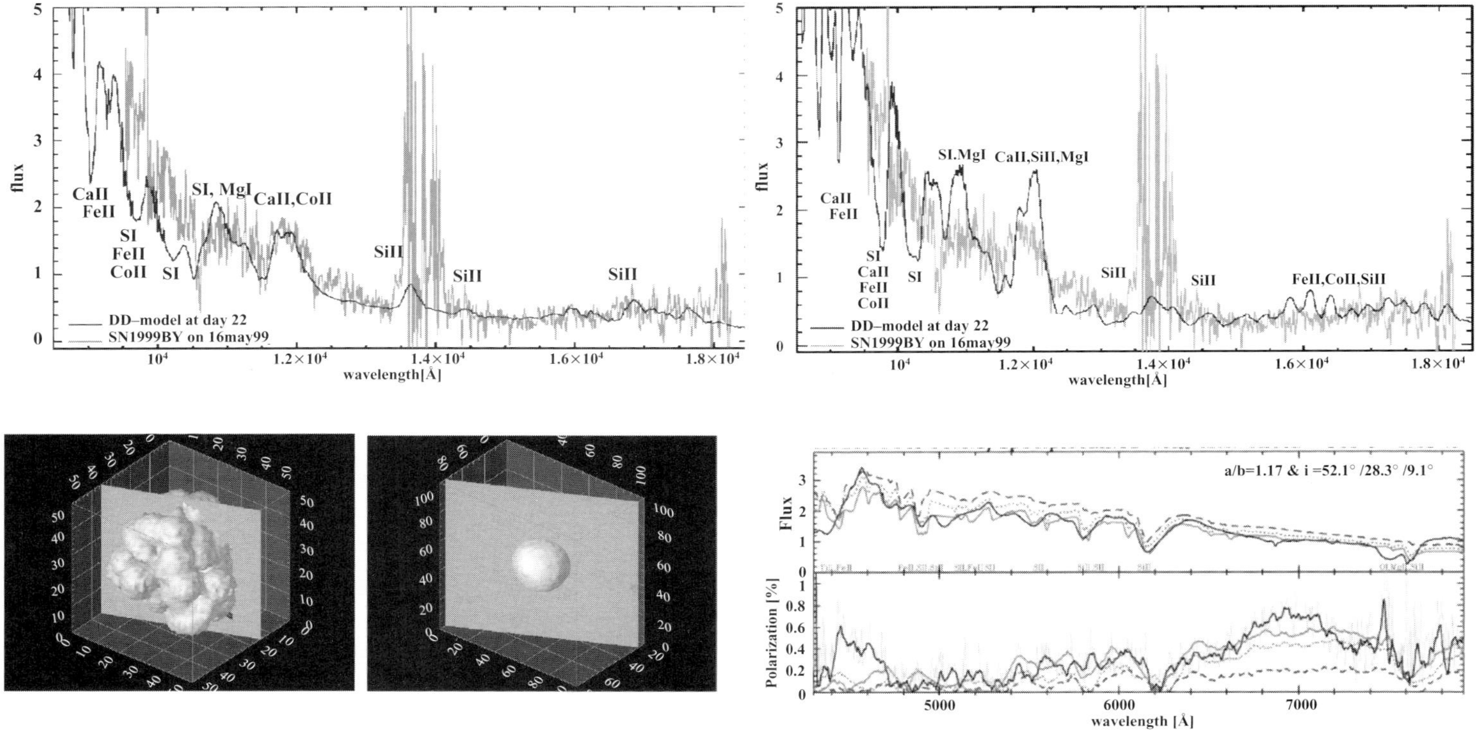
CaII
FeII
SI
FeII
CoII
SI
SI, MgI
CaII,CoII
SiII
SiII
SiII
DD–model at day 22
SN1999BY on 16may99
flux
wavelength[Å]
SI,MgI
CaII,SiII,MgI
FeII,CoII,SiII
a/b=1.17 & i =52.1° /28.3° /9.1°
Flux
Polarization [%]
wavelength [Å]

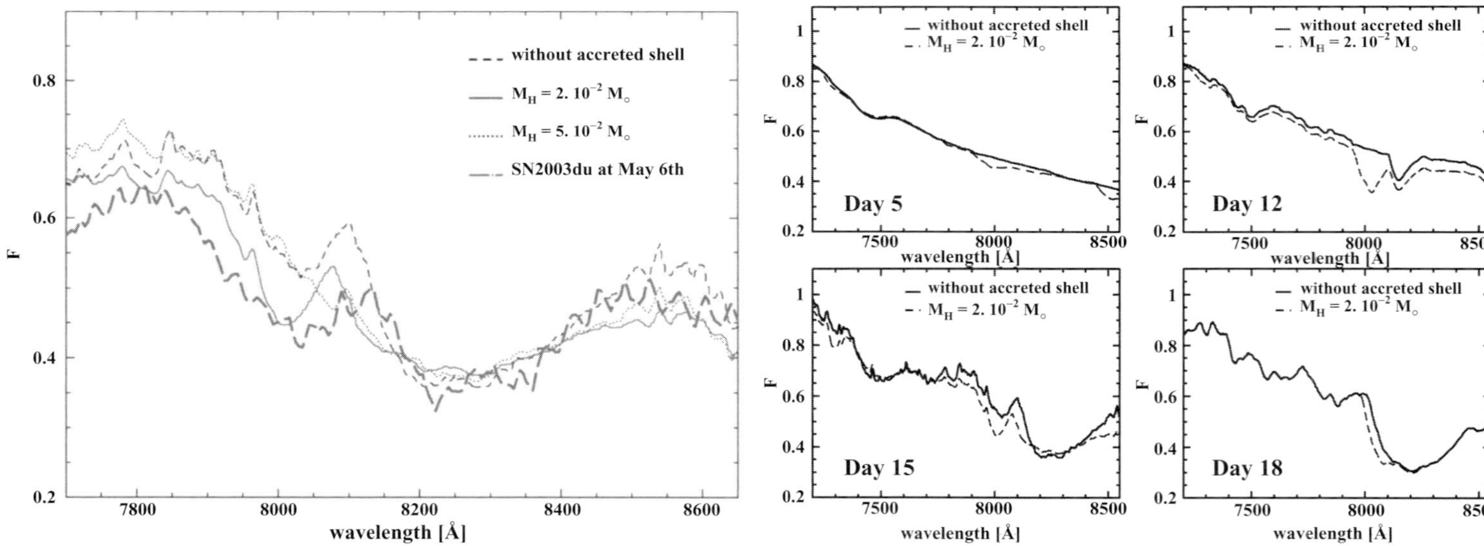

Fig. 16.5. CaII IR feature observed in SN 2003du on May 6rd in comparison with theoretical models at about 15 days after the explosion, and its evolution with time. The models are based on a delayed detonation model which interacted with a H-rich shell of 0.02 and 0.05 $M_\odot$ during the early phase of the explosion. The dominant signature of this interaction is the appearance of a secondary, high velocity Ca II feature or, for high shell masses, a persistent high velocity component in a broad Ca II line. Without ongoing interaction, no H or He lines are detectable. Note that, even without a shell, a secondary Ca II feature can be seen for a period of 2 to 3 days during the phase when Ca III recombines to Ca II emphasizing the importance of a good time coverage for the observations.

phase of homologous expansion. In the upper, right panel, the density distribution is given at about 5 days after the explosion. The steep gradients in the density in the upper right and left panels are located at the interface between the He-core and the H-rich mantel. In the lower, left panel, the resulting bolometric LCs are given for explosion energies of 2E51erg (dotted line) and 1E51erg, respectively. Based on full 3-D calculations for the radiation & γ-ray transport, we have calculated the location of the recombination front (in NLTE) as a function of time. The resulting shape of the photosphere is always prolate. The corresponding axis ratio and the polarization seen from the equator are shown (lower, right panel). Note the strong increase of the asphericity after the onset of the recombination phase between day 30 to 40 (Höflich *et al.* 2002). For the polarization in a massive, H-rich envelope, P seems to be directly linked to the recombination process and asymmetric excitation.

16.4 Thermonuclear explosions

For thermonuclear explosions, polarization turned out to be an important tool to probe for A) the global asymmetry caused by the WD, i.e. the continuum component, B) for the signatures of the progenitor system, and C) possible remains of the deflagration pattern produced during the early phase of nuclear burning.

Case A: In general, the maximum, continuum polarization is about an order of magnitude smaller than in core collapse SNe and it is decreasing with time but, again, with a well defined axis of symmetry (e.g. Wang *et al.* 2003). This decrease occurs despite a significant contribution of electron scattering to the opacity till about 1 to 2 weeks after the explosion. Overall, the objects are rather spherical as could be expected for thermonuclear explosions of a WD. For the subluminous SN1999by, the continuum was polarized up to about 0.7% (Howell *et al.* 2001). The well defined axis can be understood in the framework of rotating WDs which may be a consequence the accretion process in a binary system or the merging of two WDs. For the strongly subluminous SN 1999by, the accretion on an accreting WD is clearly favored from the analysis of light curves and spectra (Fig. 16.4). Its unusually high continuum polarization may suggest a correlation between the subluminosity and its low Ni production, i.e. the propagation of the nuclear burning front.

Case B: Interaction with the circumstellar environment may be detected by the appearance of a high velocity component in Ca II (Fig. 16.5). In SN1994D, it may be understood as an ionization effect when Ca III recombines to Ca II and, thus, forming two, radially separated features (Höflich *et al.* 1998, Hatano *et al.* 2001). Alternatively, double features of Ca II may be attributed to abundance pattern (Fisher *et al.* 1997). The observed polarization in SN2001el was high in Ca II clearly identifying this feature as a morphological distinct pattern but its origin remained open (Wang *et al.* 2003, Kasen *et al.* 2003). Based on a detailed analysis of SN2003du, Gerardy *et al.* (2003) showed that this feature can be understood in the framework of the interaction between the SN eject and its H-rich nearby surroundings.

Case C: Wang *et al.* (1998) showed that the observed polarization pattern may consistent with chemical inhomogeneities at the Si/Ni interface as can be expected from 3-D deflagration models. However, this detection was on a 1σ level, and needs to be confirmed in other objects.

References

Cropper M., Bailey J., McCowage J., Cannon R., Couch W. 1988, MNRAS 231, 685
Fesen, R. A. & Gunderson, K. S. 1996, ApJ, 470, 967
Fisher A., Branch D., Nugent P., Baron E. 1997, 481L, 89
Gerardy C., Höflich P., Quimby R., Wang L. + the HET-SN team 2003, ApJ, submitted
Hatano K., Branch D., Lentz E. J., Baron E., Filippenko A. V., Garnavich P. 2000, ApJ 543, L94
Hillebrandt, W., Niemeyer, J. 2000, ARAA 38, 191
Höflich, P. 1991 A&A 246, 481
Höflich, P. 1995, ApJ 443, 89
Höflich P., Wheeler, J. C., Hines, D., Trammell S. 1995, ApJ 459, 307
Höflich, P., Khokhlov A., Wang L., 2002, AIP-Publ. 586, p. 459 & astro-ph/0104025
Höflich, P., Gerardy, C., Linder, E., & Marion, H. 2003, in: Stellar Candles, eds. Gieren *et al.*, Lecture Notes in Physics, Springer Press, in press & astro-ph/0301334
Howell A., Höflich P., Wang L., Wheeler J. C. 2001, ApJ 556, 302
Jeffrey D. J., 1991, ApJ, 375, 264
Kasen, D., Nugent, P., Wang, L., Howell, A., Wheeler, J. C., Höflich, P., Baade, D., Baron, E., Hauschildt, P. 2003, ApJ, in press & astro-ph/0301312
Khokhlov, A. 2001, astro-ph/0008463
Leonard D. C., Filippenko, A. V., Barth A. J., Matheson T. 2000, ApJ 536, 239
Mendez R. H. *et al.* 1977, ApJ 334, 295
Scheck L., Plewa T., Janka H.-T., Kifonidis K., Müller E. 2003, Phys.Rev.Let, submitted
Strom R., Johnston H. M., Verbunt F., Aschenbach B. 1995, Nature, 373, 587
Trammell S., Hines D., Wheeler J. C. 1993, ApJ 414, 21
Wang L., Wheeler J. C., Li Z., Clocchiatti A., 1996, ApJ 467, 435
Wang L., Howell A., Höflich P., Wheeler C. 2001, ApJ 550, 1030
Wang L., Baade D., Höflich P., Wheeler C., Fransson C., Lundqvist P. 2002, ApJ, submitted
Wang L., *et al.* 2002b, ApJ Let., in press & astro-ph/0205337
Wang, L., Baade, D., Höflich, P., Khokhlov, A, Wheeler, J. C., Kasen, D., Nugent P., Perlmutter S., Fransson C., Lundqvist P. 2003, ApJ 591, 1110
Wang, L., Baade, D., Höflich, P., Wheeler, J. C., Kawabata, K., Nomoto, K. 2003, ApJ, submitted

17

Broad lightcurve SNe Ia: asymmetry or something else?

D. A. Howell

Department of Astronomy, University of Toronto, M58 2L4

P. Nugent

Lawrence Berkeley National Laboratory, Berkeley CA, 94720

Abstract

It is the conventional wisdom that overluminous Type Ia supernovae have an overproduction of their elemental powerhouse, ^{56}Ni, leading to broader light curves, higher temperatures, higher ionization states, and peculiar spectra similar to that of SN1991T. However, this simple picture is incomplete: we show that a broad lightcurve width does not necessarily predict spectroscopic peculiarity, nor does a spectrum resembling SN1991T guarantee a broad lightcurve. There is circumstantial evidence that asymmetry may play a role in the explanation of the diverse properties of broad lightcurve and SN1991T-like SNe Ia.

As an illustrative example, we present optical and NIR light curves, and Lick 3m and HST STIS spectra of the SN Ia with the broadest light curve observed to date, SN 2001ay. SN 2001ay has $\Delta m_{15}(B) = 0.6$ and stretch $s = 1.6$, yet at maximum light is fairly spectroscopically normal. The exception is an extremely high Si velocity, $v = 15{,}000$ km s^{-1}. The secondary peak in the *I*-band lightcurve is higher than the primary peak, and the J_s and H lightcurves remain flat over the entire 55 days of observation. SN 2001ay also does not appear to obey lightcurve shape-luminosity relationships, at least as they are currently formulated. Despite its broad lightcurve, the SN has normal absolute magnitudes after correction for Milky Way and host galaxy extinction. Thus, if a stretch or $\Delta m_{15}(B)$ correction is applied, the resulting magnitude would be overcorrected by $\sim$1 mag.

17.1 Introduction

The year 1991 saw the discovery of two of the most famous Type Ia supernovae (SNe Ia) ever observed: the narrow lightcurve, underluminous, spectroscopically peculiar SN1991bg (Filippenko *et al.* 1992b, Liebundgut *et al.* 1992), and the broad lightcurve, overluminous and spectroscopically peculiar SN1991T (Filippenko *et al.* 1992a, Phillips *et al.* 1992, Ruiz-Lapuente 1992). The opposite nature of these SNe Ia proved that not all have the same luminosity, but it also led Phillips (1993) to the discovery that the lightcurve width could be used to "correct" the absolute magnitudes of a SNe Ia, rendering them standardized candles. This allowed the discovery that the universe is accelerating, propelled by some unknown Dark Energy (Riess *et al.* 1998, Perlmutter *et al.* 1999).

Nugent *et al.* (1995) compared a range of SNe Ia at various lightcurve widths to a range of theoretical models and showed that the variation in spectroscopic features appears to be correlated with temperature differences. Since the light curves of SNe Ia are ultimately powered by the decay of ^{56}Ni, it follows that SNe with large amounts of synthesized ^{56}Ni are brighter, broader, and have hotter atmospheres, leading to higher ionization states like those seen in SN1991T-like SNe: Si III, Fe III, Co III, and thus a "peculiar" spectrum.

If this simple picture tells the whole story, then it follows that the SNe Ia with broader lightcurves should also be the brightest and most spectroscopically extreme. However, we present SN 2001ay, which confounds our expectations. Other authors have pointed out that not all broad lightcurve SNe are similar to SN1991T (see e.g. Lira *et al.* 1998), but this fact still goes unnoticed by many. In an effort to make as complete accounting of the known properties of these SNe as possible, here we examine all SNe available in the literature with either broad lightcurves or spectra similar to SN1991T.

A note on the nomeclature used: We use $\Delta m_{15}(B)$ (lower numbers mean broader lightcurves; Phillips 1993) and stretch, s, (higher numbers mean broader lightcurves; Perlmutter *et al.* 1997) to parameterize the widths of lightcurves in this paper. Also, we refer to all SNe with strong lines of Fe III and weak Si II near maximum light as SN1991T-like SNe. We acknowledge the difference between SN 1991T (no Ca) and SN1999aa (has Ca), but choose to call even SN 1999aa-like SNe "SN1991T-like" here to avoid the cumbersome label "SN1991T/1999aa-like".

17.2 SN 2001ay

SN 2001ay was discovered on Apr. 18.4 (UT) by LOTOSS (Swift 2001) in the nearby ($z = 0.03$) galaxy IC 4423. Our collaboration began obtaining photometry (Fig. 17.1) and spectroscopy (Fig. 17.2) near maximum light. Full details of the data analysis will be presented in an upcoming paper.

It is apparent from Fig. 17.1 that SN 2001ay has an unusual collection of lightcurves. In the B-band, $\Delta m_{15}(B) = 0.6^m$ and stretch $s = 1.6$, making SN 2001ay the SN Ia with the broadest lightcurve observed to date. The V-band lightcurve is even broader, while the R lightcurve declines only $\sim 0.5^m$ over the first 30 days of observation. The I lightcurve has a secondary peak higher than its primary

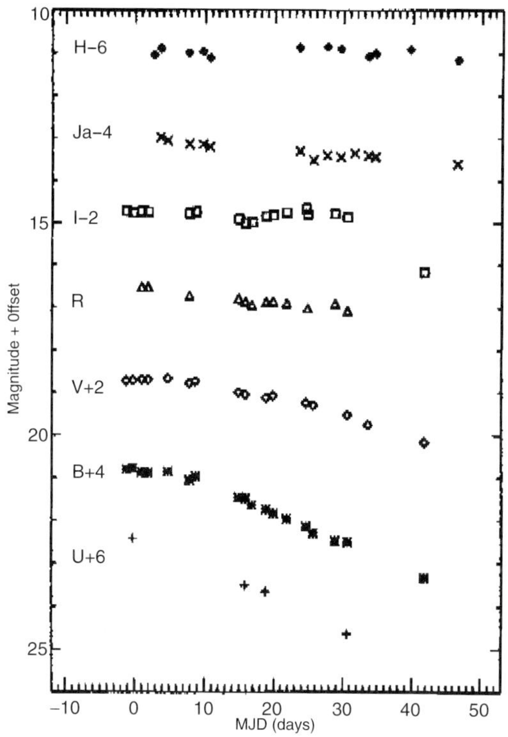

Fig. 17.1. Lightcurves of SN 2001ay. Optical data is from the Lick 1m, LCO 2.5m, and the CTIO 0.9m and 1.5m telescopes. IR data was obtained with the LCO 1m and 2.5m telescopes. We do not overplot a comparison SN Ia lightcurve, because no known SN lightcurve is as broad as SN 2001ay. Here we take April 22 as the date of maximum, obtained from a stretch fit to the B-band lightcurve. The error on the date of maximum is unusually large, ±2 days. This is due to The flatness of the lightcurves, and the lack of constraining premaximum data. Note that a later date of maximum would decrease the measured stretch somewhat.

maximum. The IR lightcurves are unusual as well, with only a shallow or no decline in J_s and H over 55 days of observation.

The Galactic reddening in the direction of SN2001ay is $A_V = 0.062$ mag (Schlegel *et al.* 1998). From NaI D lines in our high resolution spectrum (not shown in the smoothed version of the spectrum in Fig. 17.2), we place a preliminary upper limit on the reddening from the host galaxy of $A_V = 0.12$ mag using the method of Ho & Filippenko. (1995). We acknowledge that this method may not be entirely reliable, but it is difficult to estimate the reddening from the colors of this SN, due to its peculiar nature. Taking our reddening numbers at face value, we estimate an absolute magnitude for SN2001ay of $M_B = -19.2 \pm 0.1$. Current lightcurve shapeluminosity relationships would overcorrect the magnitude of SN2001ay by $\sim 1^m$.

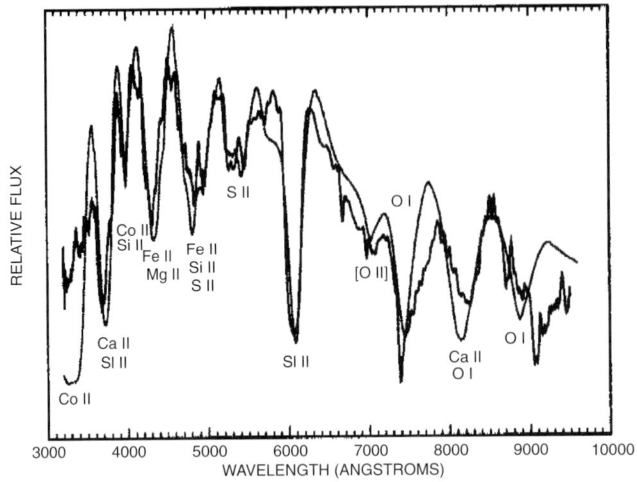

Fig. 17.2. Smoothed Keck ESI Spectrum of SN 2001ay at maximum light, compared to a SYNOW model prepared by David Branch. This spectrum is in F_ν to highlight features at the red end.

At maximum light, spectra of SN1991T-like SNe show weak Si II, and lines of SiIII and Fe III dominate. However, despite its much broader lightcurve, SN 2001ay does not show the same behavior. Its spectrum resembles that of normal SNe Ia, with one notable exception: the Si II features at 4000Å and 6150Å are extremely strong and located at high velocities, $v = 15{,}000$ km s^{-1}. This may indicate a higher than average density in the outer layers of the SN (Branch, private communication). Such a phenomenon was inferred for SN1984A (Lentz *et al.* 2001), but it remains to be seen whether this can reproduce the extreme features of SN 2001ay. In his talk at this conference, and a recent paper (2003), Hamuy argues that interaction with circumstellar material can explain the unusually broad light curve for SN 2002ic. In that case, the spectrum showed evidence that one would expect from such an interaction – narrow Hydrogen, and weak overall features, as if the spectrum of a normal SN Ia had been diluted by light from the interaction. This hypothesis cannot account for the broad lightcurve of SN 2001ay because the spectrum does not appear diluted. On the contrary, the Si absorption is one of the strongest ever seen in a SNIa.

17.3 Broad-lightcurve SNe

SN 2001ay provides dramatic evidence that not all broad lightcurve supernovae are spectroscopically peculiar at maximum light. Is this merely a special case, or do other such supernovae exist? To answer this question, we compiled a list of SNe

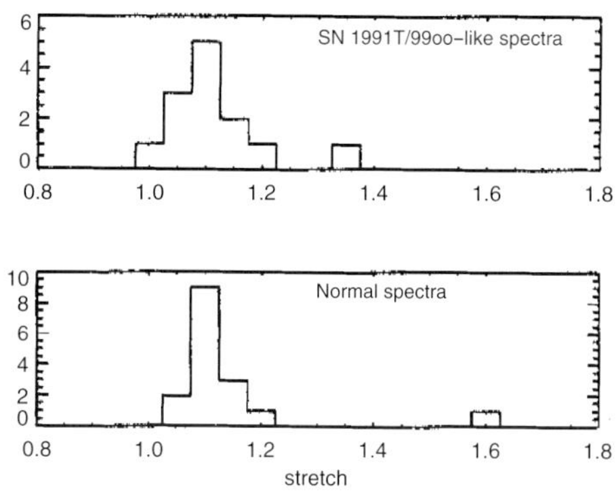

Fig. 17.3. Histogram of stretch values of SNe Ia with SN1991T-like spectra (upper panel). We compare these to the histogram of broad-lightcurve SNe ($s > 1.05$) with normal spectra (lower panel). The stretch distribution for all SNe Ia is roughly gaussian, peaked at $s = 1$, with $\sigma = 0.12$ (Perlmutter *et al.* 1999; Alex Conley, private communication), so SN1991T-like SNe are in general about 1σ broader than average.

with broad lightcurves, defined as $\Delta m_{15}(B) < 1.05$ or $s > 1.05$. SN photometery was taken from: Hamuy *et al.* (1996), Riess *et al.* (1999), Phillips *et al.* (1999), and Krisciunas *et al.* (2001), and the data were fit using the stretch method to determine the stretch parameter, s, in the B-band. If the supernovae were known to have normal spectra, or if spectra were taken, but no peculiarities were listed in the IAU circulars, then the SNe were classified as 'spectroscopically normal'. A histogram of the stretches of these SNe is plotted in the lower panel of Fig. 17.3.

We also searched all IAU circulars for any mention of supernovae with spectra similar to SN1991T or SN 1999aa. We kept all of these SNe whose stretches could be fit, even if they had $s < 1.05$. In Figure 17.3 we show the stretch distribution of the SN1991T-like SNe compared with the distribution of the broad lightcurve SNe with normal spectra mentioned above. Little difference is apparent between the two stretch distributions. It is clear that many broad-lightcurve SNe Ia are spectroscopically normal at maximum light. Indeed, it appears that among broad lightcurve SNe, there are more with normal spectra than with spectra like SN1991T or SN 1999aa.

Overall, there appears to be a tendency for SN1991T-like SNe to have broader than average lightcurves, but this is not always the case. For example, SN1997br had nearly normal light curves, with $\Delta m_{15}(B) = 1.0$ and $s = 1.04$. Peculiarity does not seem to increase with the broadness of the lightcurve. SN1991T has $\Delta m_{15}(B) = 0.94$, while the more normal SN 1999aa has $\Delta m_{15}(B) = 0.746 \pm 0.024$ (Krisciunas *et al.* 2000), though this may be somewhat of an underestimate (Garavini 2003).

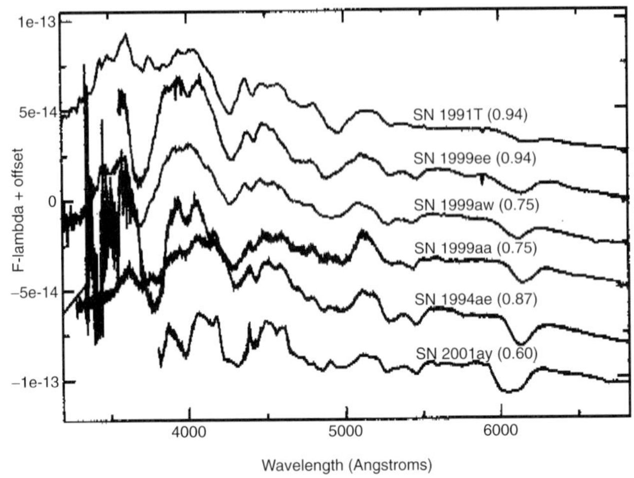

Fig. 17.4. Spectra of broad-lightcurve SNe Ia near maximum light, arranged in order of increasing Si II 4000 and 6150 line strength. $\Delta m_{15}(B)$ values are indicated in parentheses. Note that Si line strength does not appear to correlate with $\Delta m_{15}(B)$, indicating that SNe Ia are not a one-parameter family.

One trait that many broad lightcurve SNe share is unusual Si II 6150 Å features. If we take this as some measure of "degree of peculiarity," then we might hope to find some trend that correlates with lightcurve shape. Figure 17.4 shows spectra of several broad lightcurve SNe Ia near maximum light, arranged in order of increasing Si II line strength. In parentheses, the value of $\Delta m_{15}(B)$ is noted for each SN. Si line strength seems to be uncorrelated with $\Delta m_{15}(B)$. **Thus, for broad-lightcurve SNe Ia, the broadness of the lightcurve does not predict spectroscopic peculiarity.** SNe are not a one-parameter family. Our simple picture of increasing lightcurve width, luminosity, and temperature with ^{56}Ni mass cannot explain all of the diversity of SN features.

17.4 Asymmetry?

Could asymmetry of the SN ejecta explain some of the observed properties of broad lightcurve or SN1991T-like objects? While there is no smoking gun, the available data are all consistent with the predictions one would make if SNe Ia displayed a moderate amount of asymmetry:

- **A dispersion in luminosity.** For SNe Ia that deviate from spherical symmetry by 20%, a dispersion in luminosity of 0.2 mag is expected, depending on the viewing angle (Howell *et al.* 2001b).
- **Diverse spectra.** If we see different layers of ejecta depending on the viewing angle, it is possible that some broad lightcurve SNe show more ionized elements, such as Fe III when

we are seeing down into the hotter layers of the SN. This could happen, for example, if a hole is left in the ejecta though interaction with the companion star. This idea is explored by Dan Kasen in his talk at this meeting.

- **Same progenitor population.** One alternative to asymmetry as an explanation for the properties of these SNe is that they come from different progenitor systems. This does not appear to be the case. Howell (2001a) showed that both SN1991T-like SNe and broad lightcurve SNe come from late-type galaxies, and thus a younger stellar population.
- **Polarization** If broad-lightcurve SNe are asymmetric, then sometimes their spectra should be polarized. Thompson scattering polarizes light in the atmospheres of SNe, but the polarization vectors cancel if it is spherically symmetric. While spectropolarimetry is sparse for these SNe, there is a claim in the IAU circulars that one is polarized, SN1997bp (Wang *et al.* 1998), though the data remain unpublished. Some other SNe Ia also show polarization, such as the "normal" SN 2001el (Wang *et al.* 2003, Kasen *et al.* 2003), and the subluminous SN1999by (Howell *et al.* 2001b).

17.5 Concluding remarks

We are preparing two papers based on the work presented here – one on the full data set obtained for SN 2001ay, and one on the properties of broad lightcurve SNe. Anyone with unpublished data on broad lightcurve or SN1991T-like SNe Ia is encouraged to contact the first author to join the collaboration.

While space limitations of this conference proceeding prohibit us from listing the full author list at the beginning, we would like to thank the following collaborators who contributed to this work: Kevin Krisciunas, Weidong Li, Alex Conley, Mark Phillips, Lifan Wang, Dan Kasen, Greg Aldering, Saul Perlmutter, Nick Suntzeff, Chris Smith, Y. Qiu, S. Gonzalez, and David Branch.

References

Filippenko, A. V. *et al.* 1992a, ApJL, 384, 15
Filippenko, A. V. *et al.* 1992b, AJ, 104, 1543
Garavini, G., 2003, in preparation
Hamuy, M. *et al.* 2003, Nature, 424, 641
Hamuy, M. *et al.* 1996, AJ, 112, 2408
Ho, L., C. & Filippenko, A. V. 1995, ApJ, 444, 165
Howell, D. A., Höflich, P., Wang, L., Wheeler, J. C. 2001b, ApJ, 556, 302
Howell, D. A. 2001a, ApJL, 554, 193
Krisciunas, K. *et al.* 2001, AJ, 122, 1616
Lentz, E., Baron, E., Branch, D., Hauschildt, P. 2001, ApJ, 547, 402
Liebundgut, B. *et al.* 1993, AJ, 105, 301
Lira, P. *et al.* 1998, 115, 234
Nugent, P. *et al.* 1995, ApJL, 455, 147
Perlmutter, S. *et al.* 1997, ApJ, 483, 565

Perlmutter, S. *et al.* 1999, ApJ, 517, 565
Phillips, M. *et al.*, 1992, AJ, 103, 1632
Phillips, M. 1993, ApJL, 413, 105
Phillips, M. *et al.* 1999, AJ, 118, 1766
Riess, A. *et al.* 1998, AJ, 116, 1009
Riess, A. *et al.* 1999, AJ, 117, 707
Ruiz-Lapuente, P. *et al.* 1992, ApJ L, 387, 33
Schlegel, D. J., Finkbeiner, D. P., & Davis, M. 1998, ApJ, 500, 525
Swift, B. and Li, W. D. 2001, IAUC 7611
Wang, L., Wheeler, J. C., Höflich, P., & Howell, D. A. 1998, BAAS, 193, 4715

18

Synthetic spectrum methods for three-dimensional supernova models

R. C. Thomas

University of Oklahoma, Department of Physics and Astronomy, 440 W. Brooks Street Rm. 131, Norman, Oklahoma 73071 (Present Address: Lawrence Berkeley National Lab, 1 Cyclotron Road MS 50R5008, Berkeley, California 94720)

Abstract

Current observations stimulate the production of fully three-dimensional explosion models, which in turn motivates three-dimensional spectrum synthesis for supernova atmospheres. We briefly discuss techniques adapted to address the latter problem, and consider some fundamentals of line formation in supernovae without recourse to spherical symmetry. Direct and detailed extensions of the technique are discussed, and future work is outlined.

18.1 Introduction

Spectrum synthesis is the acid test of supernova modelling. Unless synthetic spectra calculated from a hydrodynamical stellar explosion model agree with observations, the model is not descriptive. Some explosion modellers contend that only three-dimensional (3-D) models faithfully describe the physics of the real events. If this is so, then the evaluation of those models requires solutions to the 3-D model supernova atmosphere problem. These solutions require full *detail*, the inclusion of as much radiation transfer physics as possible. Otherwise, a bad fit of a synthetic spectrum to an observed one might have less to do with the accuracy of the hydrodynamical model, and more to do with the shortcomings of the radiation transfer procedure.

On the other hand, solutions (of a sort) to the ill-posed inverse problem constrain parameter space available to hydrodynamical models. Fast, iterative, parameterized fits to observed spectra characterize the ejection velocities and identities of species found in the line forming region. Most importantly, the procedure reveals species that *cannot* be identified by simply Doppler-shifting line lists on top of observed spectra in search of feature coincidences. Generalizing this *direct* analysis technique to 3-D is key to constraining the geometries of real explosions.

This proceedings contribution briefly describes some steps toward the complimentary goals of detailed and direct analysis in 3-D, with an emphasis on pedagogy. For an in-depth application of the more detailed technique, refer to the contribution of D. Kasen.

18.2 Approach

Work underway to extend spherically symmetric non-LTE modelling codes to 3-D could take the better part of a decade. An alternative approach, which yields a direct analysis code along the way, is to begin with a simple 3-D code and augment its physics details. The ultimate goal is complete non-LTE radiation transfer in 3-D with full and realistic treatment of the boundary conditions at depth. This means that the evolution of radiation from deposition to escape is modelled without the central "light bulb" approach.

We embark on the journey toward full 3-D non-LTE modelling from the elegant and humble shores of the Sobolev method (Castor, 1970; Rybicki & Hummer, 1978). The Sobolev method greatly reduces the scale of the wavelength-domain of the problem by approximating line transfer, but does so at an accuracy cost. When faced with the alternative, some inaccuracy in the line transfer is acceptable. Full solutions to the 3-D radiation transfer problem are simply not possible yet, so the limited inaccuracies (and the awareness of them) seems a small price to pay for progress.

The particular implementation of the Sobolev method is the Monte Carlo technique, based largely on the formalism described primarily by L. Lucy in a series of papers (e.g., Lucy, 1999). The key innovation described in those papers we call the equal-energy packet (EEP) technique. In the EEP picture, individual photon trajectories are not simulated, but rather monochromatic photon packets of equal energy propagate through the model atmosphere. This paradigm obviates recursive trajectory calculations which make development and extension of transfer codes difficult. Of course, the scalability of the Monte Carlo technique is one of its greatest virtues. Provided the model atmosphere need not be split up across distributed nodes, communication proves almost nonexistent. Most importantly, this implementation micro-manages the energy conservation (flux divergence-less-ness) of the radiation field at all positions (or depths). Scattering is coherent in the comoving frame. Absorption and re-emission is accomplished through roulette-wheel selection (producing either a simple equivalent two-level atom or non-LTE source function). The Monte Carlo algorithm propagates variations in the radiation field instantaneously (provided enough packets are used), and this makes a simple Λ-iteration work.

Clearly, waiting for enough packets to exit the model atmosphere in any given direction of interest is not the most efficient use of limited computer time. Hence, at least for flux spectra, we can easily use the packets to establish the radiation field (energy density) throughout the envelope. From that, we derive the source function at all points and compute the emergent spectrum for any line of sight, a kind of "formal integral," again simplified by the Sobolev approximation. This technique greatly reduces the number of packets required to build a spectrum.

Other extensions to the method besides the formal integral for emergent spectra include polarized transfer and a way of treating energy deposition from radioactive decay in the ejecta. The propagation of Monte Carlo Stokes vectors for computing polarization spectra and progress away from the central "light bulb" approximation are exemplified in D. Kasen's contribution.

18.3 Simple constraints on nonsphericity

The starting question is a rather obvious one, but its answer is fundamental. If deviations from spherical symmetry occur in supernova atmospheres, what is the corresponding detection threshold for their evidence to appear in flux spectra? The changes to line profiles resulting from nonsphericity are easy to understand, but we seek means of quantitatively exploiting the results to constrain supernova geometry.

To find out, we conduct a very simple experiment. We generate a series of distributions of Sobolev optical depth covering a range of nonsphericity scales. For each model, we compute line profiles from a large number of lines of sight. This gives us a sense of what sort of diversity can be generated by what size of perturbation. In Figure 18.1 are line and covering fraction profiles averaged over 100 lines of sight. The scatter around the average is summarized by a $1 - \sigma$ deviation from the average. The covering fraction for a given velocity is defined as the fraction of the projected photosphere surface obscured by Sobolev optical depth exceeding 1. We deduce three facts from the results. First, nonsphericity of the kind considered here can most strongly influence absorption features. Second, these nonsphericities only weakly influence emission features. (Both of these previous points are reversed if the line source function significantly exceeds that of resonance scattering.)

Third and most importantly, the diversity trend as a function of the line of sight is directly correlated with the photospheric covering fraction. Hence, it is possible to suggest a threshold scale of clumpiness below which clumping goes undetected in flux spectra. At first this result seems purely academic, since nature presents us with only one line of sight to a given supernova.

However, we can apply the result to make a claim about models for spectroscopically normal Type Ia supernovae. Such supernovae are spectroscopically homogeneous, and we find that the depth of the Si II feature in these events is a fairly repeatable 0.7 times the local continuum. In fact, the measured scatter seems to

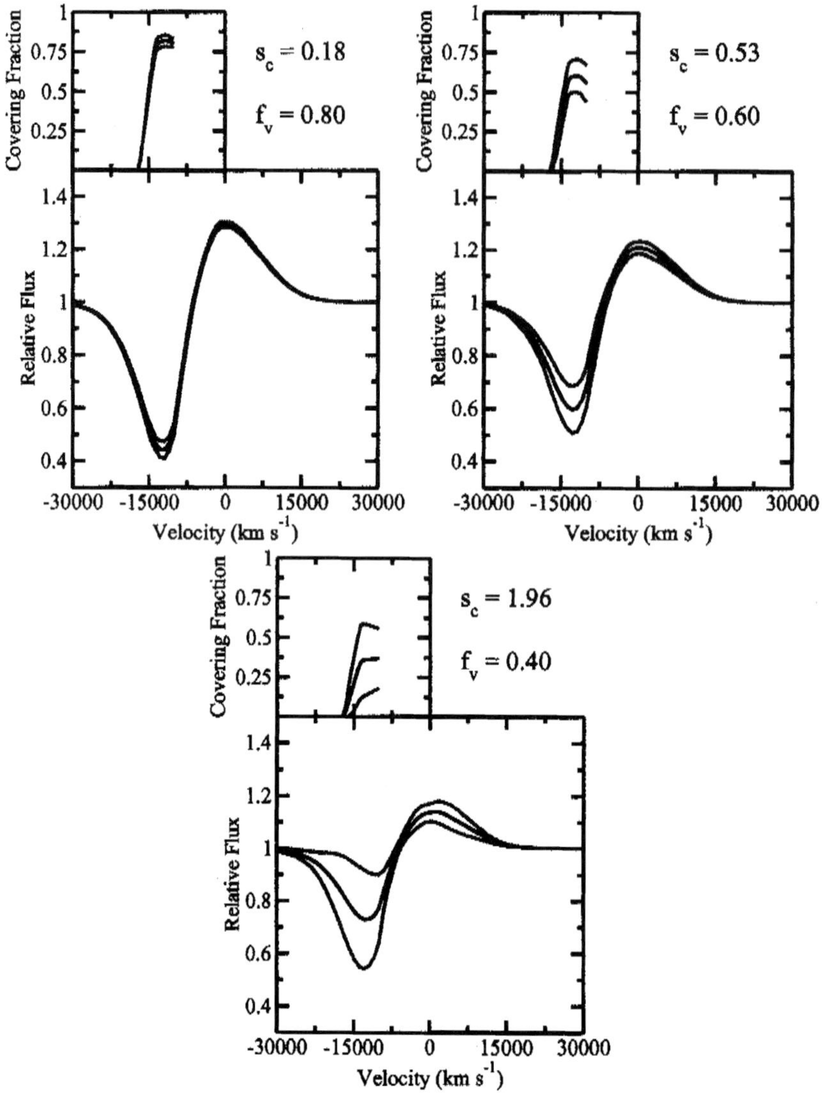

Fig. 18.1. Sample results from a simple experiment. The parameter f_V is the fraction of the line-forming region with non-zero optical depth, and S_c is the ratio of the cubical "clump" edge length to the photospheric radius. Average line and covering fraction profiles (dark curve) plus "1-σ" deviation (light curves) from the average profiles are shown.

suggest that perturbations like those explored here must be smaller than 10% the size of the photosphere area if they are indeed present. More importantly, if pure 3-D deflagrations exhibit such large scale perturbations leading to wildly fluctuating line covering fractions as a function of perspective, they cannot account for such events.

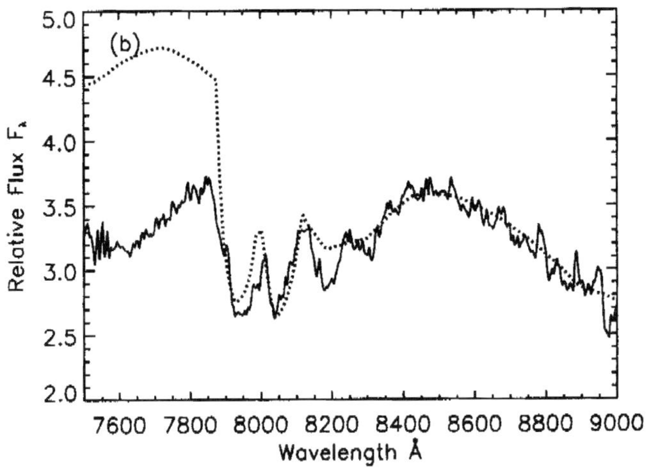

Fig. 18.2. High velocity Ca II fit using a 3-D optical depth parameterization. From Thomas *et al.* (2003).

18.4 Direct analysis in 3-D

The oft-repeated goals of spherically symmetric direct analysis are to identify lines and velocity intervals within the line forming region where the parent ions of those lines are found. In direct analysis, special attention is given to treating line blending, since this is an important feature of supernova spectra. Without this attention, we have seen that identifications are problematic, and these problems make it more difficult to narrow down the range of hydrodynamical models that are worth pursuing.

To provide the same direct analysis capabilities, but without recourse to spherical symmetry, we developed a code called Brute, based on the earlier, spherically symmetric (and non-Monte Carlo) code Synow. The basic picture is the same, except that instead of radial functions of Sobolev optical depth in a reference line of each ion included, we use a template (constructed in any fashion) that need not be spherically symmetric.

An application of this code to the unique Type Ia supernova 2000cx appears in Thomas *et al.* (2003). The maximum light spectra of this object exhibited unusual, narrow high-velocity features, particularly in the Ca II infrared triplet. Using alternate 1D and 3-D models of optical depth in Ca II, we find that partial blocking of the photosphere by a nonspherical ejecta distribution can help explain some of these features. Simultaneous fits are attempted to the corresponding Ca II UV feature, and we find that the 3-D distribution is less problematic than the 1D one. The origin of the high-velocity material in this supernova is still a mystery, but how much of the observed diversity in these objects is due to 3-D distributions of ejecta at high velocity? More modelling of more objects is needed.

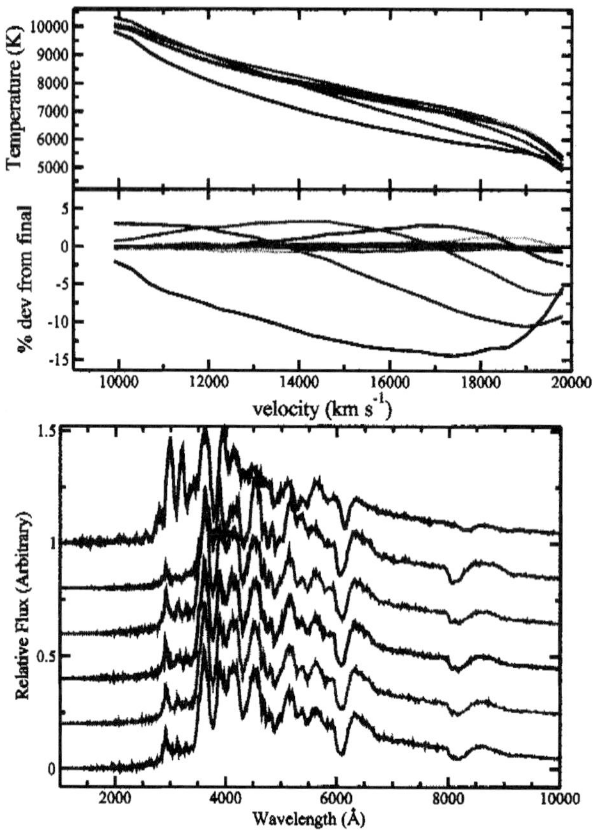

Fig. 18.3. Test results for a mixed W7 model. Rapid convergence in the temperature structure is evident. Resulting spectra from iterations 1, 3, 5, 7, 9 and 10 are shown to the right.

18.5 Detailed analysis in 3-D

A special characteristic of the Monte Carlo technique is that it allows for simple solution to the radiative equilibrium problem. This permits more self-consistent modelling of emergent spectra from real hydrodynamical models. Given a composition and density structure (in any geometry), we use the Λ-iteration procedure to construct self-consistent temperature structures.

In Figure 18.3, temperature structures and spectra are shown from 10 iterations to a converged model. Though this particular model is spherically symmetric (a W7-like model mixed above 9000 km s^{-1}), the convergence speed is striking. This provides us with hope that (at least for now) we can begin modelling real hydrodynamical explosion models, at least in LTE.

18.6 Conclusion

The path to fully detailed 3-D non-LTE spectrum synthesis for supernova models is clear. Including non-LTE and continuum transfer effects will permit us to examine core collapse supernovae more closely, to help unlock the connection between supernovae and gamma-ray bursts. The eventual goal of dispensing with the Sobolev approximation will also be reached, and progress is already underway with realistic lower boundary conditions and gamma-ray transport.

Eventually, 3-D hydrodynamical stellar explosion models will be carried to the homologous expansion phase. Those models will be converted into flux and polarization spectra to be compared with observations. Until then, there is much computer time to be burned.

This work is supported by grant HST-AR-09544-01.A, provided by NASA through the STScI, operated by the AURA, Inc., under NASA contract NAS5-26555.

References

Castor, J. 1970, MNRAS 149, 111
Lucy, L. 1999a, A&A, 345, 211
Rybicki, G., & Hummer 1978, ApJ, 219, 654
Thomas, R., *et al.* 2003, ApJ in press, astro-ph/0302260

19

A hole in Ia? Spectroscopic and polarimetric signatures of SN Ia asymmetry due to a companion star

D. Kasen

Lawrence Berkeley Lab, University of California
Berkeley Berkeley, California USA

Abstract

In the popular progenitor scenario, Type Ia supernova are the result of a white dwarf exploding in a binary system. The presence of a nearby companion star could cause a substantial asymmetry in the supernova ejecta – according to the models of Marietta et. al. (2000), the companion carves out an hole in the ejecta. The opening angle of the hole is as large as 40°. Such an asymmetry would leave signatures in the supernova flux and polarization spectra. We explore this possibility using a three-dimensional Monte-Carlo LTE radiative transfer code which includes gamma ray transport and a temperature correction procedure. We calculate synthetic spectra and polarization levels from multiple lines of sight to see how an ejecta hole model compares to observations.

19.1 Introduction

While some Type Ia supernovae (SNe Ia) are known to be aspherical, the exact nature of the asymmetry is unknown. The direct evidence of the asphericity is the detection of non-zero intrinsic polarization in, for example, SN 1999by [4] and SN 2001el [17]. In both cases, the polarization level was rather low (~0.7% for SN 1999by, ~0.4% for SN 2001el), which indicates a mild asymmetry along the line of sight. In addition, the polarization angle was fairly constant across the majority of line features, indicating that the bulk of the ejecta obeyed a near axial symmetry. The exact shape of the supernova ejecta is an important question, as it must be closely tied to the explosion processes and progenitor systems of SNe Ia. Most theoretical attempts at modeling the spectropolarimetry have so far assumed the ejecta was ellipsoidal [3, 6, 4].

One potential cause of asymmetry in SNe Ia is the binary nature of the progenitor system. In the favored progenitor scenario, SNe Ia arise from a white dwarf accreting material from a non-degenerate companion star. As the companion star is close enough to be in Roche-lobe overflow, it is a rather large nearby presence from the perspective of the white dwarf. Soon after the white dwarf explodes, the ejected supernova material runs over the companion star. In the impact, it would not be surprising if a substantial asymmetry was imprinted on the supernova ejecta.

The ejecta-companion interaction has been studied with 2-D hydrodynamical models [1, 10, 11]. In the most recent and extensive models, Marietta *et. al.* (2000) find that the impact with the companion star carves out a conical hole in the supernova ejecta. The opening angle of the hole is 30°–40°, and because the ejecta is moving supersonically, the hole does not close with time. With the advance of spectropolarimetric observations of SNe Ia, this ejecta asphericity becomes a relevant signature of the progenitor scenario.

Here we describe radiative transfer calculations that address the possibility of SNe Ia having an ejecta hole asymmetry. We calculate the line of sight variation of the spectrum, luminosity, and polarization for the aspherical SN Ia near maximum light. In contrast to the ellipsoidal models, the line of sight variations in an ejecta-hole geometry can be rather extreme, especially when one looks near the hole itself. These variations would necessarily introduce some diversity into the observed properties of SNe Ia.

19.2 The ejecta-hole model

We use an ejecta model based upon the spherical w7 explosion model [14]. The ejecta hole was introduced by hand, using an analytic function that resembles the density structures seen in Marietta *et. al.* (2000). In the radial direction the density follows an exponential profile with an e-folding length of 2,500 km s^{-1}. The conical hole has an opening angle of $\theta_H = 40^\circ$ and the density in the hole is a factor $f = 0.05$ less then that outside. The material that is displaced from the hole gets piled up into a density peak just outside the hole edge, with angular size $\theta_p = 20^\circ$. The composition used was that of w7, except that one adjustment was necessary. In order to fit the depth and width of the CaII H&K feature in a spherical model, we need to increase the calcium abundance by a factor of 10 in the outer, unburnt layers ($v > 15,000$ km s^{-1}).

Our models are calculated using a 3-D Monte Carlo radiative transfer code. Because of the unusual geometry, a spherical inner boundary surface was not used, rather photon packets were allowed to propagate throughout the entire supernova envelope, including the optically thick center. The initial source of photon packets was determined by performing a multi-dimensional gamma ray transfer calculation to see where radioactive energy is deposited in the supernova envelope. The electron scattering and bound-bound opacities were computed assuming LTE and a two

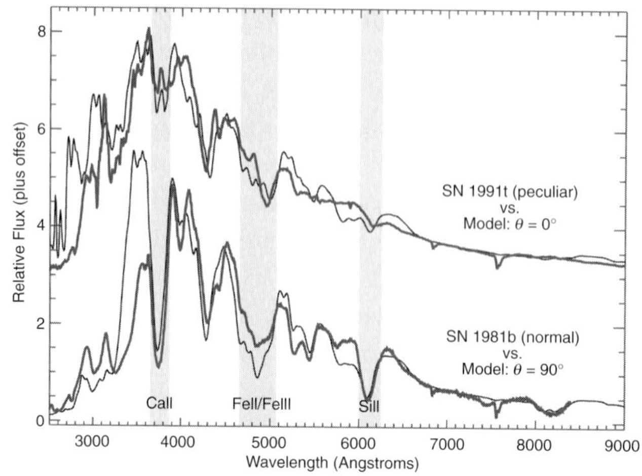

Fig. 19.1. The maximum light spectra of the ejecta-hole model (thin lines) from two different inclinations are compared to two observed supernovae (thick lines). Bottom: the view from the side ($\theta = 90°$) compared to the normal SN Ia 1981B. Top: the view down the hole ($\theta = 0°$) compared to the peculiar SN Ia 1991T.

level equivalent atom formulation with $\epsilon = 0.05$. The temperature structure was computed self-consistently, by solving the equation of radiative-equilibrium.

19.3 Spectral signatures

Line opacity in a spherical, expanding SN atmosphere gives rise to the well known P-Cygni profile – i.e. a blueshifted absorption trough with a redshifted emission peak. An ejecta-hole asymmetry dramatically alters the line profile from some lines of sight. In the typical P-Cygni formation, material in front of the photosphere obscures the light below and gives rise to the blueshifted absorption feature. When one looks down the ejecta hole ($\theta < \theta_H$) there is much less of this obscuring material and the line absorption will therefore be much weaker. In addition, because one is seeing deeper into the ejecta, the minimum of the absorption will be less blueshifted. In fact the hole uncovers deeper layers of ejecta that would otherwise be buried beneath the photosphere. By revealing the inner layers of ejecta, the hole allows for line formation from the hotter, more ionized material at depth.

In sum, the spectrum in the ejecta-hole model will look the same as in a spherical model for all lines of sight *except* when one looks almost directly down the hole ($\theta < \theta_H$). In the later case one sees a peculiar spectrum characterized by more highly ionized species, weaker absorption features, and lower absorption velocities. Figure 19.1 compares the synthetic spectra of the ejecta model to two well known SNe Ia. While the view away from the hole ($\theta = 90°$) resembles the normal SN Ia 1981B, the spectrum down the hole ($\theta = 0°$) is clearly very different than a normal SN Ia. We compare it to the peculiar SN 1991T, which it resembles in the following ways: (1) the SiII absorption near 6150 Å is weak and has an unusually low velocity; in addition, the SiII absorption near 4000 Å is absent. (2) The CaII H&K feature is weak and the CaII IR triplet absorption is absent. (3) In the iron blend near 5000 Å, the broad FeII absorption is weak while the sharper FeIII blend is prominent (4) The UV portion of the spectrum (2500 Å $< \lambda <$ 3500 Å) is much brighter down the hole, due to the decreased line blocking.

The comparison of Figure 19.1 is meant to demonstrate that the spectrum emanating from the hole is characterized by so-called SN 1991T-like peculiarities. What connection, if any, the hole asymmetry may have to SN 1991T itself will be discussed further in the conclusion. Note there are also apparent differences between SN 1991T and the model, among them: (1) The SII "W-feature" near 5500 Å is weak but visible in the model, whereas no clear feature is seen in SN 1991T; (2) The model has too much emission in the SiII 6150 and CaII IR triplet features. (3) The velocities of the FeIII lines are too low in the model, by about 2000 km s^{-1}.

The observed luminosity of the supernova also depends upon line of sight. When viewed near or down the hole ($\theta < \theta_H$) the supernova is brighter than average by up to 0.3 mag in B. This is because photons preferentially escape out the hole due to the lower opacities. For those lines of sight where the spectrum looks normal ($\theta > \theta_H$) the ejecta-hole asymmetry still causes some dispersion in peak magnitude. For example, the supernova is dimmer when viewed from the side ($\theta = 90°$) then from the back ($\theta = 180°$), because when viewed from the side the supernova is missing a "wedge" of emitting material. The total dispersion about the mean is $\sim$0.2 mag in B. For a hole with a smaller opening angle ($\theta_H = 30°$) the B-band dispersion is only $\sim$0.1 mag.

19.4 Polarization signatures

The polarization is the most direct indication of asymmetry in the ejecta. Light becomes polarized in supernova atmospheres due to electron scattering; other sources of opacity, such as bound-bound line transitions, are usually considered to be depolarizing. We define the continuum polarization as the polarization computed using only electron scattering opacity – at maximum light this is most closely realized in the red end of a supernova spectrum (near 7000 Å), where there is the least amount of line opacity. In an axially-symmetric geometry like the ejecta-hole model, the net polarization can be non-zero and will align either parallel or perpendicular to the axis of symmetry. We use the convention that positive (negative) polarization designates a polarization oriented parallel (perpendicular) to the axis of symmetry.

Figure 19.2 shows the ejecta-hole model continuum polarization as a function of inclination. When viewed directly down the hole ($\theta = 0°$) the projection of the

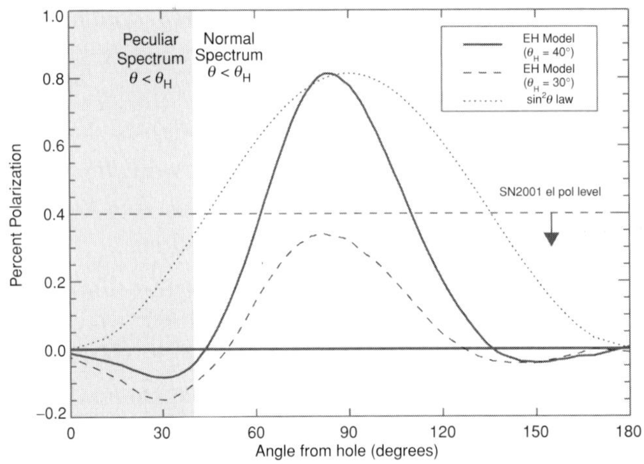

Fig. 19.2. Continuum polarization of the ejecta-hole model near maximum light as a function of inclination. The solid line is the model used throughout the paper, while the dashed line is a model where the hole opening angle has been reduced to 30°. We also over-plot $\sin^2\theta$ (multiplied by 0.8%), which is the inclination dependence of an optically thin scattering envelope.

supernova atmosphere is circularly symmetric and the polarization cancels. As the line of sight is inclined, the polarization increases, reaching a maximum of 0.8% when the supernova is viewed nearly side-on ($\theta \approx 90°$). This maximum value is some-what higher than that seen for example in SN 2001el, however two points should be stressed: (1) the number of high quality polarization observation of SNe Ia is still relatively small, so it is possible that more highly polarized SNe Ia (those observed near 90°) have so far been missed; (2) the polarization level is sensitive to the particular density structure of the ejecta hole. To demonstrate this we have over-plotted the continuum polarization of a model with a smaller hole (opening angle $\theta_H = 30°$). This tames the asymmetry and decrease the continuum polarization by more than a factor of two. In general, the more extreme the asymmetry of the hole (i.e. the larger and more evacuated it is) the higher the average polarization level. A larger sample of SN Ia polarization observations could therefore put constraints on the size of the hole.

One correlation to keep in mind is that the continuum polarization in the ejecta hole geometry is always relatively small ($<0.1\%$) for views near the hole where the spectrum looks peculiar. For views away from the hole, the continuum polarization may be either small or large. However the continuum polarization is not the whole story, and the polarization over the line features is substantial for $\theta < \theta_H$. This is because the line opacity partially obscures the underlying light, creating a less effective polarization cancellation over the line absorption features [7]. The polarization spectrum for viewing angles just off the hole axis ($\theta \approx 20°$) will therefore look "line peak dominated" – i.e. the continuum polarization will be low, but large polarization peaks will be associated with the blueshifted line absorption features (in particular the SiII 6150 feature and the CaII IR triplet). Because no such polarization spectrum occurs in ellipsoidal geometries, the large line peaks are an important signature of the hole geometry. The failure to see such signatures of the hole in future spectropolarimetric observations could have interesting consequences for the progenitors of SNe Ia.

19.5 Conclusion

An ejecta hole asymmetry may be one source of diversity in SNe Ia, but it would not be the only one. The dominant source of diversity in SNe Ia is thought to be variations in the amount of ^{56}Ni synthesized in the explosion. The peculiar spectra of SN 1991T-like supernovae are usually explained in this context, as supernovae with high envelope temperatures due to a relatively large ^{56}Ni mass [15, 13, 5]. As the models in this paper demonstrate, if an ejecta-hole asymmetry exists, there is a physically very different route to the same sort of spectral peculiarities. Our models show that the spectrum should show some level of peculiarity for $\theta < \theta_H$ or about 12% of the time. The observed rate of SN 1991T/SN 1999aa like supernovae is $\sim$20% according to Li *et. al.* (2001) – therefore it is possible that a substantial percentage of the class (but likely not all) is the result of an ejecta-hole asymmetry.

In this paper we have chosen to compare the spectra emanating from the hole with SN 1991T only because it is the well-known prototype of a certain kind of spectral peculiarity. Whether SN 1991T itself was an example of looking down the ejecta hole is debatable. There are several indications are that SN 1991T had a large ^{56}Ni mass, including the broad light curve, and the high velocities of nebular iron lines [12]. Confusing this conclusion, however, is the fact that SN 1991T had an unusually low velocity SiII 6150 line [2] and that, according to new distances measurements, the over-luminosity of the supernova was only moderate ($\sim$0.3 mag [16]). In any case, among *other* supernovae with SN 1991T-like peculiarities there is a good deal of diversity, and the large ^{56}Ni mass explanation will not apply in all cases. The most obvious case in point is SN 2002cx [8]. The spectrum of SN 2002cx resembled SN 1991T in having prominent FeIII lines and weak SiII, CaII and FeII lines, but the supernova was *underluminous* by $\sim$2 mag. The singularity of the supernova may indicate that we are seeing multiple channels of diversity operating at once – one scenario to entertain now is that we are looking down the ejecta hole of a supernova with a small ^{56}Ni mass. The fact that SN 2002cx had unusually low line velocities fits well into the picture of peering through a hole into the inner layers of a supernova.

In any case, SN 2002cx highlights the fact that the diversity of SNe Ia is more complicated than a one-parameter sequence based upon ^{56}Ni. Other channels of diversity therefore need to be identified, and the non-zero polarization measurements suggest an asymmetry of some sort likely plays an important role. However, because we recognize that an asymmetry like an ejecta hole is only one of several possible sources of diversity in SNe Ia, it will be difficult to isolate the geometrical effects from the other variations that may be operating. The only hope is to collect a large sample of supernovae with well observed light curves, spectra and polarization, so that one might try to pull out the different trends.

References

Fryxell, B. A. & Arnett, W. D. 1981, ApJ, 243, 994
Hatano, K., Branch, D., Qiu, Y. L., Baron, E., Thielemann, F.-K., & Fisher, A. 2002, New Astronomy, 7, 441
Höflich, P., Wheeler, J. C., Hines, D. C., & Trammell, S. R. 1996, ApJ, 459, 307+
Howell, D. A., Höflich, P., Wang, L., & Wheeler, J. C. 2001, ApJ, 556, 302
Jeffery, D. J., Leibundgut, B., Kirshner, R. P., Benetti, S., Branch, D., & Sonneborn, G. 1992, ApJ, 397, 304
Jeffrey, D. J. 1991, ApJ, 375, 264
Kasen, D., Nugent, P., Wang, L., Howell, D. A., Wheeler, J. C., Höflich, P., Baade, D., Baron, E., & Hauschildt, P. H. 2003, ApJ, 593, 788
Li, W., Filippenko, A. V., Chornock, R., Berger, E., Berlind, P., Calkins, M. L., Challis, P., Fassnacht, C., Jha, S., Kirshner, R. P., Matheson, T., Sargent, W. L. W., Simcoe, R. A., Smith, G. H., & Squires, G. 2003, PASP, 115, 453
Li, W., Filippenko, A. V., Treffers, R. R., Riess, A. G., Hu, J., & Qiu, Y. 2001, ApJ, 546, 734
Livne, E., Tuchman, Y., & Wheeler, J. C. 1992, ApJ, 399, 665
Marietta, E., Burrows, A., & Fryxell, B. 2000, ApJs, 128, 615
Mazzali, P. A., Cappellaro, E., Danziger, I. J., Turatto, M., & Benetti, S. 1998, ApJl, 499, L49+
Mazzali, P. A., Danziger, I. J., & Turatto, M. 1995, AA, 297, 509
Nomoto, K., Thielemann, F., & Yokoi, K. 1984, ApJ, 286, 644
Nugent, P., Phillips, M., Baron, E., Branch, D., & Hauschildt, P. 1995, ApJl, 455, L147+
Saha, A., Sandage, A., Thim, F., Labhardt, L., Tammann, G. A., Christensen, J., Panagia, N., & Macchetto, F. D. 2001, ApJ, 551, 973
Wang, L. et al. 2003, ApJ, in press

20

Hunting for the signatures of 3-D explosions with 1-D synthetic spectra

E. J. Lentz

Department of Physics and Astronomy, University of Georgia, Athens, GA 30602

E. Baron

Department of Physics and Astronomy, University of Oklahoma, Norman, OK 73025

P. H. Hauschildt

Hamburger Sternwarte, Gojenbergsweg 112, 21029 Hamburg, Germany

Abstract

Multi-dimensional models of supernovae show radial and non-radial variation in the density and composition not seen in one-dimensional models. Many of the questions about the flow of radiation through the expanding, multi-dimensional atmosphere will require multi-dimensional radiation transport calculations, but some may be tested with existing one-dimensional transport codes. So far, tests with models of Type Ia supernovae have shown that the unburned fuel (C+O) mixed down into deeper layers in multi-dimensional models have only a simple (C II lines) and modest signature in the spectra. This places only light constraints on the mixing of C+O into the lower layers of Type Ia supernovae.

20.1 3-D effects on supernova spectra

The proliferation of multi-dimensional explosion models for supernovae has made it clear that using one dimensional models for spectrum synthesis is not fully adequate. How adequate are the old 1-D spectral models? What are the multidimensional signals present in the light received from distant supernovae? Can we calculate spectra with 1-D codes that can explore the multi-dimensional effects?

The clearest multi-dimensional signal in spectrum of supernova is polarization. The number of polarization observations of supernovae have increased dramatically in the last decade, and polarization has now been detected in all types and sub-types of supernovae. Since it is impossible to calculate polarization in the 1-D models we calculate, we will not discuss polarization further, but Kasen discusses polarization elsewhere in this volume.

Small scale variations in composition or density due to hydrodynamic instabilities are another 3-D effect that can affect the spectra. For 1-D spectrum modeling, we must assume that such variations are smoothed. These effects on the spectrum are discussed by Thomas elsewhere in this volume. There are also related and unanswered questions about the flow of radiative energy through a gas with alternating small regions of higher and lower opacity. These need to be studied with 3-D models, either using mean opacities or Monte Carlo techniques.

The two other potential effects involve hydrodynamic effects introduced in multidimensional models and large scale asymmetries. These are discussed in more detail in the following two sections.

20.2 Effects of 3-D hydro on the spectra of Type Ia supernovae

To represent a typical 1-D explosion model for SNe Ia we look at the structure of W7 (Nomoto *et al.* 1984; Thielemann *et al.* 1986). In the inner region, $v < 8000$ km s^{-1}, we find Fe-peak material. The next region, $8000 < v < 15000$ km s^{-1}, consists of the intermediate mass elements, Ti, Ca, S, Si, and Mg. Each of those two regions typically contain about half of the 1.4 $M_\odot$ ejecta. A small quantity of unburned carbon and oxygen is often found at higher velocities, $v > 15000$ km s^{-1}.

In contrast the 3-D models computed by Gamezo *et al.* (2003; this volume) and by others have shown large plumes of burnt material moving upward surrounded by sinking columns of unburnt C+O 'fuel'. Even though these multi-dimensional hydrodynamic calculations have not reached the free expansion epoch yet, they do present a significant challenge to the previously calculated spectra based on 1-D explosion models. What effect does the unburnt material sinking to the low velocity inner regions have on the spectrum and how much C+O is permissible before delayed detonation models must be invoked to burn it after it has sunk toward the center?

To address this question we are in the midst of a series of numerical experiments using the PHOENIX, version 12 (Hauschildt & Baron 1999), stellar atmospheres and radiative transport code. We include a large database of line and continuum opacities. The model atmosphere is computed in spherical symmetry, including relativistic effects, in energy balance with gamma-ray deposition. The density and composition structure is a homologously expanded hydrodynamic model, in this case W7. We have computed the model with the most important species treated in non-LTE: C II, O II, Na I, Mg II, Si II, S II, Ca II, Fe II, Co II, and Ni II. We have previously shown that W7 fits normal Type Ia spectra like SN 1994D (Lentz *et al.* 2001). For these calculations, the 'base' model has the same parameters we found to fit SN 1994D.

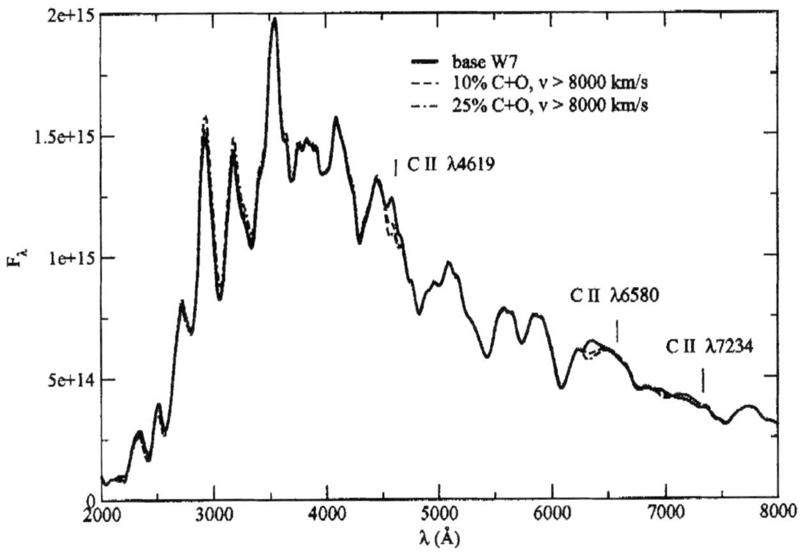

Fig. 20.1. W7 at 9 days comparison of models with different amounts of C + O mixed into the layers outside the Fe core.

In our first comparison (Figure 20.1), we show the effect of replacing 10% and 25% of the mass of the layers outside the Fe-rich core region, $v > 8000$ km s^{-1}, with a mixture of C+O to represent the downward streams of unburnt material passing through the region in the 3-D models. This calculation is for 9 days after the explosion, or about 10 days before maximum brightness. Two features show a modest effect. The C II λ4619 line is in a crowded region of the spectrum, and the C II λ6580 line is easily confused with the Hβ line of hydrogen. To separate the two, a non-ambiguous addition signal is required. For the identification of hydrogen we found that the Paα should be strong if hydrogen is present (Lentz *et al.* 2002).

Figure 20.2 shows the C+O mixing signature 18 days after explosion, or 1–2 days before maximum brightness. Again the the C II λ4619 line is in a crowded region and C II λ6580 can be confused with Hβ, but now the C II λ7234 line is strong. The measured blue-shift velocities from the minima of the two carbon lines in the red region are $\sim$ 8000 km s^{-1}. This would demonstrate that the carbon is found at layers much deeper than seen in the 1-D models.

We have previously reported similar models with the C+O replacement at the center of the Fe-rich core (Baron *et al.* 2003). For these models we replaced 25% of the mass inside 4000 km s^{-1} with C+O or 50% of the mass inside 5000 km s^{-1} with C+O (Figure 20.3) at 35 days after the explosion and compared it to the observation of SN 1994D two weeks after maximum brightness. The overall difference between

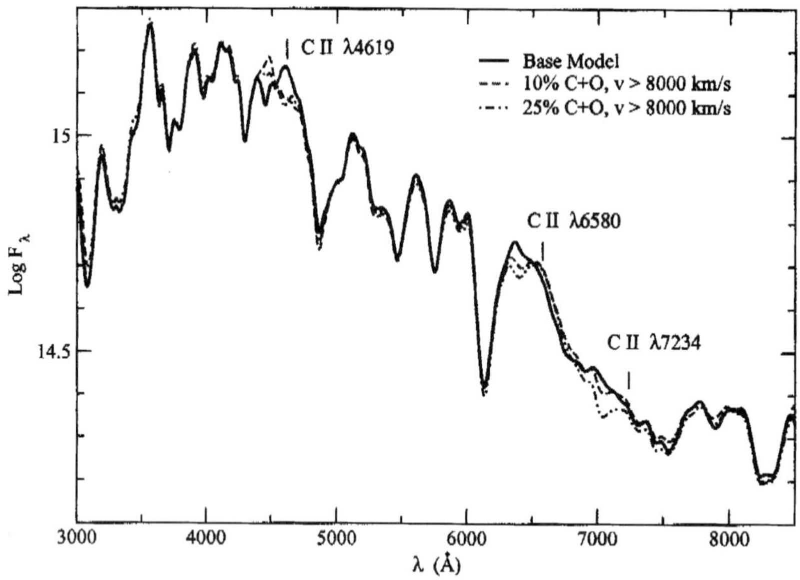

Fig. 20.2. W7 at 18 days comparison of models with different amounts of C + O mixed into the layers outside the Fe core.

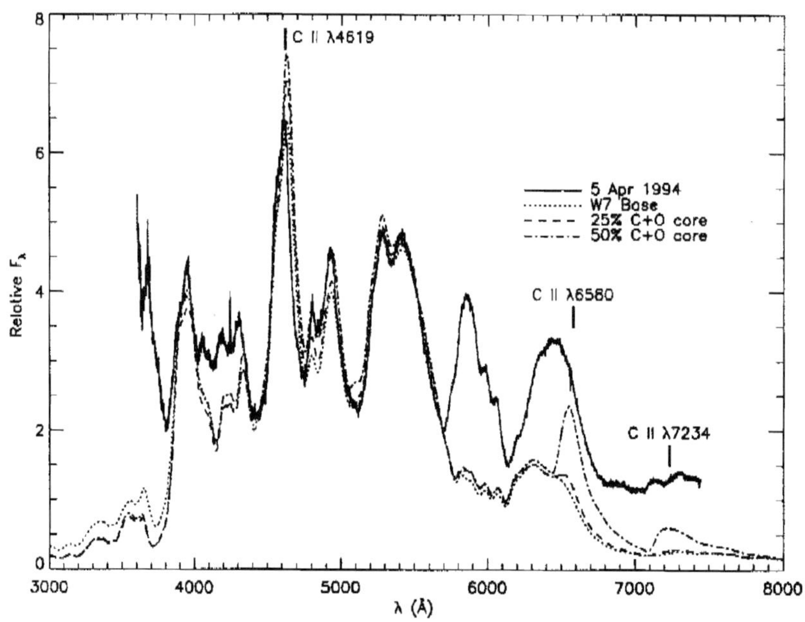

Fig. 20.3. W7 at 35 days comparison of models with different amounts of C + O mixed into the inner portion the Fe core. From Baron *et al.* 2002.

the models and the observation reflects the lack of forbidden lines in our line list for this nearly nebular supernova. The signal from the two carbon lines in the red band of the spectrum are clear for the "50% C+O core" model but not for the "25% C+O core" model. It seems that large amounts of carbon, up to about 25%, can be hidden in the deepest regions of the core.

The downward mixing of unburnt C+O into the burnt regions of the exploding white dwarf in Type Ia supernova models are not strictly excluded by the absence of low-velocity carbon or oxygen lines in the observed spectra. As much as 10% of the mass in the intermediate velocity region where the 'streamers' of 'fuel' are sinking and 25% of the inner core where it accumulates could be unburnt C+O without significantly affecting the spectrum. We are working on models that combine these calculations for the separate regions and checking the limit on the upward flow of Fe-rich material into higher and faster layers.

20.3 Modeling supernovae with large asymmetries

Some supernova models show large asymmetries like the jet powered 'hypernova' models (Maeda & Nomoto 2003; Maeda, this volume; Höflich *et al.* 1999, etc.) or the collision of a Type Ia supernova with its companion star (Marietta *et al.* 2000). In these cases there are large, distinct regions with not only different compositions, but densities as well. This should lead to diffential rates of energy flow and thus different temperature structures in each region. The question is whether each region is large enough to be treated semi-independently of the others. One method used for stars with surface variation is to treat each separate region with an independent 1-D model and assume there is no interaction. For supernovae this is somewhat more difficult since supernovae are more spherical, that is the height above the photosphere is a much larger than in stars, so the lateral distance for the supernova atmosphere to become optically thick, and therefore independent, is a much larger fraction of the sphere.

We are beginning to pursue a model of SN 1998bw using the 2-D model of Maeda & Nomoto (2003). Preliminary calculations show that it should be opaque enough to treat the polar and non-polar ejecta independently for the early spectroscopic phases. The regions will be treated as if they were separate 1-D models and then combined after convergence to give a single spectrum. One significant question without an easy answer is the luminosity. Normally, we fit spectra by adjusting the total luminosity which is tied to the temperature of the photosphere, but the two component models have two separate photospheres. Do they have the same effective temperature? This would have the flux dominated by the larger radius, faster, photosphere in the dense jet ejecta. Or, is the luminosity per solid angle a constant, which would lower the effective temperature of the faster and larger radius photosphere of the dense jet? Likely, neither of these is fully correct and extracting the difference may be difficult from the observed flux spectra. Ultimately the answer to this question will have to come from 2-D light curve modeling.

We have seen how many important effects of multi-dimensional structure in supernovae can be probed using 1-D spectrum synthesis.

This work has been supported in part by NSF grant AST-9720704, NASA grants NAG 5-8424 and NAG 5-3619 to the University of Georgia; and by NASA grant NAG 5-12127, NSF grant AST-0204771, and an IBM SUR grant to the University of Oklahoma.

References

Baron, E., Lentz, E. J., & Hauschildt, P. H. 2003, Astrophys. J. Lett., 588, L29
Gamezo, V. N., Khokhlov, A. M., Oran, E. S., Chtchelkanova, A. Y., & Rosenberg, R. O. 2003, Science, 299, 77
Hauschildt, P. H. & Baron, E. 1999, J. Comp. Appl. Math., 109, 41
Höflich, P., Wheeler, J. C., & Wang, L. 1999, Astrophys. J., 521, 179
Lentz, E. J., Baron, E., Branch, D., & Hauschildt, P. H. 2001, Astrophys. J., 557, 266
Lentz, E. J., Baron, E., Branch, D., & Hauschildt, P. H. 2002, Astrophys. J., 580, 374
Maeda, K. & Nomoto, K. 2003, Astrophys. J., in press
Marietta, E., Burrows, A., & Fryxell, B. 2000, Astrophys. J. Supp. 128, 615
Nomoto, K., Thielemann, F.-K., & Yokoi, K. 1984, Astrophys. J., 286, 644
Thielemann, F.-K., Nomoto, K., & Yokoi, K. 1986, Astron. & Astrophys., 158, 17

21

On variations in the peak luminosities of Type Ia supernovae

F. X. Timmes, E. F. Brown, J. W. Truran

Center for Astrophysical Thermonuclear Flashes and Dept. of Astronomy & Astrophysics, The University of Chicago, Chicago, IL, USA

Abstract

We explore whether the observed variations in the peak luminosities of Type Ia supernovae originate in part from a scatter in metallicity of the main-sequence stars that become white dwarfs. Previous, numerical, studies have not self-consistently explored metallicities greater than solar. One-dimensional, Chandrasekhar mass models of SNe Ia produce most of their ^{56}Ni in a burn to nuclear statistical equilibrium between the mass shells 0.2 $M_\odot$ and 0.8 $M_\odot$, for which the electron to nucleon ratio Y_e is constant during the burn. We show analytically that, under these conditions, charge and mass conservation constrain the mass of ^{56}Ni produced to depend *linearly* on the original metallicity of the white dwarf progenitor. This effect is most evident at metallicities greater than solar. Detailed post-processing of W7-like models confirms this linear dependence, and our calculations are in agreement with previous self-consistent calculations over the metallicity range common to both calculations. The observed scatter in the metallicity ($1/3\ Z_\odot - 3\ Z_\odot$) of the solar neighborhood is enough to induce a 25% variation in the mass of ^{56}Ni ejected by Type Ia supernova and is sufficient to vary the peak V-band brightness by $|\Delta M_V| \approx 0.2$. This scatter in metallicity is present out to the limiting redshifts of current observations ($z \lesssim 1$). Sedimentation of ^{22}Ne can possibly amplify the variation in ^{56}Ni mass to $\lesssim$50%. Further numerical studies can determine if other metallicity-induced effects, such as a change in the mass of the ^{56}Ni-producing region, offset or enhance the variation we identify.

21.1 Introduction

The peak luminosities of Type Ia supernovae (SNe Ia) are essential not only for understanding the explosion mechanism, but also for the use of SNe Ia as distance indicators in cosmology (Fillipenko 1997; Branch 1998; Leibundgut 2001). For the nearby SNe Ia with Cepheid-determined distances, Fig. 21.1 shows the overall

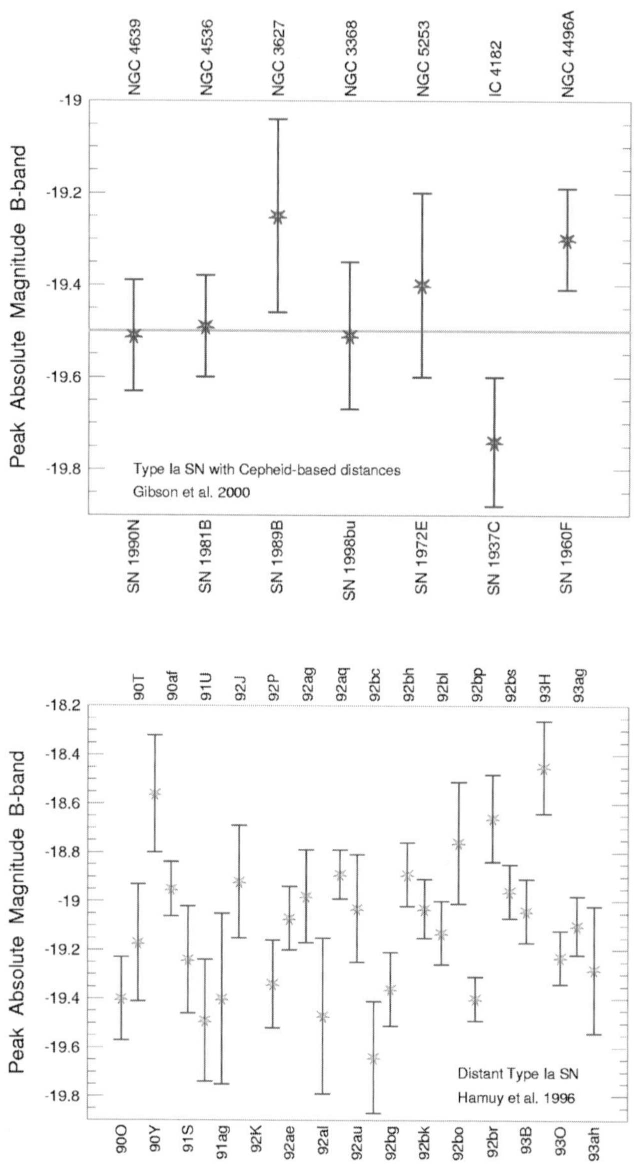

Fig. 21.1. Peak absolute B magnitudes of nearby (with Cepheid-determined distances; *left panel*; Fillipenko 1997; Saha *et al.* 1999; Gibson *et al.* 2000) and distant (*right panel*; Hamuy *et al.* 1996; Riess *et al.* 1998) SNe.

dispersion in the peak magnitude measurements is rather small, about 0.5 mag. in *B* and *V* (Fillipenko 1997; Saha *et al.* 1999; Gibson *et al.* 2000). When the sample is enlarged to include more distant SNe Ia, there are several subluminous events that broaden the variation to about 1 mag. in *B* (Hamuy *et al.* 1996; Riess *et al.* 1998; Perlmutter *et al.* 1999; Phillips *et al.* 1999; Ivanov, Hamuy, & Pinto 2000), but the bulk of the SNe Ia sample have peak brightnesses within a 0.5 magnitude range in B and V.

An interesting feature of SNe Ia is that their peak luminosity is set not by the explosion, for which the deposited energy goes into expansion, but rather by the decay of ^{56}Ni and ^{56}Co formed during the nucleosynthesis (Arnett 1982; Pinto & Eastman 2000). At the time of peak luminosity, ^{56}Co has not yet decayed, and hence the peak luminosity is a measurement of the mass of ^{56}Ni ($M(^{56}\mathrm{Ni})$) synthesized during the explosion. This amount presumably depends on the progenitor and on the details of the explosion. In particular, it has long been known that the amount of ^{56}Ni synthesized depends in part on the asymmetry between neutrons and protons in the progenitor white dwarf (Truran, Arnett, & Cameron 1967; Arnett, Truran, & Woosley 1971).

Nearly all one-dimensional Chandrasekhar mass models of SNe Ia produce most of their ^{56}Ni in a burn to nuclear statistical equilibrium (NSE) between the mass shells 0.2 $M_\odot$ and 0.8 $M_\odot$ (Nomoto, Thielemann, & Yokoi 1984; Höflich, Wheeler, & Thielemann 1998; Iwamoto *et al.* 1999; Höflich *et al.* 2000). In this region, unlike in the innermost 0.2 $M_\odot$ (Brachwitz *et al.* 2000), weak interactions operate on timescales longer than the time for the thermonuclear burning front to disrupt the white dwarf. Following this rapid burn to NSE, most of the mass is in the iron-peak nuclei ^{56}Ni, ^{58}Ni, and ^{54}Fe. First consider the case when ^{56}Ni and ^{58}Ni are the only two competing species. Mass and charge conservation then imply that the mass fraction of ^{56}Ni goes as $X(^{56}\mathrm{Ni}) = 58Y_e - 28$, where $Y_e = \Sigma X_i Z_i / A_i$ is the electron abundance and isotope i has Z_i protons, A_i nucleons, and a mass fraction X_i. In particular, $X(^{56}\mathrm{Ni})$ is linear in Y_e.

Most of a main-sequence star's initial Z comes from the CNO and ^{56}Fe nuclei inherited from its ambient interstellar medium. During helium burning in the white dwarf progenitor, the CNO catalysts are converted to ^{22}Ne. Using this to calculate Y_e in the white dwarf, Timmes, Brown, & Truran (2003) obtained a linear expression for the mass fraction of ^{56}Ni in an NSE distribution in terms of the main-sequence star's initial metallicity

$$
\begin{aligned}
X(^{56}\mathrm{Ni}) &= 1 - 58\left[\frac{X(^{12}\mathrm{C})}{12} + \frac{X(^{14}\mathrm{N})}{14} + \frac{X(^{16}\mathrm{O})}{16} + \frac{X(^{56}\mathrm{Fe})}{28}\right] \\
&= 1 - 0.057\frac{Z}{Z_\odot}. \qquad (21.1)
\end{aligned}
$$

The average peak B and V magnitudes of nearby SNe Ia (Saha *et al.* 1999; Gibson *et al.* 2000) strongly imply that a fiducial SNe Ia produces $\approx 0.6\,M_\odot$ of ^{56}Ni. Taking eq. (21.1) to represent the mass fraction of ^{56}Ni relative to this fiducial mass gives

$$
M(^{56}\mathrm{Ni}) \approx 0.6 M_\odot \left(1 - 0.057\frac{Z}{Z_\odot}\right). \qquad (21.2)
$$

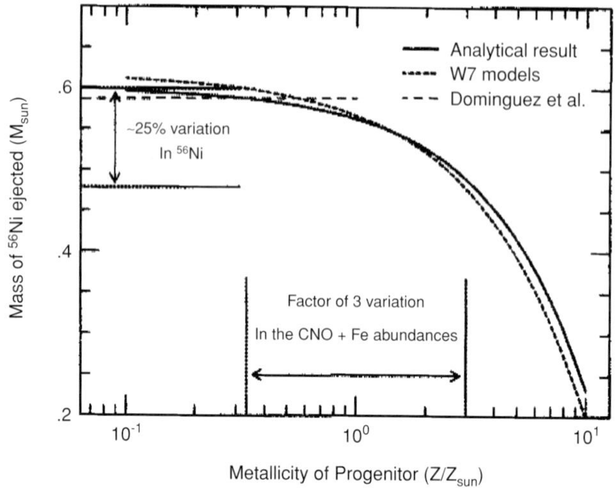

Fig. 21.2. Mass of ^{56}Ni ejected by SNe Ia as a function of the initial metallicity Z. Shown is the linear relation (*solid curve*; the curvature is from the logarithmic abscissa) of eq. (1.2) for $X(^{22}\mathrm{Ne}) = 0.024(Z/Z_{\odot})$, a sequence of W7-like models (*short-dashed curve*), and the calculation of D01 for 1.5 $M_{\odot}$ progenitors (*long-dashed curve*). From Timmes *et al.* (2003).

Here we assume that Y_e is uniform throughout the star, and that all material within the ^{56}Ni-producing mass shell passes through NSE with normal freeze-out.

The ^{56}Ni mass is affected only slightly if ^{54}Fe is produced instead of ^{58}Ni (Timmes *et al.* 2003), as the charge-to-mass of ^{58}Ni (which sets the slope in eq. [21.1]) is $28/58 = 0.483$ and differs by only 0.3% from the charge-to-mass of ^{54}Fe, $26/54 = 0.481$. The simple linear relation between Y_e and the mass fraction of ^{56}Ni in eq. (21.1) is robust, as it only relies on basic properties of NSE. Fig. 21.2 shows the $M(^{56}\mathrm{Ni})$ ejected (*short-dashed curve*) obtained by integrating a 510 isotope nuclear reaction network over the thermodynamical trajectories of W7-like models (Nomoto *et al.* 1984; Thielemann, Nomoto, & Yokoi 1986; Iwamoto *et al.* 1999) with Z allowed to vary. Also plotted are the linear relation (eq. [21.2]; *solid curve*), with $Z/Z_{\odot}$ adjusted to give the ^{22}Ne abundance used by W7, $X(^{22}\mathrm{Ne}) = 0.025\ Z/Z_{\odot}$. Most of the differences between eq. (21.2) and the detailed W7-like models are attributable to the assumption that all of the ^{56}Ni comes from the 0.2 to 0.8 $M_{\odot}$ region in the white dwarf, with an additional small correction for weak interactions that slightly decrease Y_e.

The long-dashed curve in Fig. 21.2 shows the results of Dominguez *et al.* (2001, hereafter D01) for their 1.5 $M_{\odot}$ progenitors; other progenitor masses in their survey display the same trend with Z. Our analytical result and post-processed W7-like models essentially agree with the findings of D01 over the range of metallicities common to all three calculations. As is evident in Fig. 21.2, the largest variation in $M(^{56}\mathrm{Ni})$ occurs at $Z > Z_{\odot}$. We note that the calculations of D01 evolve a main-sequence star into the SNe Ia progenitor (see also Umeda *et al.* 1999a; Umeda *et al.* 1999b), whereas our calculation and those of Iwamoto *et al.* (1999) start from a given white dwarf configuration.

As a caveat, we note that our post-processing of the W7 thermodynamic trajectories is not completely self-consistent. The reason is that the temperature and density profiles of the W7 were calculated using the energy released by burning matter of solar *Z*. While *Z* is likely to influence the flame propagation (via the change in the rate of energy production; Hix & Thielemann 1996, 1999), in the mass range under consideration Si-burning is complete, so our assumption of NSE still holds. Indeed, Dominguez & Höflich (2000) found that the mass of ^{56}Ni synthesized was also rather insensitive to the details of the flame microphysics. It is also possible the density, ρ_{tr}, at which a transition from deflagration to detonation occurs will influence the amount of ^{56}Ni produced (Höflich, Khokhlov, & Wheeler 1995, and references therein). Fits to observations seem to require, however, that $\rho_{\mathrm{tr}} \approx 10^7$ g cm^{-3}, corresponding to a Lagrangian mass coordinate of $\sim 1.0\ M_{\odot}$, which is exterior to the main ^{56}Ni producing layers.

21.2 Scatter in the initial metallicity and the induced brightness variations

The age-metallicity relationship is based on calibration of [Fe/H], since the accumulation of ^{56}Fe in the ISM increases monotonically with time (Wheeler, Sneden, & Truran 1989). The metallicity of local field stars rapidly increased about 10–13 Gyr ago during formation of the Galaxy's disk and then increased much more gradually over the last $\sim$10 Gyr (Twarog 1980; Edvardsson *et al.* 1993; Ivanov, Hamuy, & Pinto 2000; Feltzing, Holmberg, & Hurley 2001). More importantly for our purposes, however, is the relatively large scatter in stellar metallicities, Δ[Fe/H] ~ 0.5, at any given age. Feltzing *et al.* (2001) constructed an age-metallicity diagram for 5828 dwarf and sub-dwarf stars from the Hipparcos Catalog using evolutionary tracks to derive ages and Strömgren photometry to derive metallicities. They concluded that the age-metallicity diagram is well-populated at all ages, that old, metal-rich stars do exist, and that the scatter in *Z* at any given age is larger than the observational errors. Other surveys of stellar metallicities (Edvardsson *et al.* 1993; Chen *et al.* 2000) are in good agreement with these trends.

The most abundant elements in the Galaxy after H and He are CNO. Both [C/Fe] and [N/Fe] in halo and disk dwarfs are observed to be roughly solar and constant (Laird 1985; Carbon *et al.* 1987; Wheeler *et al.* 1989). The [O/Fe] ratio is larger at low metallicities – oxygen being the dominant element ejected by SNe II – and then slowly decreases due to variations in mass and *Z* (Gratton 1985; Peterson, Kurucz, &

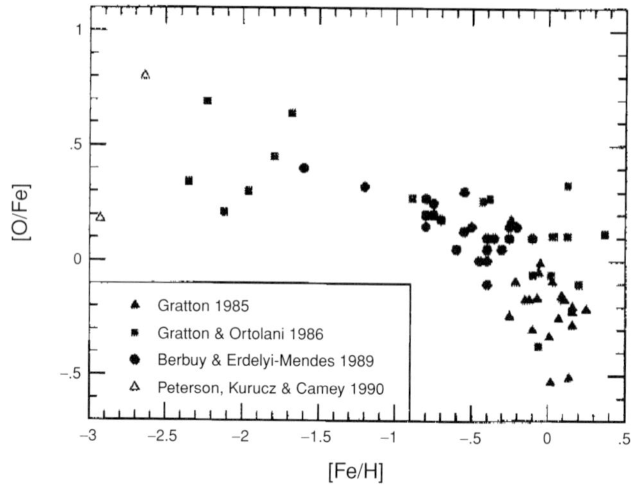

Fig. 21.3. The scatter in [O/Fe] is roughly from 1/3 to 3 times solar at any age (traced by [Fe/H]).

Carney 1990). Within these general trends is a relatively large scatter, Δ[C/Fe] $\sim$ Δ[N/Fe] $\sim$ Δ[O/Fe] $\sim$ 0.5 dex, at any given [Fe/H] (Fig. 21.3).

According to the simple analytical relation (eq. [21.1]) and the detailed W7-like models, a scatter of a factor of 3 about the mean in the initial metallicity ($1/3\ Z_\odot < Z < 3\ Z_\odot$) leads to a variation of about 25% (0.13 $M_\odot$) in $M(^{56}\mathrm{Ni})$ ejected by SNe Ia if the ^{22}Ne and ^{56}Fe are uniformly distributed within the white dwarf (Fig. 21.2, *arrows*). The minimum peak brightness variations caused by this variation in $M(^{56}\mathrm{Ni})$ are $|\Delta M_V| \approx 0.2$ (Pinto & Eastman 2001). Thus, the amplitude of this effect cannot account for all of the observed variation in peak luminosity of local SNe Ia (0.5 magnitude in B and V; Fig 21.1, *left*). The observed scatter in peak brightnesses may be even larger, as Cepheid-based distances to the host galaxies of peculiar events such as sub-luminous SN 1991bg or brighter-than-normal SN 1991T haven't been measured yet (Saha *et al.* 1999; Contardo, Leibundgut, & Vacca 2000; Leibundgut 2000). There is evidence for a larger scatter when more distant supernovae (Fig. 21.1, *right*) are included (Hamuy *et al.* 1996; Riess *et al.* 1998). It would take a scatter of about a factor of seven (Δ[Fe/H] $\sim$ 0.8 dex) in the initial Z to account for a factor of two variation in $M(^{56}\mathrm{Ni})$ and peak luminosity.

21.3 Implications

The variation from 1/3 to 3 times solar metallicity observed in dwarf stars in the Galactic disk implies that SNe Ia should have a minimum variation of 25% in the ejected $M(^{56}\mathrm{Ni})$. This conclusion rests on (1) considerations of nuclear equilibrium during the explosion, and (2) the observed scatter in Z of stars within the Galactic disk and implicitly assumes that the metallicity does not affect the mass of the white dwarf that is burned to nuclear statistical equilibrium.

There are implications for the brightness variations observed in both near and distant SNe Ia and for galactic chemical evolution. The most distant SNe Ia observed today have redshifts $z \lesssim 1$. This corresponds to a lookback time of about 4–7 Gyr depending on the cosmological model, and implies a mean [Fe/H] ratio between -0.1 and -0.3. There is, however, still a scatter of Δ[Fe/H] ~ 0.5 at these mean [Fe/H] ratios (see the discussion in § 21.2). We therefore expect that the variation in peak luminosities of SNe Ia in spiral galaxies at $z \lesssim 1$ will show the same minimum variation of $|\Delta M_V| \approx 0.2$ in peak luminosity as nearby SNe Ia. This variation is superposed on other evolutionary effects, such as that from a reduction in the C/O ratio (Höflich *et al.* 1998; D01).

For combustion to NSE with no freeze-out (see § 21.1), the ^{56}Ni mass is anticorrelated with the mass of ^{54}Fe and ^{58}Ni ejected, and subluminous SNe Ia will tend, therefore, to have larger ^{54}Fe/^{56}Fe ratios than brighter ones. Because SNe Ia produce $>1/2$ of the galactic iron-peak nuclei, the isotopic ratios among the iron group in SNe Ia ejecta should not exceed the solar ratios by about a factor of two (Wheeler *et al.* 1989; Iwamoto *et al.* 1999). There are several uncertainties with the ^{54}Fe/^{56}Fe ratio. First, some of the ^{54}Fe is produced in the core where weak interactions are important for determining the final Y_e (although this may be alleviated by the overturning of matter in the core from Rayleigh-Taylor instabilities; see Gamezo *et al.* 2003 for a recent calculation). At high densities these reaction rates are sensitive to the input nuclear physics (Brachwitz *et al.* 2000; Martinez-Pinedo, Langanke, & Dean 2000). Isotopic measurements of cosmic rays, such as with *Ulysses* (Connell & Simpson 1997; Connell 2001) and *ACE* (Wiedenbeck *et al.* 2001) can probe the evolution of ^{54}Fe/^{56}Fe over the past $\approx$5 Gyr.

Since most of the ^{56}Ni created in one-dimensional SNe Ia models lies in the $0.2\,M_\odot$–$0.8\,M_\odot$ mass shell, our assumption of a uniform ^{22}Ne distribution deserves close scrutiny. There has long been speculation about sedimentation of ^{22}Ne and its effect on the cooling of isolated white dwarfs (Hansen *et al.* 2002; Bildsten & Hall 2001; Deloye & Bildsten 2002, and references therein). If, for example, all of the ^{22}Ne from the outermost $0.6\,M_\odot$ were to settle into the shell between $0.2\,M_\odot$ and $0.8\,M_\odot$, then the effective Y_e in this shell would double. Indeed, Bildsten & Hall (2001) noted that the production of ^{54}Fe was an indirect test of the sedimentation of ^{22}Ne. For nearby SNe Ia, which presumably sample a range of progenitor ages, the variability in ^{56}Ni production would increase to 50%, or about $\Delta M_v \approx 0.3$. Since it takes about 7 Gyr for ^{22}Ne at the surface to fall through the outer $\approx 0.4\,M_\odot$ of a $1.2\,M_\odot$ CO white dwarf (Bildsten & Hall 2001), the effect of sedimentation will be diminished for those SNe Ia at $z \approx 1$.

Acknowledgements

We thank Franziska Brachwitz and Friedel Thielemann for providing the initial W7 thermodynamic trajectories. This work is supported by the Department of Energy under Grant No. B341495 to the Center for Astrophysical Thermonuclear Flashes and Grant No. DE-FG02-91ER40606 in Nuclear Physics and Astrophysics at the University of Chicago.

References

Arnett, W. D. 1982, ApJ, 253, 785
Arnett, W. D., Truran, J. W., & Woosley, S. E. 1971, ApJ, 165, 87
Bildsten, L. & Hall, D. 2001, ApJ, 549, L219
Brachwitz, F. *et al.* 2000, ApJ, 536, 934
Branch, D. 1998, ARA&A, 36, 17
Carbon, D. F., Barbuy, B., Kraft, R. P., Friel, E. D., & Suntzeff, N. B. 1987, PASP, 99, 335
Chen, Y. Q., Nissen, P. E., Zhao, G., Zhang, H. W., & Benoni, T. 2000, A&AS, 141, 491
Clifford, F. E., & Tayler, R. F. 1965, MNRAS, 129, 104
Connell, J. J. 2001, Space Sci. Rev., 99, 41
Connell, J. J., & Simpson, J. A. 1997, ApJ, 475, L61
Contardo, G., Leibundgut, B., & Vacca, W. D. 2000, A&A, 359, 876
Deloye, C. J. & Bildsten, L. 2002, ApJ, 580, 1077
Domínguez, I. & Höflich, P. 2000, ApJ, 528, 854
Domínguez, I., Höflich, P., & Straniero, O. 2001, ApJ, 557, 279 (D01)
Edvardsson, B., Andersen, J., Gustafsson, B., Lambert, D. L., Nissen, P. E., & Tomkin, J. 1993, A&A, 275, 101
Feltzing, S., Holmberg, J., & Hurley, J. R. 2001, A&A, 377, 911
Fillipenko, A. V. 1997, ARA&A, 35, 309
Gamezo, V. N., Khokhlov, A. M., Oran, E. S., Chtchelkanova, A. Y., & Rosenberg, R. O. 2003, Science, 299, 77
Gibson, B. K. *et al.* 2000, ApJ, 529, 723
Gratton, R. G. 1985, A&A, 148, 105
Hamuy, M., Phillips, M. M., Suntzeff, N. B., Schommer, R. A., Maza, J., Smith, R. C., Lira, P., & Aviles, R. 1996, AJ, 112, 2438
Hansen, B. M. S. *et al.* 2002, ApJ, 574, L155
Hix, W. R., & Thielemann, F.-K. 1996, ApJ, 460, 869
Hix, W. R., & Thielemann, F.-K. 1999, ApJ, 511, 862
Höflich, P., Khokhlov, A. M., & Wheeler, J. C. 1995, ApJ, 444, 831
Höflich, P., Wheeler, J. C., & Thielemann, F. K. 1998, ApJ, 495, 617
Höflich, P., Nomoto, K., Umeda, H., & Wheeler, J. C. 2000, ApJ, 528, 590
Ivanov, V. D., Hamuy, M., & Pinto, P. A. 2000, ApJ, 542, 588
Iwamoto, K., Brachwitz, F., Nomoto, K., Kishimoto, N., Umeda, H., Hix, W. R., & Thielemann, F. 1999, ApJS, 125, 439
Laird, J. B. 1985, ApJ, 289, 556
Leibundgut, B. 2000, A&A rev., 10, 179
Leibundgut, B. 2001, ARA&A, 39, 67
Martínez-Pinedo, G., Langanke, K., & Dean, D. J. 2000, ApJS, 126, 493
Nomoto, K., Thielemann, F.-K., & Yokoi, K. 1984, ApJ, 286, 644
Perlmutter, S., *et al.* 1999, ApJ, 517, 565
Peterson, R. C., Kurucz, R. L., & Carney, B. W. 1990, ApJ, 350, 173

Pinto, P. A. & Eastman, R. G. 2000, ApJ, 530, 744
Pinto, P. A. & Eastman, R. G. 2001, New Astronomy, 6, 307
Phillips, M. M., Lira, P., Suntzeff, N. B., Schommer, R. A., Hamuy, M., Maza, J., AJ, 118. 1766
Riess, A. G. *et al.* 1998, AJ, 116, 1009
Saha, A., Sandage, A., Tammann, G. A., Labhardt, L., Macchetto, F. D., & Panagia, N. 1999, ApJ, 522, 802
Thielemann, F.-K., Nomoto, K., & Yokoi, K. 1986, A&A, 158, 17
Timmes, F. X., Brown, E. F., & Truran, J. W. 2003, ApJ, 590, L83
Truran, J. W., Arnett, D., & Cameron, A. G. W. 1967, Canad. J. Phys., 45, 2315
Twarog, B. A. 1980, ApJ, 242, 242
Umeda, H., Nomoto, K., Kobayashi, C., Hachisu, I., & Kato, M. 1999, ApJ, 522, L43
Umeda, H., Nomoto, K., Yamaoka, H., & Wanajo, S. 1999, ApJ, 513, 861
Wheeler, J. C., Sneden, C., & Truran, J. W. 1989, ARA&A, 27, 279
Wiedenbeck, M. E., *et al.* 2001, Space Sci. Rev., 99, 15

Part IV

Theory of Core Collapse Supernovae

22

Rotation in core collapse progenitors: single and binary stars

N. Langer

Astronomical Institute, Utrecht University, Princetonplein 5, NL-3584 CC, Utrecht, The Netherlands

Abstract

Current massive single star evolution models with rotation, especially when magnetic fields are included, appear to get close in reproducing the spin rates of young neutron stars. This, however, excludes them as progenitors of gamma-ray bursts within the collapsar model. Close binary evolution models with rotation, on the other hand, suggest that the mass receiving star is spun-up appreciably and may retain enough angular momentum in its core until collapse, while the mass donor is spun-down to produce core rotation rates below those of single stars.

22.1 Introduction

The evolution of a single star can be strongly influenced by its rotation (e.g., Heger & Langer 2000; Meynet & Maeder 2000), and evolutionary models of rotating stars are now available for many masses and metallicities. While the treatment of the rotational processes in these models is not yet in a final stage (e.g., magnetic dynamo processes are just about to be included; Heger *et al.* 2003), they provide first ideas of what rotation can really do to a star.

Effects of rotation, as important as they are in single stars, can be much stronger in the components of close binary systems: Estimates of the angular momentum gain of the accreting star in mass transferring binaries show that critical rotation may be reached quickly (Packet 1981; Langer *et al.* 2000). Therefore, we need binary evolution models which include a detailed treatment of rotation in the stellar interior, as in recent single star models. However, in binaries, tidal processes as well as angular momentum accretion need to be considered at the same time. Some first such models are now available and are discussed below.

These models provide evidence for rotational processes in binaries being essential for some of the most exciting cosmic phenomena, which may occur exclusively

Table 22.1. *Magnetic Field and angular velocity evolution in a 15 $M_\odot$ star of solar composition and 200 km s^{-1} equatorial surface rotation on the ZAMS. We give typical toroidal (B_r) and radial (B_ϕ) magnetic field strengths in the inner 1–2 $M_\odot$ if radiative or outside the convective core for different evolution stages (see below), and the central angular velocity [from Heger et al. (2003)].*

evolution stage	B_ϕ(G)	B_r(G)	Ω_c(rad s^{-1})
MS	5×10^3	1	5.7×10^{-5}
TAMS	1×10^4	2	2×10^{-5}
He ignition . . .	4×10^4	30	8×10^{-5}
He depletion .	1×10^4	2	6×10^{-5}
C ignition	1×10^6	300	2.5×10^{-4}
C depletion. .	3×10^7	5×10^3	2×10^{-3}
O depletion . .	5×10^7	7×10^3	4×10^{-3}
Si depletion . .	3×10^8	2×10^5	7×10^{-3}
pre-SN	5×10^9	10^6	0.1

in binaries: Type Ia supernovae, the main producers of iron and cosmic yardsticks to measure the accelerated expansion of the universe, and gamma-ray bursts from collapsars – which, as we shall see, current stellar models with rotation preclude occurring in single stars – may provide the most powerful explosions in the universe and trace star formation to its edge.

22.2 Single stars

The pre-supernova rotation state in rotating single star models has recently been reviewed by Heger *et al.* (2003). Without the action of magnetic fields, hydrodynamic rotationally-induced instabilities and convection alone may not transport angular momentum efficiently enough to avoid forming pulsars rotating at break up (assuming that most of the angular momentum is conserved during the collapse). However, in massive single stars that lose significant amounts of mass already on the main sequence, the corresponding (non-magnetic) spin-down is very significant (Langer 1998). When the dynamo process proposed by Spruit (2002) is included in the models (Tables 22.1 and 22.2), one obtains small iron core spins for all initial masses, corresponding to pulsar periods of about 4–7 ms (Table 22.2). Assuming that these periods may still increase by 20% due to neutrino losses from the protoneutron star, they approach those of the fastest-rotating observed young pulsars (Table 22.3). In any case, the rotation rates found in the magnetic stellar models are too slow for most current gamma-ray burst models that require rapidly rotating stellar cores.

Table 22.2. *Pulsar rotation and angular momentum for different masses. We assume initial solar composition and a ZAMS equatorial surface rotation rate of* 200 km s^{-1}. *In the* 2nd *column we give the total angular momentum in the inner 1.7* $M_{\odot}$ *of the stellar core. Assuming no further loss of angular momentum and that a neutron star with moment of inertia* $I = 1.44 \times 10^{45}$ *g* cm^2 *is formed (* $R = 12$ *km,* $M = 1.4\ M_{\odot}$, $I = 0.36 M R^2$*; Lattimer & Prakash 2001) the resulting (lower limits for the) pulsar periods are given in the* 3rd *column [from Heger et al. (2003)].*

stellar mass	J (erg s)	period (ms)
15 $M_{\odot}$	1.4×10^{48}	6.7
20 $M_{\odot}$	1.8×10^{48}	5.0
25 $M_{\odot}$	2.1×10^{48}	4.3

Table 22.3. *Experimental periods (Marshall et al. 1998) and angular momentum estimates for young pulsars [from Heger et al. (2003)].*

pulsar	period (ms)	j (cm^2 s^{-1})	J (erg s)
PSR J0537-6910 (N157B, LMC)	16	2.0×10^{14}	5.67×10^{47}
PSR B0531+21 (crab)	33	9.9×10^{13}	2.75×10^{47}
PSR B0540-69 (LMC)	50	6.5×10^{13}	1.81×10^{47}
PSR B1509-58	150	2.2×10^{13}	6.05×10^{46}

22.3 Binary stars

We have constructed binary evolution models using the code of Wellstein *et al.* (2001), but including the physics of rotation as in the single star models of Heger *et al.* (2000) for both components. In addition, spin-orbit coupling according to Zahn (1977) has been added, and rotationally enhanced winds are implemented as in Langer (1998). The specific angular momentum of the accreted matter is assumed to be that of Kepler rotation at the stellar equator in the case of disk accretion, and determined by integrating the equation of motion of a test particle in the Roche potential in case the accretion stream impacts directly on the secondary star (Wellstein 2001).

Fig. 22.1 shows the evolution of the equatorial rotation velocity in a system starting out with a 16 $M_{\odot}$ and a 15 $M_{\odot}$ star in a 3 day orbit. The initial rotational velocity of both stars is unimportant since they evolve quickly into rotation that

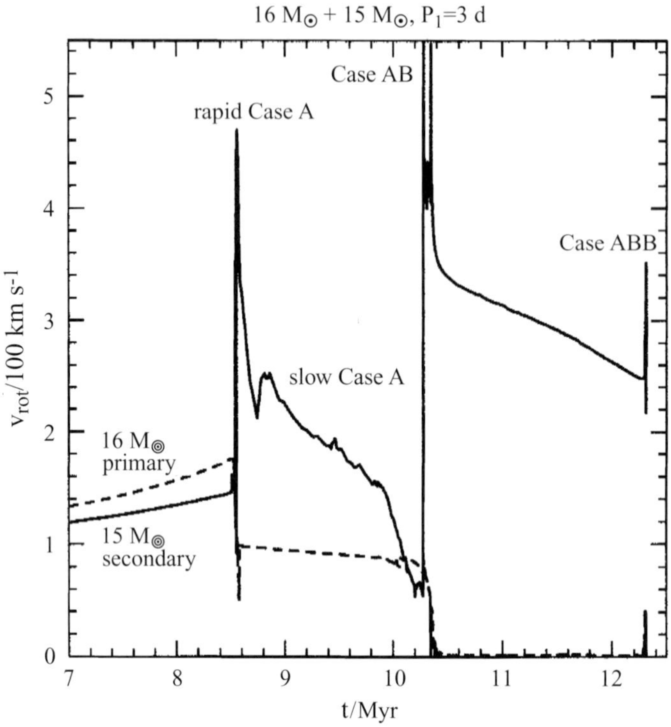

Fig. 22.1. Equatorial rotation velocity for primary (dashed line) and secondary (solid line) component of a 16 $M_\odot$ + 15 $M_\odot$ system with an initial orbital period of 3 d as function of time, starting at an age of 7 Myr, i.e., well before the onset of mass transfer, until the end of Case ABB mass transfer, which corresponds to the time of the supernova explosion of the primary. The four different mass transfer phases which occur in this system are indicated; except for the slow Case A mass transfer they occur on the thermal time scale of the primary star (see also next figure).

is synchronous with the orbital revolution, due to spin-orbit coupling. Each of the three thermal time scale mass transfer phases through which this system evolves (rapid Case A, Case AB, and Case ABB; see Wellstein *et al.* 2001) leads to a strong spin-up of the secondary star and an equally drastic spin-down of the primary (see Langer 1998, for the purely mechanical spin-down effect).

22.3.1 Spin evolution of the mass donor

From these models, we conclude that the initially more massive stars in massive close binaries are even less likely to produce a gamma-ray burst than single stars. First of all, they lose so much mass that even stars with a very large initial mass may not even form a black hole but rather a neutron star (see also Wellstein & Langer

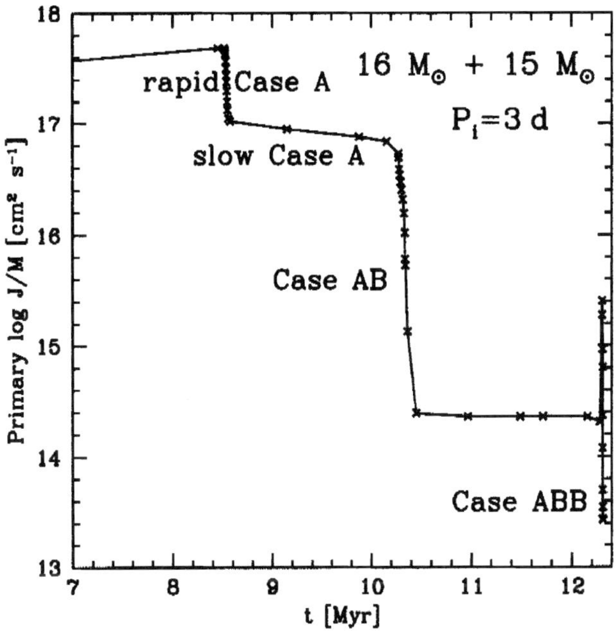

Fig. 22.2. Average specific angular momentum J/M as function of time for the mass donor in our 16 $M_\odot$ + 15 $M_\odot$ system with an initial orbital period of 3 d, starting at an age of 7 Myr, i.e., well before the onset of mass transfer, until the supernova explosion of the mass donor (see also previous figure).

1999). And secondly, Figure 22.2 shows how drastic these stars are spun down as a consequence of their heavy mass loss. The 16 $M_\odot$ star in the computed binary system is expected to produce a neutron star with an initial spin period of more than one second!

The only way to avoid both drawbacks is to employ Case C evolution, which leads to a core evolution as in single stars. However, even then the CO-core of the star needs to be spun-up significantly to produce a collapsar and a gamma-ray burst – a possibility suggested by Brown *et al.* (1999, 2000) in the context of common envelope evolution and spiral-in. No detailed models for this scenario exist at present.

22.3.2 Spin evolution of the mass gainer

Our models imply that the initially less massive star in a massive binary may accrete large amounts of angular momentum and will acquire a larger core spin than a corresponding single star (see Fig. 22.1). It is thus conceivable that accretion stars are the progenitors of asymmetric supernova explosions and rapidly spinning compact objects. Those sufficiently massive to transform into a Wolf-Rayet star

Table 22.4. *Clues on accretion efficiencies from observed binaries*

object	sp. types	orb. period	masses/ratio	accretion?
V729 Cyg[1]	O7+O7[2]	6.6 d	$q = 3.5$	YES
Wray 977[3]	BI+NS	44 d	40 $M_\odot$ + 1.4 $M_\odot$	YES
3 systems[4]	WNE+O	~8 d	$q \simeq 0.5$	NO[5]
4U 1700-37[6]	O6I+NS/BH	3.4 d	58 $M_\odot$ + 2.4 $M_\odot$	NO[5]

[1] Bohannan & Conti (1976)
[2] both components have the same visual magnitude
[3] Kaper *et al.* (1995), Wellstein & Langer (1999)
[4] Petrovic & Langer (2002)
[5] meaning: 10% or less
[6] Clark *et al.* (2002)

during core helium burning, or those which lose their envelope in a reverse Case C mass transfer, may possess all required ingredients to produce a gamma-ray burst within the collapsar model.

22.3.3 How much matter can stars accrete from a binary companion?

As mentioned above, non-magnetic accretion, i.e. accretion via a viscous disk or via ballistic impact, transports angular momentum and can lead to a strong spin-up of the mass gaining star. For disk accretion, it appears plausible that the specific angular momentum of the accreted matter corresponds to Kepler-rotation at the stellar equator. For rigid rotation, this leads to a spin-up of the whole star to critical rotation when its initial mass is increased by about 20% (e.g., Packet 1981). Can accretion continue beyond this?

Theoretically, this appears possible, as viscous processes may transport angular momentum outward through the star, the boundary layer, and the accretion disk (e.g., Paczynski 1991). However, as the star may be strongly rotating, its wind mass loss may be dramatically increased (Langer 1997, 1998), which may render the mass transfer process inefficient.

Observations of massive post-mass transfer binary systems constrain this effect. Table 22.1 lists parameters of four different kinds of massive close binary systems which give opposite answers. The two O stars in the Case A binary (mass transfer starts while both stars undergo core hydrogen burning) V 729 Cyg have a mass ratio of 3.5 but the same spectral type and visual flux. Clearly, an initial mass ratio close to 1 is required to get close to the observed current mass ratio. However, as during Case A the primary star (as we designate the initially more massive star in a binary) loses just about half of its mass, a mass ratio of at most 2 could be produced were the secondary (the initially less massive star in a binary) not allowed to accrete.

Another system showing strong evidence for accretion is the massive X-ray binary Wray 977; it would require that stars of $\gtrsim 40\ M_\odot$ form neutron stars to explain this system without accretion (Wellstein & Langer 1999).

Several Galactic short period WNE+O binaries, on the other hand, cannot be understood had the O star accreted substantial amounts from the WNE progenitor (Petrovic & Langer 2002). While those might have formed through common envelope evolution – for which little accretion is expected – the key X-ray binary 4U 1700-37 has such a short period that a major accretion phase can be excluded. However, as Case C evolution (mass transfer starts after core helium burning of the primary star) would lead to a compact object much more massive than $2.4\ M_\odot$, Case B evolution (mass transfer starts just after core hydrogen burning of the primary star) is most likely here (Clark *et al.* 2002).

We conclude from these observed binary systems that in some massive close (Case A or B) binaries, the mass transfer process is nearly conservative, while in others it is strongly non-conservative. It has been shown by Langer *et al.* (2003) that including rotational physics into binary evolution calculations allows one to recover both features within the same physical model.

22.4 Conclusions

The angular momentum evolution of single stars of various masses can be – extremely simplified – summarized in the following table (Langer 2003).

Specific angular momenta / $cm^2\ s^{-1}$

	ZAMS	end obs.	stage mod.	rigid rot.
$2 \ldots 8\ M_\odot$	10^{18}	WD: 10^{14}	10^{15}	10^{10}
$>8\ M_\odot$	10^{18}	NS: 10^{14}	$5\ 10^{14}$	10^{10}

Stars above $2\ M_\odot$ (i.e., those which do not suffer from magnetic braking) start out with $j \simeq 10^{18}\ cm^2\ s^{-1}$ on the main sequence. The observed angular momentum in white dwarfs and young neutron stars is of the order of $10^{14}\ cm^2\ s^{-1}$ (cf. Maeder & Eenens 2003). If the birth event of these compact remnants does not affect their spin, white dwarfs and young neutron stars should indeed have a similar specific angular momentum, as the angular momentum loss from the core in the rotating massive star models occurs during the early evolutionary phases. Note that $10^{14}\ cm^2\ s^{-1}$ is (logarithmically) about half way between local angular momentum conservation and completely rigid rotation. The predictions of rotating models for intermediate mass (Langer *et al.* 1999) and massive stars (see above) are still slightly above the observed values (the magnetic effects have not yet been considered in intermediate mass models). However, in view of this situation it seems difficult to assume that rotational effects can play any significant role in supernovae or gamma-ray bursts originating in single stars – which may defer the occurrence of gamma-ray bursts and jet-driven supernovae to binary stars altogether.

In close binary systems, there is indeed hope that the accreting component is significantly spun-up, and that rapid rotation prevails in the core until the end. The calculation of corresponding core collapse supernova progenitor models is currently underway. However, if accretion spins up supernova progenitors, this should hold as well for Type Ia events – which is indeed predicted by corresponding models (Yoon & Langer 2004).

References

Bohannan B., Conti P. S., 1976, ApJ 204, 797
Brown G. E., Lee C.-H., Bethe H. A., 1999, New Astron. 4, 313
Brown G. E., Lee C.-H., Wijers R. A. M. J., Lee H. K., Israelian G., Bethe H. A., 2000, New Astron. 5, 191
Clark J. S., Goodwin S. P., Crowther P. A., Kaper L., Fairbairn M., Langer N., Brocksopp C., 2002, A&A 392, 909
Heger A., Langer N., 2000, ApJ 544, 1016
Heger A., Langer N., Woosley S. E., 2000, ApJ 528, 368
Heger A., Woosley S. E., Langer N., Spruit H. C., 2003, in: *Stellar Rotation*, Proc. IAU-Symp. 215, A. Maeder & P. Eenens, eds, in press
Kaper L., Lamers H. J. G. L. M., Ruymaekers E., van den Heuvel E. P. J., Zuidervijk E. J., 1995, A&A 300, 446
Langer N., 1997, in *Luminous Blue Variables: Massive Stars in Transition*, A. Nota, H. J. G. L. M. Lamers, eds, ASP Conf. Ser., Vol. 120, p. 332
Langer N., 1998, A&A 329, 551
Langer N., Heger A., Wellstein S., Herwig F., 1999, A&A 346, L37
Langer N., Deutschmann A., Wellstein S., Höflich P., 2000, A&A 362, 1046
Langer N., Yoon S.-C., Petrovic J., Heger A., 2003, in: *Stellar Rotation*, Proc. IAU-Symp. 215, A. Maeder & P. Eenens, eds, in press
Lattimer J. M., Prakash M., 2001, ApJ 550, 426
Maeder A., Eenens, P., 2003, *Stellar Rotation*, Proc. IAU-Symp. 215, in press
Marshall F. E., Gotthelf E. V., Zhang W., Middleditch J., Wang Q. D., 1998, ApJ 499, 179
Meynet G., Maeder A., 2000, A&A 361, 101
Packet W., 1981, A&A 102, 17
Paczynski B., 1991, ApJ 370, 597
Petrovic J., Langer N., 2002, in *A massive star odyssey: from main sequence to supernova*, proc. IAU-Symp. 212, (San Francisco: ASP), K. A. van der Hucht, A. Herrero & C. Esteban, eds., p. 418
Spruit H. C., 2002, A&A 381, 923
Wellstein S., 2001, PhD thesis, Potsdam University
Wellstein S., Langer N., 1999, A&A 350, 148
Wellstein S., Langer N., Braun H., 2001, A&A 369, 939
Yoon S.-C., Langer N., 2004, this volume
Zahn J.-P., 1977, A&A 57, 383

23

Large scale convection and the convective supernova mechanism

S. A. Colgate

MS 227, Los Alamos Nat. Lab, P.O. Box 1663, Los Alamos, NM 87545; colgate@lanl.gov

M. E. Herant

Boston University BME, 44 Cummington St., Boston, MA 02215.

Abstract

It is a weird and unlikely circumstance that a collapse supernova (Type II) should explode. The peculiar mechanism that facilitates this explosion is the formation and preservation of large scale structures in a high entropy atmosphere residing on the surface of a nearly formed neutron star. The high entropy atmosphere is maintained by two sources: the gravitational energy of initial formation of the neutron star, released by diffusion and transport of neutrinos and secondly and possibly dominantly by the gravitational energy released at the suface by additional low entropy matter falling through to the neutron star surface. The preservation of this entropy contrast between up and down flows requires thermal isolation between the low entropy down flows and the high entropy up flows. This entropy contrast allows an efficient Carnot cycle to operate and thus allows the efficient conversion of thermal energy to mechanical, which in turn drives the explosion. The P-V diagram of various up and down going mass elements in the calculations demonstrates the existence of the cycle and its efficiency. Greater thermal isolation should occur in 3-D as opposed to 2-D calculations because of the difference in relative thickness or surface to mass ratio for the same mass flow in 2 and 3-D. This may explain the observed stronger explosion in 3-D calculations.

23.1 Prolog

This paper is written in honor of a long and lasting friendship between Craig Wheeler and the first author for more than half his current life. It starts with a short letter exchange which seems as valid today as ever because it is the confusion concerning large scale convection which is so central to both supernova as well as stars. It is largely based upon a paper Colgate, Herant, & Benz (1993) in honor of Willy Fowler, but with significant additions by Colgate & Fryer (1994).

From wheel@astro.as.utexas.edu Sun Nov 1 1992 To: colgate@eagle.lanl.gov Subject: hot bubble

Stirling,

I received your Fowler preprint last week and read it with great interest. Alexei Khokhlov, Itamar Lichtenstadt, a student and I will get together to talk about it next week.

I would like to clear up one conceptual point. In originally reading Herant *et al*, I got the idea that when you said "convection" you meant circulation of matter in the sense that some stuff went up and then joined a downward plume to complete the cycle. The current paper again uses the words "circulation" and "convection" in ways that seem to connote the context. But in Aspen last summer, and perhaps implicitly in some contexts of this paper, you said that down flow is heated and driven in the rising plumes, but that no up welling plume ever turns into a downward plume.

I have this feeling that there might be significant differences depending on whether matter truly circulates or whether high entropy plumes only rise, never to return. The present paper is still rather ambiguous on this point. Can you clarify it for me?

Thanks,

Craig

To: wheel@astro.as.utexas.edu Subject: Re: hot bubble

I guess I have never tried to close the loop on the argument about convection, up and down, but my comment last summer was more or less correct. If a hot bubble, plume rises, to first order it never comes back down. It is never cooled beyond its adiabatic expansion. The down flow that reaches to the neutron star surface bifurcates at the neutron star surface. The inner fraction cools i.e. deleptonizes and joins the neutron star with negligible change in volume. Another fraction is heated by the putative high energy neutrino deposition to high entropy and fills a much larger volume and becomes another plume. The ensemble of all such higher entropy plumes is then the desired hot bubble needed to make the explosion, i.e. no fall back. Thanks for the encouragement. best Stirling

23.2 Introduction

The spherically symmetric diffusion of heat from a thermonuclearly explosive fuel is extraordinarily stable as the existence of all the various stars attests. In stars the free energy of thermonuclear burn is many orders of magnitude greater than the gravitational binding energy and so even a very small runaway thermonuclear reaction should lead to explosion, but it usually does not happen, novas and SN Ia's being the exception. The lepton degenerate core of a forming neutron star on the other hand, is strongly bound and so there is no explosive free energy available, yet the supernova explodes and ejects nearly the whole star. However, although the free energy of this interior lepton degenerate gas is small compared to the binding energy at the surface of the degenerate gas, it is large compared to the binding energy of the stellar matter at larger radii of the star. The question is how can this free energy be transported to a much larger radius sufficiently rapidly so that the heat of this free energy cannot diffuse away more rapidly than the hydrodynamic equilibration time i.e., the explosion time? How also can this free energy be transported without doing excessive work against gravity, i.e., leave "its" mass behind?

The structure of any star is inherently stable because, with the heat generated in the interior, the diffusion of heat from the center is always slow compared to diffusion from the surface because of the very large density gradient. In the case of a collapse supernova, after the formation of the neutron star core and the transient trapping of the free binding energy, the neutrino flux from both this energy source and from the subsequent accretion flows lead to a high entropy, neutrino-dominated atmosphere on the surface as opposed to the interior of the neutron star. High entropy in this case means matter whose internal energy is so high that it is not gavitationally bound. (It is confined by the external pressure of accretion.) We think that the properties of convection, truncated at the large scale by diffusion, uniquely solves this problem. The large scale of the Rayleigh – Taylor instability occurs because neutrino diffusion at the neutrino sphere prevents the growth, i.e., truncates the growth of all smaller scale wave lengths than the local scale height. Convection allows the transport of heat without doing work against gravity. A large scale convective element, a plume in a high entropy, relativistic gas of specific heat ratio, $\gamma = 4/3$, allows a plume or large scale mass element to rise and survive without entrainment or mixing for many scale heights of displacement, because the expansion is homologous and faster than Helmholtz mixing. Thus the plume or mass element can reach a height where a significant change in gravitational potential occurs. In turn, provided the displacement of the mass element remains adiabatic, the Carnot efficiency for converting the high entropy or heat of the plume to useful work remains high. It is this work, we believe, that causes the supernova ejection or explosion. A fraction of the high entropy atmosphere formed adjacent to the neutron star, is transported, adiabatically, by large scale convection to a larger radius without significant loss of heat and thus to a much lower gravitational potential. In a lower gravitational potential it requires less work from the adiabatic expansion to eject the matter. It requires almost a conspiracy of all three physics, in our view, to cause a collapse supernova to explode.

The beginning physics in this scenario is the formation of a high entropy atmosphere lying on top of a nascent neutron star. This unlikely sounding circumstance is the inevitable result of continuing neutrino heat diffusion from the contracting neutron star core and as well low entropy and higher density matter falling onto a tightly bound neutron star. Furthermore the rate of release of this free energy must be great enough to establish a neutrino fire-ball above the neutrino sphere where the energy flux is so high that the radiation energy density in neutrinos becomes opaque to the neutrinos themselves. We will discuss the origin of this atmosphere first while attempting to evaluate the dynamic range of physical processes that might limit its formation. However, we consider the lack of universal understanding of the Carnot cycle in large scale convection the largest limitation to the understanding of the convective transport process.

The second law of thermodynamics establishes the limiting efficiency for accessing the free energy of two, different temperature, reservoirs. The Carnot cycle describes the sequence of deformations necessary to accomplish this efficiency. A necessary part of this cycle to access this free energy is the thermal isolation of the two parts of the cycle of compression from expansion. Similarly, the necessary heat flow isolation between the up and down flows of large scale convection determines the efficiency for transporting heat and free energy from the deep gravitational potential. In our view this isolation is absolutely necessary in order to maintain an efficient Carnot cycle and is therefore the basis for the somewhat greater energy release of the 3-D calculations (Fryer & Warren, 2002) over those performed in 2-D. In 3-D a given low entropy mass flow is cylindrical in shape and therefore thicker against neutrino heat flow transport than the corresponding entropy mass flow in the form of a sheet in 2-D, namely the surface to mass ratio is more favorable in 3-D.

23.2.1 Entropy

We summarize the discussion of entropy in Colgate, Herant, & Bentz (1993) and in Colgate & Fryer (1994). In order to describe matter during the history of the explosion, we use an entropy, S, (in units of the Boltzman factor per nucleon) that is approximated from several discrete events in the equation of state in the spirit of Bethe and Wilson (1985) and later Bethe (1990). We identify first the radiation contribution to the entropy where $\Delta S_{rad} = \int dQ/T$. As they pointed out when $Q = aT^4$, or $= (11/4)a\,T^4$ when pairs are important, then $S_{rad} = (4/3)(11/4)aT^3/\rho = 2W_{rad}/W_{nuc}$.

Thus $S_{rad} > 2$ corresponds to the transition from matter dominated energy to a radiation dominated gas. We will be concerned with the conditions when $S_{rad} \gg 2$, because when the radiation flux (including the neutrino component) from a neutron star surface, the neutrino sphere, into a gedanken vacuum, becomes large enough to make a neutrino fire ball, the entropy of the fire ball matter becomes near infinity and $2W_{rad} \gg nmc^2 W_{nuc}$. Importantly, such high entropy matter is unbound,

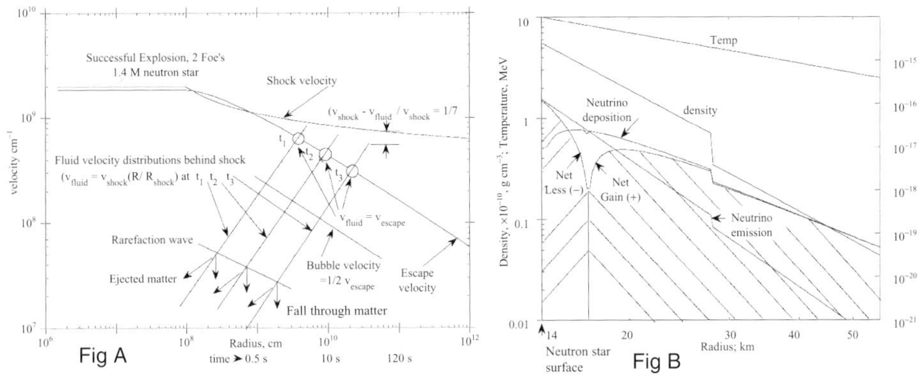

Fig. 23.1. Fig A shows a typical explosion shock history, starting at a velocity of 2×10^9 cm s^{-1} and decreasing, beyond 2×10^8 cm radius, to 1×10^9cm s^{-1} after roughly a second of time. The velocity distribution of the matter behind the shock is also shown for subsonic flow extending from behind the shock back to the neutron star. As can be seen much of this matter has velocities significantly less than escape and therefore may possibly fall back onto the NS. The bubbles are shown as rising relative to the shocked matter at a limiting velocity of roughly $1/2$ the local free-fall or escape speed. The ejected matter of the bubbles is depicted with a flatter slope and some of the heavier, low entropy matter will "fall through" the rising bubbles and is depicted with a steeper slope. The trace of the trajectories of the fall-through matter is the rarefaction wave that overtakes the fall-back mass fraction. Fig. B shows the temperature, density, neutrino emission rate, and deposition rate of a two layer equilibrium atmosphere resting on a 1.4 $M_\odot$ neutron star. The entropy of each layer and radius of the transition is chosen such that T = 10 MeV at the NS surface where $S_{rad} = 9.3$ for R < 2 R_{NS} and $S_{rad} = 18.6$ for R > 2 R_{NS}. The neutrinos are emitted in the semi-transparent limit. One observes that the neutrino heating rate exceeds the cooling rate at a gain radius of 1.2 R_{ns}. Outside of this radius, heating exceeds cooling. (from Colgate, S. A., Herant, M, and Benz, W. 1993).

$MG/R \simeq 0.2c^2$, and would expand to infinity if not confined by the accretion pressure. This is the condition we seek for the neutrino fire ball so that the equation of state is relativistic or the specific heat ratio, $\gamma = 4/3$. The pressure equilibrium atmosphere associated with this relativistic equation of state is that particular distribution such that a bubble, rising in such an atmosphere, is shape preserving and thus allows isentropic transport of heat over many scale heights in radius or change in gravitational potential.

The entropy of matter created by the explosion shock, including a nuclear thermal decomposition component of $\Delta S_{nuc} \simeq 8.6$ is shown in Fig. 23.1A starting at a velocity of 2×10^9 cm s^{-1} and decreasing, beyond 1×10^8 cm radius, to 1×10^9 cm s^{-1} after roughly a second of time.

The entropy of the matter behind the shock is small, $S_{rad} < 3$ out to a radius $R > 3 \times 10^9$ cm and in non-rarefied matter. This means that this mass of matter behind the explosion shock and external to the neutron star with low entropy will be Rayleigh – Taylor unstable when supported by the high entropy neutrino heated atmosphere. Only a small fraction of this low entropy mass can be accreted onto the neutron star without creating an extra massive neutron star or black hole. In other words the down flow of this low entropy matter must be shut off by some means that we presume and have calculated is the formation of extra large high entropy bubbles. These bubbles or plumes must be in near pressure equilibrium with the NS surface, and so we will consider equilibrium atmospheres where the pressure is dominated by radiation and where S_{rad} is the major fraction of the entropy.

23.3 Equilibrium atmospheres

The assumptions of an equilibrium atmosphere are:

1. The pressure gradient is in equilibrium with the density distribution and the gravity of the neutron star.
2. The entropy is large compared to unity so that the pressure and internal energy are dominated by photons and pairs ($(11/4)aT^4 \gg (3/2)n_{nuc}(kT)$), and as a consequence the mass of the atmosphere is small compared to that of the neutron star.
3. The entropy distribution is either an increasing function of radius, or uniform with radius, because a negative entropy gradient drives convection until the entropy gradient is zero. A positive entropy gradient is stable (see the relevant chapter from Bethe 1990). Here we assume that the entropy is a constant within a given region of the atmosphere and make our total atmosphere of finite regions of uniform entropy.

23.3.1 A constant entropy atmosphere

At pressure equilibrium and next to a near constant mass neutron star, M_{NS},

$$dP/dR = -g\rho = -M_{NS}G_{\rho}/R^2, \tag{23.1}$$

with S independent of R. Then, if we integrate from an outer boundary, R_1 with pressure P_1, to a radius R with pressure P, we have

$$P = \left[(M_{NS}G/4)(S_{rad}/S_o)^{-1}(1/R - 1/R_1) + P_1^{1/4}\right]^4 \tag{23.2}$$

In the limit where the outer boundary (subscript "1") is at a large radius relative to that of the neutron star, as is the case of the explosion shock for times longer than $1/10$ s, then we neglect $1/R_1$ and $P_1{}^{1/4}$ compared to $1/R$ and $P^{1/4}$. Then in terms of the radiation entropy per nucleon and for a cold neutron star of 1.4 $M_\odot$ we express the radius of the neutron star in units of 10^6 cm. The atmosphere can then be expressed as

$$P_{NS} = 1.16 \times 10^{35} S_{rad}^{-4} R_6^{-4} \text{dyne cm}^{-2} \tag{23.3}$$

$$T_{NS} = 174\ S_{rad}^{-1} R_6^{-4}\ \text{MeV} \tag{23.4}$$

$$\rho_{NS} = 2.8 \times 10^{15}\ S_{rad}^{-4} R_6^{-3} \text{g cm}^{-3} \tag{23.5}$$

and the mass of the atmosphere becomes

$$M_{atm} = 17.4\ S_{rad}^{-4}\ ln(R_{max}/R_{NS}) M_{\odot} \tag{23.6}$$

These numbers are so large, because we neglect nuclear and degeneracy pressure, which become comparable to P_{rad} for $S_{rad} \sim 5$. However, the temperature next to the neutron star of an equilibrium atmosphere with an entropy as large as $S_{rad} = 16$, would be exceedingly high, ~11 MeV, and the neutrino emission from such an atmosphere will approach the black body limit. The mass of the atmosphere, on the other hand, is small, ~1/2% of $M_{\odot}$, so that only continuing down-flows of low entropy matter can significantly add to the mass of the neutron star. We are interested in what happens to such an atmosphere, depicted in Fig. 23.1B, due to neutrino emission.

23.4 Neutrino heating and the thermal cycle

Such a high entropy atmosphere is maintained by neutrino heating as shown analytically in Colgate, Herant, and Benz (1993), by numerical calculations of Herant, Benz, and Colgate (1992), in 2-D in Herant *et al.* (1994) and in very many calculations some before and some since, but particularly with sophisticated Boltzman neutrino transport in Rampp & Janka (2002a,b) and in 3-D in Fryer and Warren (2002). An entropy at the gain radius of $S_{rad} = 55$ ensures a powerful unstable drive of the R-T instability compared to the initial down flow entropy of $S_{rad} \simeq 2$. The problem is understanding the scale of the primary unstable mass elements and the degree to which the up and down flows remain unmixed and the degree to which such a flow can perform work by the high entropy element, Fig. 23.2.

In Fig. 23.3A we show the idealization of the Carnot cycle as a heat engine where the up and down flows are thermally separated. Fig. 23.3B shows the pressure-volume plot following many zones of the computation (Herant *et al.* 1994). The enclosed or loop area is the useful work of the cycle. The degree to which heat is exchanged between the up and down flows determines the thermodynamic efficiency of the cycle.

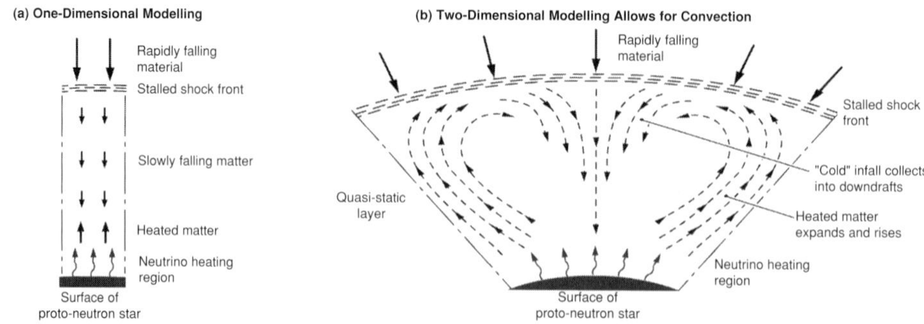

Fig. 23.2. shows on the right an idealized large scale, rising, high entropy buoyant mass element with a narrow low entropy down flow. The transverse scale ratio is just the inverse of the entropy ratio at pressure equilibrium. On the left is the usual 1-D representation of a diffusion atmosphere where there is no differentiation between the up and down flows and, because of diffusion equilibrium, no possibility of obtaining useful work from the entropy or temperature contrast. (from Herant, M, Colgate, S. A., and Benz, W. 1996).

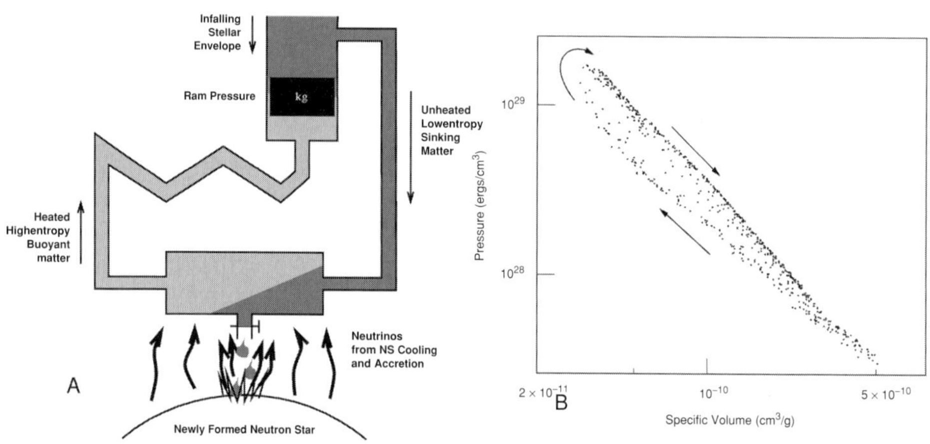

Fig. 23.3. Fig. 23.3A shows the idealization of the Carnot cycle as a heat engine where the up and down flows are thermally separated. The second law of thermodynamics predicts a limiting efficiency $\Delta T/T$. This efficiency and hence, useful work would approach zero if the two flows were in thermal contact. Fig. 23.3B shows the pressure-volume plot following many zones of the computation. The enclosed or loop area is the useful work of the cycle. An integration of this loop area corresponds closely to the explosion energy. If heat were fully exchanged between the up and down flows, this loop would shrink to a straight line. (From Herant *et al.* 1994).

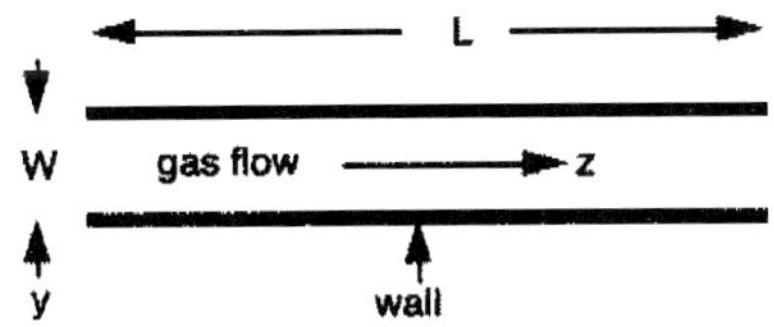

Fig. 23.4. shows an idealized channel of uniform width W and length L. The gas flows in the z direction with an average velocity v, and the heat flows both transversely in the y direction to and from the walls as well as in the z direction. The extent of the channel into the paper is B. The temperature is T and the change in temperature is ΔT. The mean velocity, half cycle average is v, the density ρ, and so the mass throughput is $BW_{v\rho} = dM/dt$. (From Colgate & Petschek 1993).

23.5 Heat exchange loss

Since one of the purposes of this paper is to point out the affect of the geometric difference between 2 and 3-D representations of the large scale convective flow, we calculate the entropy generation and hence the loss by heat and viscous exchange between the up and down flows.

Heat flow generates entropy. Here, we consider a steady, average value during each half of the cycle. For a temperature gradient $\nabla(T)$ we have a heat flux $\kappa\nabla(T)A$ through an area A and over a distance Δx. The temperature goes from T to $(T - \nabla(T)\ \Delta x)$ so that the entropy change rate in a volume $\Delta V = A\Delta x$ is:

$$\Delta\dot{S} = \kappa(\nabla T)A\left(\frac{1}{T - \nabla T\Delta x} - \frac{1}{T}\right) = \kappa\left(\frac{\Delta T}{T}\right)^2 \Delta V \tag{23.7}$$

Assuming ∇T is independent of the coordinates in the B and L directions (Fig. 23.4), we have

$$\frac{\dot{S}}{\dot{M}} = \frac{BL}{BWV\rho}\int_{W/2}^{W/2} \kappa\left(\frac{\nabla T}{T}\right) dy \tag{23.8}$$

To determine the entropy gain or loss of efficiency of the convective flows this integral must be evaluated over the rising plumes and collimated down flows. The point is that the heat flow from hot to cold flows results in a direct increase in entropy and therefore a loss in efficiency of the convective work. To the extent that the down flows in 3-D should be more isolated than in 2-D, we expect the thermal mixing or conduction loss to be smaller in 3-D and therefore lead to a more efficient convective work and explosion.

We leave to another paper a discussion of the scale of the buoyant elements as determined by the heat diffusion at the neutrinosphere and the expected homologous expansion of the buoyant element in the $\gamma = 4/3$ atmosphere.

This research was performed under the auspices of the Department of Energy.

References

Bethe, H. A., 1990, Rev. of Mod. Phys., **62**, 801.
Bethe, H. A., & Wilson, J. R. 1985 Ap J, **295**, 14.
Colgate, S. A., Herant, M, and Benz, W. 1993, *Physics Reports*, **227**, 157
Colgate, S. A. & Fryer, C. L., 1994, *Physics Reports*, **256**, 5
Colgate, S. A. & Petschek, A. G. 1993, Advances in Cryogenic Engineering, *CEC/ICMC* **39** Albuquerque, NM,
Fryer, C. L. & Warren, M. S., 2002 *Ap.J. (Letters)* **541**, 1033 *Mon. Not. R. astr. Soc.*, **196**, 731–745.
Herant, M., Benz, W., Hix, W. R., Fryer, C., and Colgate, S. A., 1994, ApJ, **435**, 339
Herant, M., Benz, W., and Colgate, S. A. 1992, Ap J **395**, 642.
Herant, M, Colgate, S. A., and Benz, W. 1996 LAScience
Rampp, M. & Janka, H.-Th 2002a, *Ap.J. (Letters)*, **432**, L119)
Rampp, M. & Janka, H.-Th 2002b, *Astr. Ap.*, **396**, 361

24

Topics in core-collapse supernova theory

A. Burrows

Department of Astronomy and Steward Observatory,
The University of Arizona, Tucson, AZ 85721

C. D. Ott

Institut für Theoretische Astrophysik, Universität Heidelberg

C. Meakin

Department of Astronomy and Steward Observatory,
The University of Arizona, Tucson, AZ 85721

Abstract

There are many interesting topics at the intersection of physics and astrophysics we call Supernova Theory. A small subset of them include the origin of pulsar kicks, gravitational radiation signatures of core bounce, and the possible roles of neutrinos and rotation in the mechanism of explosion. In this brief communication we summarize various recent ideas and calculations that bear on these themes.

24.1 What is the mechanism of pulsar kicks?

Radio pulsars are observed to have large proper motions that average $\sim$400–500 km s^{-1} (Lyne & Lorimer 1994) and whose velocity distribution might be bimodal (Fryer, Burrows, and Benz 1998; Arzomanian, Chernoff, & Cordes 2002). If bimodal, the slow peak would have a mean speed near $\sim$100 km s^{-1} and the fast peak would have a mean speed near 500–600 km s^{-1}. A bimodal distribution implies different populations and different mechanisms, but what these populations could be remains highly speculative.

Many arguments suggest that pulsars are given "kicks" at birth (Lai 2000; Lai, Chernoff, and Cordes 2001), and are not accelerated over periods of years or centuries. The best explanation is that these kicks are imparted during the supernova explosion itself. We think that this view is compelling. The two suggested modes of acceleration and impulse are via net neutrino anisotropy during the neutrino emission phase (which lasts seconds) and anisotropic mass motions and aspherical explosion which impart momentum to the residual core. The former requires but a

$\sim$1% *net* anisotropy in the neutrino angular distribution to provide a $\sim$300 km s^{-1} kick. However, anisotropies in the neutrino radiation field are more easily smoothed than matter anisotropies due to convection, rotation, aspherical collapse, etc. and relativistic particles such as neutrinos are not as efficient as non-relativistic matter at converting a given amount of energy into recoil (momentum). To achieve the requisite neutrino anisotropies people have generally invoked large ($\sim 10^{15-16}$ gauss) magnetic fields, which may not obtain generically (see Lai 2000 for a summary). Furthermore, all multi-D calculations to date imply that convective motions between the inner core and the shock result in significant jostling of the protoneutron star. Velocities of $\sim$100–200 km s^{-1} arise quite naturally by dint of the basic hydrodynamics of the convective mantle of the iron core after bounce ("Brownian Motion"; Burrows, Hayes, and Fryxell 1995 (BHF); Burrows and Hayes 1996a; Janka and Müller 1994; Scheck *et al.* 2003). This process *must* be a stochastic contributor to pulsar proper motions. In addition, due to the associated torques, modest spins can be imparted (Burrows, Hayes, and Fryxell 1995).

However, the average recoil speeds obtained theoretically by Burrows, Hayes, and Fryxell (1995), Burrows and Hayes (1996a), and Scheck *et al.* (2003) due to the Brownian motion of the core are only $\sim$200 km s^{-1}; this is not sufficient to explain either the average pulsar speed or the high-speed peak of a bimodal distribution. To do that might require an initial mild anisotropy ($\sim$percents) in the density or velocity profiles of the collapsing Chandrasekhar core (Lai 2000; Lai and Goldreich 2000). Such small anisotropies have been shown to result in significant impulses and implied kicks of 550 to 800 km s^{-1} (Burrows and Hayes 1996b). It may be that whatever determines whether the initial core is anisotropic results in the high-speed peak of the bimodal distribution, while the low-speed peak is due to the natural jostling by convective plumes and the resultant Brownian motion of the core. The latter is stochastic and not deterministic, but has been a robust prediction of the collapse theory for many years.

An instability in the pre-collapse structure that might result in aspherical collapse (particularly relevant in the supersonic region of the collapse since it can not smooth itself out by pressure forces) might be progenitor mass dependent; high-mass progenitors might result in high-velocity pulsars, while low-mass progenitors might result in low-velocity pulsars (on average) (or vice versa!). Whatever the origin of pulsar kicks and their apparent bimodality may be, new calculations are desperately needed. No hydrodynamic calculation to date has actually freed the very inner core to respond to the pressure impulses in a consistent fashion. All calculations have anchored the core and recoils have been inferred due to the integrated anisotropic pressure distributions seen. Freeing the core to respond to pressure and gravity effects and allowing the associated feedback processes will be crucial for obtaining self-consistent and credible results.

24.2 Gravitational waves from core collapse

Gravitational radiation signatures can in principle provide a dramatic potential constraint on core-collapse supernovae. Massive stars (ZAMS mass $\gtrsim$ 8 $M_\odot$) develop degenerate cores in the final stages of nuclear burning and achieve the Chandrasekhar mass. Gravitational collapse ensues, leading to dynamical compression to nuclear densities, subsequent core bounce, and hydrodynamical shock wave generation. These phenomena involve large masses at high velocities ($\sim c/4$) and great accelerations. Such dynamics, if only slightly aspherical, will lead to copious gravitational wave emission and, arguably, to one of the most distinctive features of core-collapse supernovae. The gravitational waveforms and associated spectra bear the direct stamp of the hydrodynamics and rotation of the core and speak volumes about internal supernova evolution. Furthermore, they provide data that complement (temporally and spectrally) those from the neutrino pulse (which also originates from the core), enhancing the diagnostic potential of each.

Most stars rotate and rotation can result in large asphericity at and around bounce. This provides hope that the emission of gravitational radiation from stellar core collapse can be significant. Furthermore, Rayleigh-Taylor-like convection in the protoneutron star, the aspherical emission of neutrinos, and post-bounce triaxial rotational instabilities are also potential sources of gravitational radiation. Together these phenomena, with their characteristic spectral and temporal signatures, make core-collapse supernovae promising and interesting generators of gravitational radiation.

Ott *et al.* (2004) use the 2D hydro code VULCAN/2D (Livne 1993) and follow Zwerger & Müller (1997) in forcing the one-dimensional initial models to rotate with constant angular velocity on cylinders according to the rotation law

$$\Omega(r) = \Omega_0 \left[1 + \left(\frac{r}{\mathrm{A}} \right)^2 \right]^{-1}, \tag{24.1}$$

where $\Omega(r)$ is the angular velocity, r is the distance from the rotation axis, and Ω_0 and A are free parameters that determine the rotational speed/energy of the model and the scale of the distribution of angular momentum. The rotation parameter β is defined by

$$\beta = \frac{E_{rot}}{|E_{grav}|}, \tag{24.2}$$

where E_{rot} is the total rotational kinetic energy and E_{grav} is the total gravitational energy. We (Ott *et al.* 2004) name our runs according to the following convention: [initial model name]A[in km]β_i[in %]. For example, s11A1000β0.3 is a Woosley and Weaver (1995) 11 $M_\odot$ model with A = 1000 km and an initial β_i of 0.3%.

Representative results are those found for model s15A1000β0.2 (Ott *et al.* 2003). The spectrum of s15A1000β0.2 is dominated by frequencies between 300 Hz and 600 Hz and peaks at 460 Hz. Most of the smaller peaks are connected to the first spike in the waveform during which 94% of the total gravitational wave energy of this model is radiated. There is, however, a contribution by the radial and non-radial ring-down pulsations that have characteristic periods of 2–2.5 ms in this model, translating into frequencies of 400–500 Hz. The peak is at 700 Hz and there are higher harmonics around 1400 Hz. With increasing β_i the spectrum shifts to lower frequencies and lower absolute values, peaking at 152 Hz ($\beta_i = 0.40\%$), 91 Hz ($\beta_i = 0.60\%$), and 38 Hz ($\beta_i = 0.80\%$). Furthermore, a prominent peak at low frequencies can be directly associated with the oscillation frequency of the post bounce cycles.

The models of Ott *et al.* (2004) yield absolute values of the dimensionless maximum gravitational wave strain in the interval $2.0 \times 10^{-23} \leq h^{TT}_{max} \leq 1.25 \times 10^{-20}$ at a distance of 10 kpc. The total energy radiated (E_{GW}) lies in the range 1.4×10^{-11} $M_\odot$ $c^2 \leq E_{GW} \leq 2.21 \times 10^{-8}$ $M_\odot$ c^2 and the energy spectra peak (with the exception of a very few models) in the frequency interval 20 Hz $\lesssim f_{peak} \lesssim$ 600 Hz.

Ott *et al.* (2004) find that at a distance of 10 kpc, i.e. for galactic distances, the 1st-generation LIGO, once it has reached its design sensitivity level, will be able to detect more than 80% of our core collapse models under optimal conditions and orientations. Assuming random polarizations and angles of incidence, this reduces to 10%. Advanced LIGO, however, should be able to detect virtually all models at galactic distances. Figure 24.1 presents peak h_{char} (the points), the maxima of the characteristic gravitational wave strain spectrum, but it also includes the actual h_{char} spectra of selected models (see Ott *et al.* 2004 for details). These h_{char} serve to put the issues of detectability in the LIGO detector into sharp relief.

24.3 Rotational effects

The evolution of the rotation parameter β and of the angular velocity is of particular interest, since they are connected to two still unanswered questions in core-collapse supernovae physics: What are the periods of newborn neutron stars? What is the role of rotation in the mechanism of core-collapse supernovae? As a prelude, Ott *et al.* (2004) addressed two related points:

1) There exists a maximum value of β at bounce for a given progenitor model and value of A. Interestingly, the maximum β is not reached by the model with the maximum β_i, but by a model with some intermediate value of β_i. β at bounce is determined by the subtle interplay between initial angular momentum distribution, the equation of

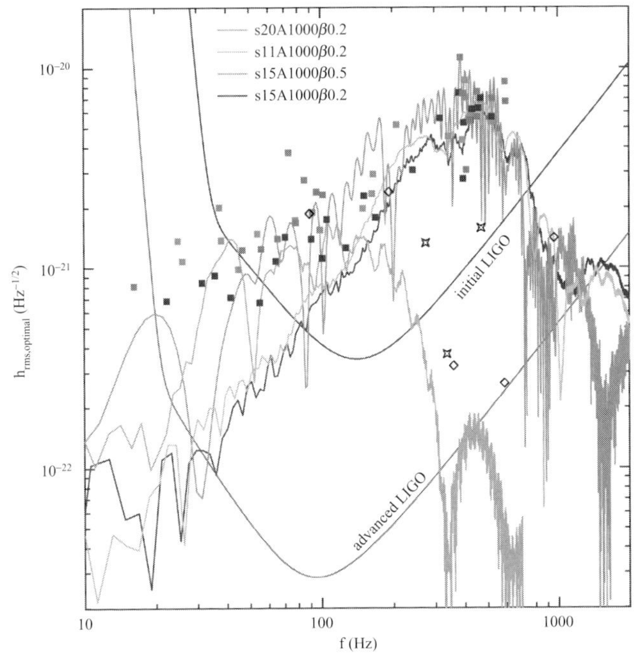

Fig. 24.1. LIGO sensitivity plot. Plotted are the optimal root-mean-square noise strain amplitudes $h_{rms} = \sqrt{f s(f)}$ of the initial and advanced LIGO interferometer designs. Optimal means that the gravitational waves are incident at an optimal angle and optimal polarization for detection and that there are coincident measurements of gravitational waves by multiple detectors. For gravitational waves from burst sources incident at random times from a random direction and a signal-to-noise ratio (SNR) of 5, the rms noise level h_{rms} is approximately a factor of 11 above the one plotted here (Abramovici *et al.* 1992) We have plotted solid squares at the maxima of the characteristic gravitational wave strain spectrum (h_{char} (f)) of our s11, s15, and s20 models from Woosley and Weaver (1995) that were artificially put into rotation. Our nonrotating models are marked with stars; diamonds stand for models from Heger *et al.* (2000, 2003). The distance to Earth was set to 10 kpc for all models. Most of our models lie above the optimal design sensitivity limit of LIGO I. Hence, the prospects for detection are good. Those models that are not detectable by the 1st-generation LIGO are those that rotate most slowly and those which are the fastest rotators. See Ott *et al.* (2004) for details.

state, centrifugal forces and gravity. The "optimal" configuration leads to the overall maximum β at bounce for a given β_i. Generally, β increases during collapse by a factor of ~10–40.

2) As with β, overall the angular velocity increases with increasing β_i until a maximum is reached. It subsequently decreases with the further increase of β_i. The initially more rigidly rotating models actually yield larger post-bounce angular velocity gradients inside 30 km. The equatorial velocity profile peaks off center for moderate β_i at radii between 6 and 10 km. An initially more differentially rotating model (at a given β_i) leads to the

highest central values of the angular velocity, while its angular velocity profile quickly drops to low values and near rigid rotation for $\beta_i \geq 0.3\%$. Model s15A500β0.2 in Ott *et al.* (2004) (see § 1.2) results in the shortest rotation period near the center ($\sim$1.5 ms). Model s15A50000β0.5 yields the shortest period of the A=50000 km model series ($\sim$1.85 ms).

In sum, the amplification of the angular velocity (frequency) due to collapse is generally large, from a factor of $\sim$25 to $\sim$1000. An initial period of 2 seconds in the iron core can translate into a period at bounce of $\sim$5 *milliseconds,* depending upon the initial rotational profile. The angular velocity shear exterior to the peak at 6–10 km exhibited by these models has also been identified in the one-dimensional study of Akiyama *et al.* (2003). These authors consider such shear a possible driver for the magneto-rotational instability (MRI), which could be a generator of strong magnetic fields.

24.3.1 Rotation and explosion

The large amplification of the angular velocity during bounce implies that rotation may be a factor in core collapse phenomenology and in the explosion mechanism. Though the latter remains to be demonstrated, there are a few aspects of rotating collapse that bear mentioning and that distinguish it from spherical collapse: 1) Rotation lowers the effective gravity in the core, increasing the radius of the stalled shock and the size of the gain region. Since ejection is inhibited by the deep potential well, rotation might in this manner facilitate explosion. 2) Rotation generates vortices that might dredge up heat from below the neutrinospheres and thereby enhance the driving neutrino luminosities. 3) Rotation lowers (slightly) the optical depth of a given mass shell, thereby increasing the ν_e neutrino luminosity. (However, as Fryer and Heger (2000) have shown the $\bar{\nu}_e$ luminosity is at the same time decreased due to the lower temperatures achieved.) 4) Importantly, rotation results in a pronounced anisotropy in the mass accretion flux after bounce. In fact, rotation can create large pole-to-equator differences in the density profiles of the infalling matter, due to the centrifugal barrier along the poles. The actual magnitude and evolution of this barrier is a function of the degree of rotation and its profile, but can be quite pronounced. Very approximately, in the equatorial region the distance from the axis (ρ, in cylindrical coordinates) of the barrier is given by $j^2/(GM)$, where M is the interior mass and j is the specific angular momentum at that mass. If the slope of j with r is positive, then as matter from further and further out accretes onto the protoneutron star the centrifugal barrier is expected to grow in extent. Even if the j profile is flat, $j^2/(GM)$ might be an interesting ($\sim$10–300 kilometers?) number (Heger *et al.* 2000,2003).

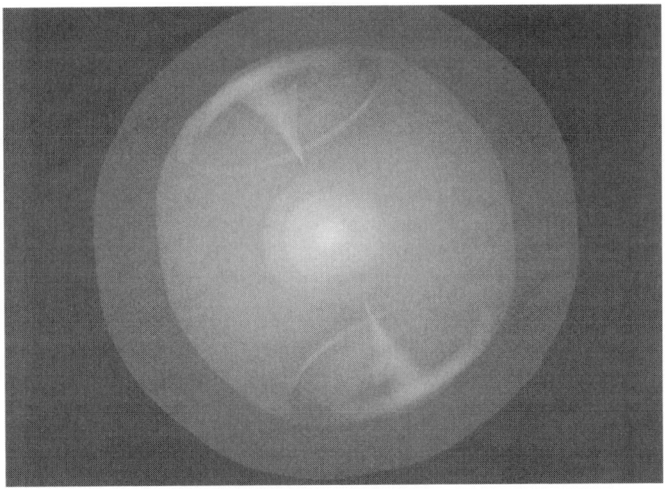

Fig. 24.2. A snapshot of a rotating collapse and bounce simulation in 2D, rendered in 3-D with nested layers of isodensity contours from 10^8 gm cm^{-3} to 10^{13} gm cm^{-3}. The funnel along the poles due to the centrifugal barrier created after bounce by the rotation of the collapsing Chandrasekhar core is clearly seen.

Figure 24.2 depicts a snapshot after bounce of various nested isodensity contours for a rapidly rotating initial progenitor ($\Omega_0 \sim 2\pi/2$ rad s^{-1}; eq. 24.1). The fact that outer (lower-density) contours pierce into the inner regions along the poles implies that a partially evacuated region is carved out along these poles. Since the total accretion rate is set by the initial mass density profile and since accretion powers the early post-bounce neutrino luminosity, this luminosity is not much changed. However, due to significant rotation the mass accretion flux at the poles is small after bounce. This suggests that the neutrino-driven mechanism is naturally facilitated along the poles (Burrows & Goshy 1993). A bipolar explosion would result (Fryer & Heger 2000, though see Buras *et al.* 2003). Thus, such bipolarity (and the consequent optical polarization of the debris) is not an exclusive signature of MHD driven explosions and may be a natural consequence of the neutrino-driven mechanism with rotation. However, for rotation to be pivotal in the mechanism, it might need to be rapid. This would beg the question of how the excess angular momentum is ejected to leave only the modestly rotating pulsars observed. Clearly, much works remains to be done.

24.4 Reprise on supernova energetics, made simple

The discussion concerning the rudiments of supernova energetics included in Burrows and Thompson (2002) summarizes our thoughts on the origin of the supernova energy scale and contains a useful perspective on the true efficiency of the neutrino-driven mechanism. It is reprised here without shame due to the continuing interest outside the supernova community in simple arguments that are not as opaque as numerical simulations can be (Thompson, Burrows, and Pinto 2003; Liebendörfer *et al.* 2001; Rampp & Janka 2000).

24.4.1 Supernova energetics made simple (?)

"It is important to note that one is not obliged to unbind the inner core ($\sim$10 kilometers) as well; the explosion is a phenomenon of the outer mantle at ten times the radius (50–200 kilometers). One consequence of this goes to the heart of a general confusion concerning supernova physics. Though the binding energy of a cold neutron star is $\sim 3 \times 10^{53}$ ergs and the supernova explosion energy is near 10^{51} ergs, a comparison of these two numbers and the large ratio that results are not very relevant. More germane are the binding energy of the mantle (interior to the shock or, perhaps, exterior to the neutrinospheres) and the neutrino energy radiated during the delayed phase. These are both at most a few $\times 10^{52}$ ergs, not $\sim 3 \times 10^{53}$ ergs, and the relevant ratio that illuminates the neutrino-driven supernova phenomenon is $\sim 10^{51}$ ergs divided by a few $\times 10^{52}$ ergs. This is $\sim 5-10\%$, not the oft-quoted 1%, a number which tends to overemphasize the sensitivity of the neutrino mechanism to neutrino and numerical details.

Furthermore, there is general confusion concerning what determines the supernova explosion energy. While a detailed understanding of the supernova mechanism is required to answer this question, one can still proffer a few observations. First is the simple discussion above. Five to ten percent of the neutrino energy coursing through the semi-transparent region is required, not one percent. Importantly, the optical depth to neutrino absorption in the gain region is of order ~ 0.1. The product of the sum of the ν_e and $\bar{\nu}_e$ neutrino energy emissions in the first 100's of milliseconds and this optical depth gives a number near 10^{51} ergs. Furthermore, the binding energy of the progenitor mantle exterior to the iron core is of order a few $\times 10^{50}$ to a few $\times 10^{51}$ ergs and it is very approximately this binding energy, not that of a cold neutron star, that is relevant in setting the scale of the core-collapse supernova explosion energy. Given the power-law nature of the progenitor envelope structure, it is clear that this binding energy is related to the binding energy of the pre-collapse iron core (note that they both have a boundary given by the same *GM/R*), which at collapse is that of the Chandrasekhar core. The binding energy of the Chandrasekhar core is easily shown to be zero, modulo the rest mass of the electron times the number of baryons in a $\sim 1.4\ M_\odot$ Chandrasekhar mass. (The Chandrasekhar mass/instability is tied to the onset of relativity for the electrons, itself contingent upon the electron rest mass). The result is $\sim 10^{51}$ ergs.

The core-collapse explosion energy is near the explosion energy for a Type Ia supernovae because in a thermonuclear explosion the total energy yield is approximately the 0.5 MeV/baryon derived from carbon/oxygen burning to iron times the number of baryons burned in the explosion. The latter is $\geq$ half the number of baryons in a Chandrasekhar mass. The result is $\sim 10^{51}$ ergs. This is the same number as for corecollapse supernovae because 1) in both cases we are dealing with the Chandrasekhar mass (corrected for electron captures, entropy, general relativity, and Coulomb effects) and 2) the electron mass and the per-baryon thermonuclear yield are each about 0.5 MeV.

While more detailed calculations are clearly necessary to do this correctly, the essential elements of supernova energetics are not terribly esoteric (if neutrino-driven), at least to within a factor of 5, and should not be viewed as such."

Acknowledgements

Support for this work is provided in part by the Scientific Discovery through Advanced Computing (SciDAC) program of the DOE, grant number DE-FC02-01ER41184, a NASA GSRP program fellowship, and by NASA through Hubble Fellowship grant #HST-HF-01157.01-A awarded by the Space Telescope Science Institute, which is operated by the Association of Universities for Research in Astronomy, Inc., for NASA, under contract NAS 5-26555.

References

Abramovici, A. *et al.*, 1992. *Science*, **256**, 325.
Akiyama, S., Wheeler, J. C., Meier, D., & Lichtenstadt, I., 2003. *Astrophys. J.*, **584**, 954.
Arzoumanian, Z., Chernoff, D. F., & Cordes, J., 2002. *Astrophys. J.*, **568**, 289–301.
Buras, R., Rampp, M., Janka, H.-Th., & Kifonidis, K. 2003, *Phys. Rev. Lett.*, **90**, 241101.
Burrows, A., Hayes, J., & Fryxell, B. A., 1995. *Astrophys. J.*, **450**, 830.
Burrows, A & and Hayes, J., 1996a. "An Origin for Pulsar Kicks in Supernova Hydrodynamics," p. 25, in the Proceedings of the conference, *High-Velocity Neutron Stars and Gamma-Ray Bursts*, eds. R. E. Rothschild & R. E. Lingenfelter, A.I.P. Press, no. 366.
Burrows, A. & Hayes, J., 1996b. "Pulsar Recoil and Gravitational Radiation due to Asymmetrical Stellar Collapse and Explosion," *Phys. Rev. Letters*, **76**, 352.
Burrows, A. & Goshy, J., 1993. *Astrophys. /J.*, **416**, L75.
Burrows, A. & Thompson, T. A., 2002. "The Mechanism of Core-Collapse Supernova Explosions: A Status Report," in the proceedings of the ESO/MPA/MPE Workshop *From Twilight to Highlight: The Physics of Supernovae*, p. 53, eds. Bruno Leibundgut and Wolfgang Hillebrandt (Springer-Verlag).
Fryer, C. L., Burrows, A., & Benz, W., 1998. *Astrophys. J.*, **496**, 333.
Fryer, C. L., & Heger, A., 2000. *Astrophys. J.*, **541**, 1033.
Heger, A., Langer, N., and Woosley, S. E., 2000. *Astrophys. J.*, **528**, 368.
Heger, A., Woosley, S. E., Langer, N., & Spruit, H. C., 2003. Stellar Rotation, Proceedings IAU Symposium No. 215.

Janka, H.-Th. & Müller, E., 1994. *Astron. and Astrophys.*, **290**, 496–502.
Lai, d. & Goldreich, P., 2000. *Astrophys. J.*, **535**, 402.
Lai, D., 2000. in Stellar Astrophysics, p. 127, ed. K. S. Cheng (Dordrecht: Kluwer).
Lai, D., Chernoff, D. F., & Cordes, J. M., 2001. *Astrophys. J.*, **549**, 1111.
Liebendörfer, M., Mezzacappa, A., Thielemann, F.-K., 2001. *Phys. Rev. D*, **63**, 104003.
Livne, E., 1993. *Astrophys. J.*, **412**, 634.
Lyne, A. G., & Lorimer, D. R., 1994. *Nature*, **369**, 127.
Ott, C. D., Burrows, A., Livne, E., & Walder, R., 2003. "Gravitational Waves from Axisymmetric, Rotating Stellar Core Collapse," *Astrophys. J.*, **600**, 834.
Rampp, M. & Janka, H.-Th., 2000. *Astrophys. J.*, **539**, 33.
Scheck, L., Plewa, T., Janka, H.-Th., Kifonidis, K., & Müller, E., 2003. astro-ph/0307352.
Thompson, T. A., Burrows, A., & Pinto, P. A., 2003. *Astrophys. J.*, **592**, 434.
Woosley, S. E. & Weaver, T. A., 1995. *Astrophys. J. Suppl.*, **101**, 181.
Zwerger, T. and Müller, E., 1997. *Astron. and Astrophys.*, **320**, 209.

25

MHD supernova jets: the missing link

D. L. Meier and M. Nakamura

Jet Propulsion Laboratory, California Institute of Technology, Pasadena, CA 91109

Abstract

We review recent progress in the theory of jet production, with particular emphasis on the possibility of 1) powerful jets being produced in the first few seconds after collapse of a supernova core and 2) those jets being responsible for the asymmetric explosion itself. The presently favored jet-production mechanism is an electrodynamic one, in which charged plasma is accelerated by electric fields that are generated by a rotating magnetic field anchored in the protopulsar. Recent observations of Galactic jet sources provide important clues to how *all* such sources may be related, both in the physical mechanism that drives the jet and in the astrophysical mechanisms that create conditions conducive to jet formation. We propose a grand evolutionary scheme that attempts to unify these sources on this basis, with MHD supernovae providing the missing link. We also discuss several important issues that must be resolved before this (or another scheme) can be adopted.

25.1 Introduction: a cosmic zoo of galactic jet sources

The last few decades have seen the discovery of a large number of different types of Galactic sources that produce jets. The purpose of this talk is to show that all of these jets sources are related, in both a physical sense and an astrophysical sense. Furthermore, Craig Wheeler's idea that most core collapse supernovae (SNe) are driven by MHD jets from a protopulsar provides the missing link in an attractive unified scheme of all stellar jet sources. Below is a list of Galactic jet sources that have been identified so far:

(1) Jets from Stars Being Born: Protoplanetary systems. These are jetted and bipolar outflows from young stars or star-forming regions. They are associated with protostars that have protoplanetary disks and are ejected at approximately the escape velocity of the central star ($v_{jet} \sim v_{esc} \sim 200$ km s^{-1}).

(2) Jets from Dying Stars:

- Planetary Nebula Systems. Stars with initial mass less than a few $M_\odot$ end their lives as a planetary nebula, leaving a white dwarf remnant. Many PN have bipolar shapes; some even have highly-collimated outflows of up to $v_{jet} \sim 1000$ km s^{-1}, the escape speed from the surface of a rather distended white dwarf.
- Core-collapse Supernovae: Stars with masses between roughly 10 and 30 $M_\odot$ are destroyed in an explosion that is triggered by the collapse of their iron cores. They leave a neutron star/pulsar remnant. There is growing evidence (see below) that core-collapse SNe also produce jets with a power comparable to the explosion itself. Expected speeds are $v_{jet} \sim 0.25$–$0.5c$ – the escape speed from the new protoneutron star.
- Isolated young pulsars (Crab, Vela, etc.): Jets now have been detected by Chandra in these objects and have speeds $v_{jet} \sim 0.5c$. These may be the remaining ghosts of core-collapse jets that occurred thousands of years ago.
- Gamma-ray bursts (GRBs): Believed to be black holes in formation, these produce jets with $\Gamma_{jet} \sim 100$–300 that point toward us. Long-duration GRBs are closely associated with powerful SNe and may represent the death of particularly massive ($>30\ M_\odot$) stars.

(3) Jets from Re-kindled Dead Stars in Binary Systems

- Symbiotic stars (*e.g.*, R Aquarii): These tend to be accreting white dwarfs where the companion star has recently left the main sequence and expanded into a red giant, transferring mass to the compact object via a strong wind or Roche lobe overflow. Jets are produced by the white dwarf and have speeds up to $v_{jet} \sim 6000$ km s^{-1}, the escape speed from a white dwarf.
- Neutron star X-ray Binaries: These occur in both low-mass and high-mass systems. Jets tend to appear at lower accretion rates. Typical speeds are $\sim 0.25c$, but can reach speeds approaching c.
- SS433-type objects: Observed properties suggest a super-Eddington accreting, magnetized neutron star in a high-mass binary system. Jets have speed $v_{jet} \sim 0.25c$.
- Classical microquasars (GRS 1915 + 105, GRO J1655-40, GX 339-4, etc.): These produce jets with $v_{jet} > 0.6 - 0.95c$ ($\Gamma_{jet} \equiv [1 - v_{jet}^2/c^2]^{-1/2} > 1.25 - 3$) and up. Virtually all are black hole candidates in low-mass X-ray binary systems.

The inclusion of core-collapse SNe above is the key to the unified model presented below. In the mid 1990s it was discovered that such SNe emit polarized light in the optical band, caused by electron scattering by an asymmetrically-expanding explosion [Wang *et al.* 2001, Wang *et al.* 2003, Leonard *et al.* 2001]. The variation of polarization properties with time and with different SN types gives important clues to its nature. In a given SN, the degree of polarization Π increases with time but the polarization direction often remains constant in time and wavelength, indicating that the asymmetry is global and maintains a fixed direction. For Type IIa SN (ones with a large hydrogen-rich envelope), the $\Pi \sim 1\%$, indicating only a 2:1 or less asymmetry. For Type IIb SN (ones that have lost their hydrogen envelope prior to the explosion, leaving only the helium envelope), Π is higher ($\sim$2%), indicating a 2.5:1 axial ratio. For SN Type Ib/Ic (ones that have lost most or all of their envelope, leaving a compact blue Wolf-Rayet star that then explodes), Π is quite high (4–7%), indicating a 3:1 axial ratio or better. Clearly, the deeper one sees into the explosion, the more elongated the exploding object appears. These observers have concluded that core-collapse SN have a global prolate shape that appears to be associated with the central engine producing the explosion. A jet, with energy comparable to that of the explosion itself, significantly alters the shape of the envelope, creating the elongated, polarized central source. In the paper below we will show that all of the above sources may be intimately related, both in the physical origins of the jet itself and in their astrophysical origins as well.

25.2 Basic principles of magnetohydrodynamic jet production

The basic principles of MHD jet production have been described elsewhere [Meier *et al.* 2001]. The reader is referred to that paper for a more detailed description, and more comprehensive figures, than in the short review given below.

25.2.1 Launching of the jet outflow: the jet engine itself

25.2.1.1 Jet production in accreting systems and pulsars

Several mechanisms for producing bipolar outflows have been suggested (explosions in the center of a rotationally-flattened cloud, radiation-pressure-driven outflows from a disk, etc.), but none of these is able to produce outflows approaching the highly-relativistic speeds observed in the fastest jet sources. The currently-favored mechanism is an electro-magneto-hydrodynamic (EMHD) one, somewhat similar to terrestrial accelerators of particle beams. Indeed, electromagnetic acceleration of relativistic pulsar winds has been a leading model for these objects since the 1960s. EMHD jet production was first suggested in 1976 [Blandford 1976, Lovelace 1976] and has been applied to rotating black holes [Blandford & Znajek 1977] (BZ) and to magnetized accretion disks [Blandford & Payne 1982] (BP). This mechanism has now been simulated and is sometimes called the "sweeping pinch" mechanism [Shibata & Uchida 1985, Kudoh *et al.* 1999, Nakamura *et al.* 2001].

The most important ingredient in the EMHD mechanism is a magnetic field that is anchored in a rotating object and extends to large distances where the rotational speed of the field is considerably slower. Plasma trapped in the magnetic field lines is subject to the Lorentz ($\boldsymbol{J} \times \boldsymbol{B}$) force, which, under conditions of high conductivity (the MHD assumption), splits into two vector components: a magnetic pressure gradient ($-\nabla B^2/8\pi$) and a magnetic tension ($B \cdot \nabla B/4\pi$). Differential rotation between the inner and outer regions winds up the field, creating a strong toroidal component (B_ϕ in cylindrical $[R, Z, \phi]$ coordinates). The magnetic pressure gradient up the rotation axis ($-dB_\phi^2/dZ$) accelerates plasma up and out of the system while the magnetic tension or "hoop" stress ($-B_\phi^2/R$) pinches and collimates the outflow into a jet along the rotation axis.

This basic configuration of differential rotation and twisted magnetic field accelerating a collimated wind can be achieved in all objects identified in Sect. 1.1. For protostars, white dwarfs, X-ray binaries, classical microquasars and GRBs, the field will be anchored in the accreting plasma, which may lie in a rotating disk (BP) and/or may be trapped in the rotating spacetime of the spinning central black hole itself (BZ). In the case of SS433-type objects, isolated pulsars, and core-collapse SN the rotating field is anchored in the pulsar (or protopulsar). In SS433 and core-collapse SN the source of the accelerated plasma is, once again, accretion, but in isolated pulsars it is believed to be particles created in spark gaps by the high (10^{12} G) field.

25.2.1.2 Jets from Kerr black holes: direct & indirect magnetic coupling

The jet-production mechanism envisioned by BZ generally involved direct magnetic coupling of the accelerated plasma to the rotating horizon. That is, magnetic field lines thread the horizon, and angular momentum is transferred along those field lines to the external plasma via magnetic tension. However, another, indirect, coupling is possible. This mechanism, suggested by Punsly & Coroniti [1990] (PC) and recently simulated by us [Koide *et al.* 2002], has the same effect as the BZ mechanism (extraction of angular momentum from the rotating black hole by the magnetic field), but the field lines do not have to thread the horizon itself. Instead, they are anchored in the accreting plasma. When this plasma sinks into the ergosphere near the black hole ($R < 2\ GM/c^2$), frame dragging causes the plasma to rotate with respect to the exterior, twisting up the field lines in a manner similar to the situation when the field is anchored in a disk or pulsar. (This occurs even if the accreting plasma has no angular momentum with respect to the rotating spacetime.) The twisted field lines then have two effects:

(1) Electromagnetic power is ejected along the rotation axis in the form of a torsional Alfvén wave. Eventually the output Poynting flux power should be dissipated in the production and acceleration of particles and a fast jet.

(2) The back-reaction of the magnetic field accelerates the ergospheric plasma (in which it is anchored) to relativistic speeds *against* the rotation of the black hole. The counter-rotating ergospheric plasma now formally has negative angular momentum and negative energy (negative mass); that is, it has given up more than its rest mass in energy to the external environment. It is on orbits that must intersect the black hole horizon, and, when it does, the mass of the black hole decreases by a value equal to that negative energy.

This process is the magnetic equivalent of the Penrose process, but instead of extracting black hole rotational energy by particle scattering, the energy is extracted by scattering of an Alfvén wave off the ergospheric plasma particles. Determining whether the BZ or PC process occurs in certain systems is an important question for future study.

25.2.2 Acceleration and collimation (A & C)

25.2.2.1 *Slow A & C is probably the norm*

There are both theoretical and observational reasons for believing that slow acceleration and collimation is probably the norm for jet outflows in these sources. Non-relativistic [Krasnopolsky *et al.* 1999] and relativistic [Vlahakis & Konigl 2001] models of MHD wind outflows attain solutions where the wind opening angle is wide near the accretion disk and then narrows slowly over several orders of magnitude in distance from the disk. Because the dynamical time scale is of order 0.1 ms or less in these objects, a steady state is set up fairly quickly in jet ejection events that last even only a few seconds. In a steady state, the wind accelerates as it expands vertically away from the rotator. A jet is not fully formed until its speed exceeds the local wave propagation speed, *i.e.*, the total Alfvén speed $V_A = [(B_R^2 + B_Z^2 + B_\phi^2)/(4\pi\rho)]^{1/2}$, where ρ is the mass density in the outflowing material. The place where this occurs, often called the Alfvén point or Alfvén surface, generally is well above the rotating object producing the accelerating torsional Alfvén wave. Analytic [Blandford & Payne 1982, Li *et al.* 1992] and numerical [Krasnopolsky *et al.* 1999] studies of this steady state show that the outflow is rather broad at the base, and it slowly focuses as it is accelerated. At a height $Z_A > 10R_0$ above the disk, the total Alfvén speed is exceeded, the flow is focused into a narrow cylindrical or conical flow, and little more acceleration and collimation takes place. The terminal jet speed $v_{\rm jet}$ is of order $V_A(Z_A)$, and this speed is usually of order the escape speed from the central rotator $V_{esc}(R_0)$.

Furthermore, there now is observational evidence that, in at least some *extragalactic* systems, the steady-state picture of slow acceleration and collimation is correct. Very high resolution VLBA radio images of the M87 jet [Junor *et al.* 1999] show a broad 60 deg opening angle at the base that narrows to only a few degrees after a few hundred Schwarzschild radii. In addition, it has been argued [Sikora & Madejski 2001] that most quasar jets must be broad at the base: they lack soft, Comptonized and relativistically-boosted X-ray emission that would be expected from a narrow, relativistic jet flow near the black hole.

25.2.2.2 Stability of highly magnetized flows during A & C

Because the acceleration and collimation is expected to be slow, the terminal velocity and final state of the outflow will be reached only far from the central engine. The character of the outflow, therefore, will depend crucially on how it interacts with the external medium that surrounds it in the acceleration region. It is important, therefore, to consider the effects of the ambient "weather" surrounding the jet.

Highly-magnetized flows are characterized by a high ratio of Alfvén to sound speed c_s

$$\frac{\dot{\varepsilon}_{fields}}{\dot{\varepsilon}_{particles}} = \frac{v_{jet} B^2/4\pi}{v_{jet} \rho c_s^2} \approx \left(\frac{V_A}{c_s}\right)^2 \gg 1 \tag{25.1}$$

This means that, early in the flow the velocity becomes supersonic, but continues to accelerate toward the Alfvén speed. In this region, where $c_s < v_{jet} < V_A$, the ratio of Poynting flux to kinetic energy flux is

$$\frac{\dot{\varepsilon}_{Poynting}}{\dot{\varepsilon}_{kinetic}} = \frac{\frac{c}{8\pi}|\boldsymbol{E} \times \boldsymbol{B}|}{v_{jet}\frac{1}{2}\rho v_{jet}{}^2} = \frac{B^2}{4\pi\rho v_{jet}{}^2} = \left(\frac{V_A}{v_{jet}}\right)^2 > 1 \tag{25.2}$$

That is, the flow is "Poynting-flux-dominated" as long as the flow remains supersonic but sub-Alfvénic.

While semi-analytic models of such "Poynting-flux-dominated" jets have been built [Li *et al.* 1992, Lovelace *et al.* 2002], no numerical simulations of highly *relativistic* jets have been performed yet. The best numerical results so far are from *non*-relativistic simulations [Nakamura *et al.* 2001, Nakamura & Meier 2003], which compute the behavior of a jet in a decreasing density (increasing v_A) atmosphere. They show that the electromagnetic power is carried by a fairly coherent torsional Alfvén wave that encompasses the jet in a twisting spiral pattern. This wave can transport energy and momentum along the flow, causing further acceleration far from the central engine.

Simulations of both stable and unstable jets are shown in Figure 25.1. We find that the stability of Poynting flux-dominated flows is critically dependent on how severe the mass entrainment in the jet is – specifically on the *gradient* of the plasma parameter $\beta_{plasma} \equiv p_{gas}/(B^2/8\pi)$. (In the following, remember that β_{plasma} is always

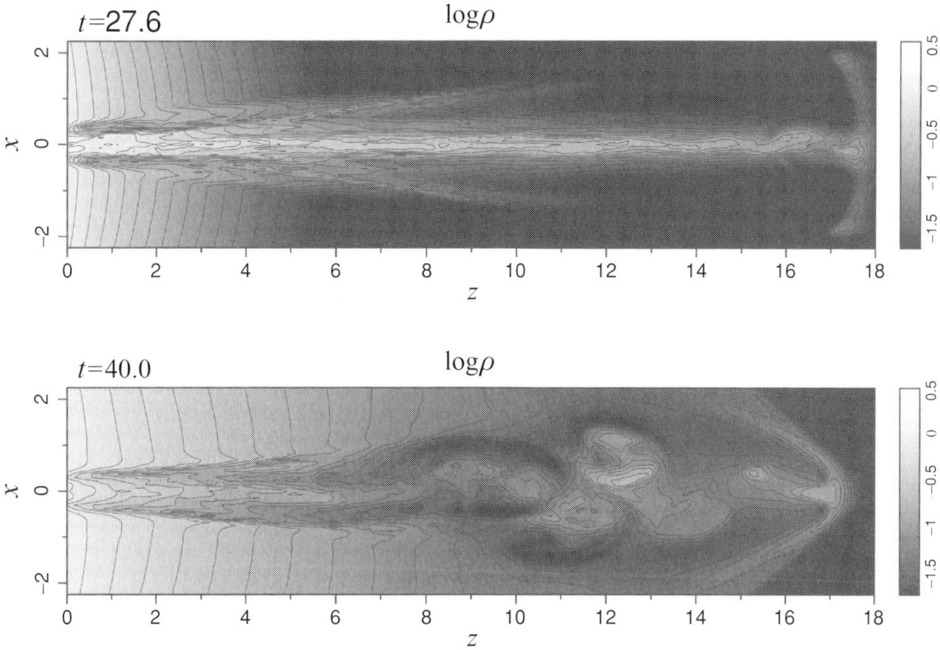

Fig. 25.1. Three-dimensional simulations of jets that are stable (top) and unstable (bottom) to the helical kink instability. Both have $\beta_{plasma} < 1$ throughout the domain, but the stable model has $d \ln \beta_{plasma}/d \ln \mathrm{r} \lesssim 0$ ($-d \ln \rho/d \ln \mathrm{r} \sim 3$, vs. 2 for the unstable model). (After [Nakamura & Meier 2003].)

less than unity for PFD jets, if the plasma is reasonably cold $P \leq \rho c^2$.) If β_{plasma} decreases or remains small as the jet propagates outward (mass loading becomes even less or stays the same), then we find that the PFD jet remains stable. However, if β_{plasma} *increases* (entrains significantly more thermal material), then we find that the jet is likely to be unstable to the helical kink instability, *even if the jet still remains magnetically dominated throughout the simulation.* Apparently even a small amount of pressure in the flow builds up over large distances, triggering a helical kink and, therefore, turbulence in the jet.

Collimated MHD outflows that are strongly dominated by magnetic stresses, and which remain that way or increase in strength, therefore propagate as straight and stable jets. Those in which the relative strength of the field decreases along the jet become unstable to the helical kink instability. In the latter case the outflow decelerates and deposits its momentum in a broad cone rather than punching through the ambient medium in a narrow jet. The restriction $\beta_{plasma} \ll 1$ is, therefore, a necessary, but not sufficient condition for jet stability. We also must have $d \ln \beta_{plasma}/d \ln r < \sim 0$. This may have important implications for MHD outflows from newly-formed pulsars in the centers of core-collapse SNe.

25.3 MHD-jet-powered supernovae

25.3.1 Basic MHD supernova model

Several authors have suggested in the past that MHD phenomena may power supernovae [LeBlanc & Wilson 1970, Bisnovatyi-Kogan 1971]. The recent discovery that SN ejecta are elongated by an asymmetric jet-like flow has stimulated renewed interest in these models and, in particular, in the possibility that an MHD jet produced by the protopulsar may be the source of the explosion energy in all core-collapse SN. We have proposed [Wheeler *et al.* 2002] an explosion mechanism that is consistent with the above properties of MHD jets. The jet is produced in the iron mantle, just outside the protoneutron star. An object with a 10^{15} G field and a rotation period of $\sim$1 ms can produce a jet power of $\sim 3 \times 10^{51}$erg s^{-1} and a total energy of $\sim 2 \times 10^{52}$ erg – more than enough to eject the outer envelope and account for the observed explosion energy. The ejecta from the vicinity of the core is nickel-iron rich, with a mass of $\sim 10^{32}$g, an initial velocity of $\sim 0.5c$, and a total momentum of $\sim 1.5 \times 10^{42}$g cm s^{-1}, decelerating somewhat as it passes through the stellar envelope.

The protoneutron star spins down to more respectable rotation periods (>10 ms) in about $\sim 10B_{15}^{-2}$ seconds. (The model is similar to that of Ostriker & Gunn [Ostriker & Gunn 1971], with their 10^{12} G pulsar fields replaced by 10^{15} G protopulsar fields.) The jet outflow is composed of iron-rich material and is initially broad at the base. Furthermore, as even these field strengths do not satisfy the necessary condition for jet stability, the jet likely will be subject to helical kink instabilities, broadening further into a wide bipolar outflow at large distances ($>10^{7-8}$ cm) from the core. It therefore can couple well to the outer envelope and eject it, imparting an elongated shape to the supernova explosion.

25.3.2 The pulsar rocket

The Crab pulsar currently has a proper motion of $\sim$120 km s^{-1} along the pointing direction of one of its twin jets. This motion has an energy of $\sim 2 \times 10^{47}$ erg and momentum of $\sim 3 \times 10^{40}$g cm s^{-1}. However, while the current Crab pulsar jets have sufficient energy to have powered this motion ($\sim 10^{49}$erg, if they had operated continuously over the past 950 yr), the present Crab jet does not produce enough momentum to accelerate the pulsar to this speed, *even if the jet had been continuously and completely one-sided* (100% asymmetric, or 10^{39}g cm s^{-1}). On the other hand, the MHD jet inferred from the MHD supernova model above would have been quite sufficient to supply both the energy ($\sim 10^{52}$erg) and momentum with only a 2% jet asymmetry. The fact that the Crab jet and proper motion are aligned, indicates that the current observed Crab pulsar jet (with a speed of $\sim$0.5c) may be a vestige of the original $\sim$0.5c jet that exploded the supernova and accelerated the pulsar.

25.3.3 A gamma-ray burst trigger

This model also includes a gamma-ray burst trigger. In very rare instances, the field can be dynamically strong in the iron mantle ($B > 10^{16}$ G), leading to satisfaction of the necessary (but not sufficient) jet stability condition $\beta_{plasma} \ll 1$. If the density gradient then is also steep ($d \ln \beta_{plasma}/d \ln r < \sim 0$) then the jet becomes narrow and very fast, coupling poorly to the mantle, punching through the outer envelope [Khokhlov & Höflich 2001], escaping the star, and producing a heavy iron "lobe" outside it, traveling at a speed of 0.05–0.3c. Because of the poor coupling to the mantle, the explosion fails, and much of the mantle falls back onto the protoneutron star, putting the system into a state very similar to that at the beginning: the "failed SN" GRB model of MacFayden & Woosely [1999]. When the mantle fallback accretes enough material onto the protoneutron star (after several minutes to hours), the neutron star is crushed to a black hole, and a new very fast ($\Gamma_{jet} \gg 1$) jet is produced via the BZ or PC mechanisms discussed above. The relativistic jet catches up with the slow, iron-rich lobe at a distance of $d \sim v_{jet}\tau_{fallback} \sim 10^{12-13}$ cm, and the interaction of jet and lobe produces gamma-rays, optical afterglow, and an iron-rich spectrum.

25.3.4 Unresolved issues

The 10^{14-15} G magnetic field strengths needed in SN cores are the real key to the success of the MHD SN model. Magnetars are believed to have surface field strengths of this order, but pulsars typically have fields of order 10^{12-13} G. Fields of this strength would have produced a slow supernova explosion lasting several months. The prediction, then, is that fast core-collapse SNe produce magnetars and slow SNe produce pulsars, but there is no observational evidence for (or against) this prediction, neither direct nor statistical. On the other hand, if stronger fields *did* exist in pulsars at the time they were formed, they must have been dissipated either during the SN process or shortly thereafter, but it is not known how that dissipation may have taken place. Secondly, there also may be competing jet mechanisms (neutrino radiation pressure, etc.) which we have not discussed here. Thirdly, while we have suggested a possible SN failure mechanism, much more detailed theoretical work will be needed before we will be able to perform the simulations necessary to test this and other such mechanisms. Finally, there is a problem that needs to be addressed by *all* SN models. The iron mantle in the progenitor star is very neutron rich. If much of it is ejected (and the $1.4M_{\odot}$ protoneutron star is left), then the predicted amount of r-process material may be much larger than that observed. It is a general problem for all core-collapse SN models to produce a neutron star remnant while still not over-producing the r-process elements.

25.4 A grand unified scheme for all galactic jet sources

SN and long-duration GRBs, therefore, potentially can be unified astrophysically as being different possible outcomes in the final stages of the death of a massive star. Also, if the above model is applicable, they both can be unified *physically* as being powered by MHD jets. This strongly suggests, therefore, a possible *grand scheme* that unifies all Galactic jet sources discussed in Sect. 25.1, both physically and astrophysically, in a similar manner.

The properties of all Galactic jet sources are similar in several ways. They all appear to produce jets when there is accretion, shrinkage, or collapse of plasma in a gravitational field. Because of that shrinkage or collapse, they occur in systems that are probably in a rapid rotation state: conservation of angular momentum implies that even a modest amount of rotation before the collapse would be amplified greatly. They also are associated with systems that have strong magnetic fields. Some directly reveal these magnetic fields in their radio synchrotron emission. Others are produced by stars that are believed to have strong fields for other reasons (protostars, accretion disks, newly-formed protoneutron stars). The grand unified scheme, therefore, asserts that all Galactic jet sources represent objects in which an excess of angular momentum has built up because of accretion or collapse. And the production of the jet itself represents the *expulsion* of that excess angular momentum by electro-magneto-hydrodynamic processes. The unifying evolutionary sequence for all Galactic jet sources is shown in Figure 25.2.

25.4.1 Protostellar and white dwarf galactic jets

The unified sequence begins with the formation of a protostar in a collapsing interstellar cloud. A jet is formed during the accretion phase, and is responsible for spinning down the star to the relatively low rotation rates seen on the main sequence. In low-mass stars jet production does not resume until shrinkage of the central stellar proto-white-dwarf core produces a bipolar planetary nebula outflow. Symbiotic star-type jets are expected in binary systems, but isolated white dwarfs may have neither the accretion fuel nor the rapid spin to do so.

25.4.2 Neutron star and black hole jets

The MHD supernova model discussed earlier provides the missing link to GRB, X-ray binary, and microquasar jets. In the grand unified scheme, most massive

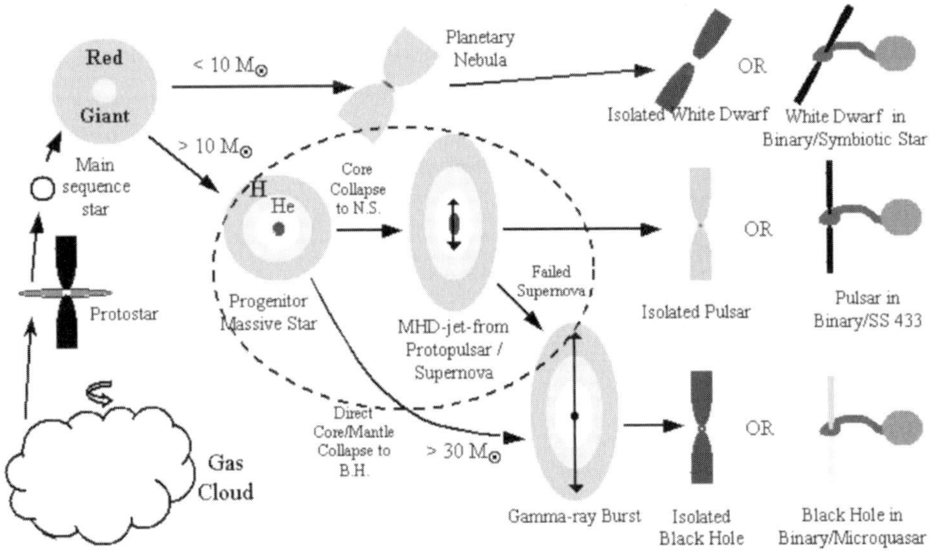

Fig. 25.2. The proposed grand unified model for Galactic jet sources. (After [Meier 2002].)

star cores collapse to a neutron star, ejecting a broad MHD jet in the process that drives the SN explosion and produces the observed asymmetry. After the envelope dissipates, if the pulsar is an isolated object, residual rotation of the magnetized remnant still drives an MHD outflow and a moderately relativistic jet like that seen in the Crab and Vela pulsars. On the other hand, if the pulsar resides in a binary system, it may accrete material from its companion star in a super-Eddington phase ($\dot{M}_{acc} \gg 10^{18}$g s^{-1}) and appear like SS433 for a brief time. Cessation of the accretion, angular momentum evolution of the pulsar, or possible collapse to a black hole all could serve to alter SS433's present state.

In rare circumstances, the MHD SN jet will fail to eject the envelope, or perhaps the progenitor core will collapse directly to form a black hole. In either case a GRB event should be generated in a manner similar to that in the failed SN model. The GRB jet event will not spin down the black hole completely (although a significant amount of rotational energy may be extracted from the black hole by another means – gravitational waves). After the envelope dissipates, if the black hole is an isolated object, it may emit little radiation and be difficult to detect. On the other hand, if the newly-formed hole is in a binary system, it also can accrete plasma and field from its companion, thereby producing a strong jet and classical microquasar. Again, changes in the accretion rate and angular momentum evolution of the black hole will alter the microquasar's observational state.

The key element of this unified model is that the evolutionary outcome of a star is ultimately determined by the magnitude and direction of the angular momentum and magnetic field of that star's core. We therefore should consider the jets seen in young pulsars and SS433-type objects to be vestiges of the mechanism that exploded the massive stars from which they came and, in a similar manner, consider the jets in microquasars to be the remnants of the gamma-ray burst that triggered the black hole's formation eons ago. These relatively modest jets are the echoes of violent events of the distant past.

References

Bisnovatyi-Kogan, G., 1971, *Soviet Astron. AJ*, **14**, 652.
Blandford, R., 1976, *M.N.R.A.S.*, **176**, 465.
Blandford, R. & Znajek, R., 1977, *M.N.R.A.S.*, **179**, 433 (BZ).
Blandford, R. & Payne, D., 1982, *M.N.R.A.S.*, **199**, 883 (BP).
Junor, W. *et al.*, 1999, *Nature*, **401**, 891.
Khokhlov, A. & Höflich, P., 2001, *Explosive Phenomena in Astrophysical Compact Objects,* AIP, 301.
Koide, S. *et al.*, 2002, *Science*, **295**, 1688.
Krasnopolsky, R. *et al.*, 1999, *Astroph. J.*, **526**, 631.
Kudoh, T. *et al.*, 1999, *Numerical Astrophysics,* Kluwer, 203.
LeBlanc, J. & Wilson, J., 1970, *Astroph. J.*, **161**, 541.
Leonard, D., *et al.*, 2001, *Astroph. J.*, **553**, 861.
Li, Z.-Y. *et al.*, 1992, *Astroph. J.*, **394**, 459.
Lovelace, R., 1976, *Nature*, **262**, 649.
Lovelace, R. *et al.*, 2002, *Astroph. J.*, **572**, 445.
MacFadyen, A. & Woosley, S., 1999, *Astroph. J.*, **524**, 62.
Meier, D., *et al.*, 2001, *Science*, **291**, 84.
D. L. Meier, in 4th Microquasar Workshop, Center for Space Physics, Kolkata, India, 165.
Nakamura, M. *et al.*, 2001, *New Astronomy*, **6**, 61.
Nakamura, M. & Meier, D. L., 2003, in preparation.
Punsly, B. & Coroniti, F., 1990, *Astroph. J.*, **354**, 583 (PC).
Ostriker, J. & Gunn, J., 1971, *Astroph. J.*, **164**, L95.
Shibata, K. & Uchida, Y., 1985, *Pub. Astron. Soc. Japan*, **37**, 31.
Sikora, M. & Madejski, 2001, astro-ph/0112231.
N. Vlahakis & A. Konigl, Astrophys. J., 563, L129 (2001).
Wang, L., *et al.*, 2001, *Astroph. J.*, **550**, 1030.
Wang, L., *et al.*, 2003, this conference.
Wheeler, J. *et al.*, 2002, *Astroph. J.*, **568**, 807.

26

Effects of super-strong magnetic fields in a core collapse supernova

S. Akiyama

University of Texas at Austin TX USA

J. C. Wheeler

University of Texas at Austin TX USA

R. C. Duncan

University of Texas at Austin TX USA

D. L. Meier

Jet Propulsion Laboratory CA USA

Abstract

Polarization and other observations indicate that supernova explosions are aspherical and often axisymmetric, implying a necessary departure from spherical models. Akiyama *et al.* investigated the effects of the magneto-rotational instability (MRI) on collapsing and rotating cores. Their results indicate that the MRI dynamo generates magnetic fields of greater than the Q.E.D. limit (4.4×10^{13} G). We present preliminary results of the effects of the super-strong magnetic field on degenerate electron pressure in core collapse.

26.1 Introduction

Although core collapse cannot be observed directly, except with neutrinos, observations of explosion ejecta can provide us with information about the explosion mechanism itself. Such observations indicate that explosions of core collapse supernovae are aspherical and often bipolar. HST observations clearly show that 1987A has aspherical ejecta for which the axis aligns roughly with the small axis of the rings (Pun *et al.* 2001; Wang *et al.* 2002). Spectropolarimetry is a powerful tool for probing ejecta asphericity, and it reveals that most, if not all, core collapse supernovae possesses asphericity and often times bipolar structure (Wang *et al.* 1996, 2001). Explosions of Type Ib and Ic are more strongly aspherical, while the asphericity of Type II supernovae increases with time as the ejecta expand and the photosphere recedes (Wang *et al.* 2001; Leonard *et al.* 2000, 2001). The indication is that it is the core collapse mechanism itself that is responsible for the asphericity.

The observational evidence of asphericity motivates the inclusion of rotation in core collapse physics. There have been studies with rotation for several decades since LeBlanc & Wilson (1970) (Müller & Hillebrandt 1981; Symbalisty 1984; Möchmeyer & Müller 1989; Yamada & Sato 1994; Fryer & Heger 2000; Ott *et al.* 2003; Kotake *et al.* 2003; Fryer & Warren 2004). Recent calculations of rotating core collapse (Yamada & Sato 1994; Fryer & Heger 2000; Ott *et al.* 2004; Kotake *et al.* 2003; Fryer & Warren 2003) agree that rotation tends to weaken bounce and to affect convection in a fashion to restrict it to the polar direction that favors neutrino emission in polar direction. This in turn results in asphericity similar to that of observed. Rotation of plasma, however, induces magnetic fields, and the role that magnetic fields play in core collapse should be examined simultaneously. Previous simulation studies (LeBlanc & Wilson 1970; Symbalisty 1984) have shown that buoyant MHD outflows form in the collapse of a magnetized rotating iron core. Wheeler *et al.* (2000, 2002) considered possible physical mechanisms for inducing axial flows, asymmetric supernovae, and related phenomena driven by magneto-rotational effects. While Wheeler *et al.* (2000) focused on the effect of the dipole field, Wheeler *et al.* (2002) found that the production of a strong toroidal field ($>10^{12}G$) is possible, and they explored the capacity of the toroidal field to directly generate axial jets by analogy with magneto-centrifugal models of jets in active galactic nuclei (Koide *et al.* 2000 and references therein).

The magnetorotational instability (MRI) is very generic because its instability criterion is basically only that the gradient in angular velocity be negative, which is satisfied by the core collapse environment as well as accretion disk systems. Akiyama *et al.* (2003) investigated the role of the MRI in core collapse and found that a magnetic field of order $10^{16}G$ can be achieved around the boundary of a protoneutron star (PNS). This strong magnetic field may induce the formation of bipolar jets. On the other hand, strongly magnetized plasma above the Q.E.D. limit behaves differently from weakly-magnetized or non-magnetized plasma, and these effects should be investigated. One of the effects is altered neutrino transport, and another is altered equation of state. We present preliminary results on the effects of super-strong magnetic fields on the equation of state.

26.2 Background

The motion of an electron in a uniform magnetic field is helical with a frequency of rotation eB/mc, the cyclotron or Larmor frequency. When the Larmor radius equals the Compton wavelength, the corresponding magnetic field reaches the Q.E.D. limit value $B_q = 4.4 \times 10^{13}$ G, above which the orbital motions of electrons are quantized into Landau levels (Canuto & Ventura 1977). The discrete Landau energy is given by

$$E = mc^2 \left[1 + x^2 + \frac{B}{B_q}(2n + 1 + s) \right]^{\frac{1}{2}}, \tag{26.1}$$

where $x = p_{||}/mc$ is the relativistic parameter (Shapiro & Teukolsky 1983), n is the Landau level, and $s = \pm 1$ corresponds to spin up or down. The relativistic parameter is the momentum parallel to the magnetic field and is continuous, while the term with the magnetic field represents the discrete perpendicular component of the momentum. The zero magnetic limit (B $\rightarrow$ 0) for the perpendicular term is obtained by letting $n \rightarrow \infty$. As a consequence of these quantum effects, the thermodynamic properties of the electron gas are modified (Canuto & Ventura 1977). The effect is strongest when only the ground level is occupied ($n = 0$) and when spin is down ($s = -1$); the perpendicular component becomes zero, thus there is no electron pressure perpendicular to the magnetic field in this case.

For a completely degenerate electron gas, the pressure in the perpendicular and parallel directions when up to the first Landau level is occupied are given by where

$$\frac{P_\perp}{P_0} = \left(\frac{B}{B_q} \right)^2 \ln \frac{\mu + \left(\mu^2 - 1 - 2\frac{B}{B_q}\right)^{\frac{1}{2}}}{\left(1 + 2\frac{B}{B_q}\right)^{\frac{1}{2}}}, \tag{26.2}$$

$$\frac{P_{||}}{P_0} = \left(\frac{B}{B_q} \right) \left[\frac{1}{4}\mu(\mu^2 - 1)^{\frac{1}{2}} - \frac{1}{4}\ln[\mu + (\mu^2 - 1)^{\frac{1}{2}}] + \frac{1}{2}\mu \left(\mu^2 - 1 - 2\frac{B}{B_q} \right)^{\frac{1}{2}} \right.$$
$$\left. - \frac{1}{2}\left(1 + 2\frac{B}{B_q}\right) \ln \frac{\mu + \left(\mu^2 - 1 - 2\frac{B}{B_q}\right)^{\frac{1}{2}}}{\left(1 + 2\frac{B}{B_q}\right)^{\frac{1}{2}}} \right], \tag{26.3}$$

$$P_0 = \frac{1}{\pi^2}\frac{mc^2}{\bar{\lambda}^3} = 1.44 \times 10^{24} dyne\ cm^{-2}, \tag{26.4}$$

$B_q = 4.4 \times 10^{13} G$, μ is the chemical potential of electron, and $\bar{\lambda}$ is the electron Compton wavelength.

26.3 Calculations

We took the magnetic field profile calculated in Akiyama *et al.* (2003) as in Fig. 26.1 to calculate the electron pressure anisotropy due to the super-strong magnetic fields. Fig. 26.2 shows the Fermi energy, the first Landau energy (n = 1), and the thermal energy as functions of radius. We used relativistic and partially degenerate formulae

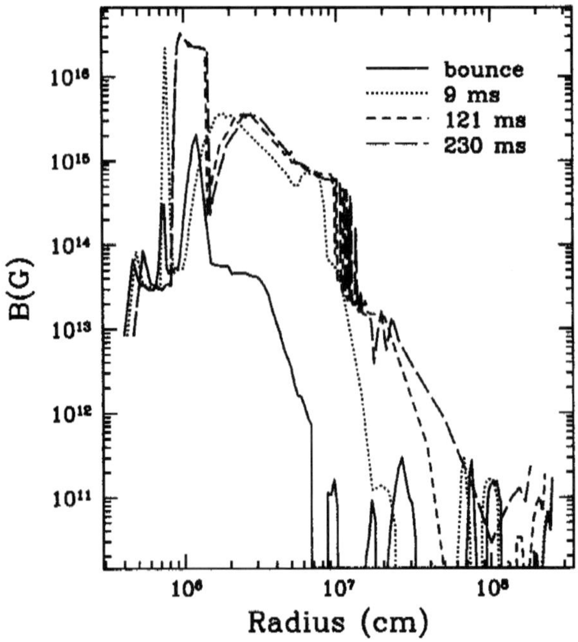

Fig. 26.1. The magnetic field profile calculated in Akiyama *et al.* (2003). At 121 ms after bounce, the peak magnetic field achieves of order $10^{16}G$ in this model.

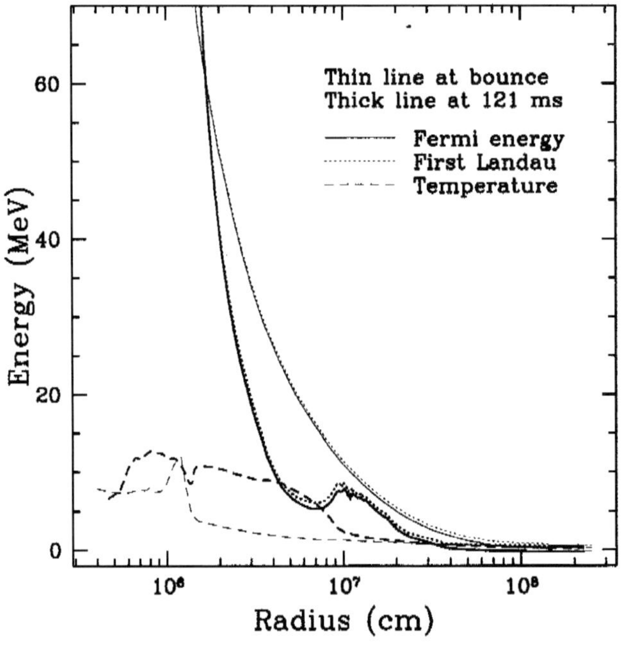

Fig. 26.2. The Fermi energy and the predicted energy of the first Landau level at various epochs is shown for the calculations of Akiyama *et al.* (2003). The first Landau level exceeds the Fermi energy at 121 ms at r $\sim 5 \times 10^6 - 10^7$ cm, but finite temperature effects also enter.

for the non-magnetized electron gas (Wheeler & Hansen 1971) and completely degenerate formulae for the magnetized electron gas (Canuto & Ventura 1977), but at large radii and in the region $4-8 \times 10^6$ cm the electron gas is no longer strongly degenerate. The electron pressure calculated is bigger than it is in the actual supernova environment (see Fig. 26.3) because the equation of state used for the non-magnetized electron gas omits Coulomb effects and other corrections.

We expect strong quantum effects in the region where the Fermi energy is less than the energy of the first Landau level. At 121 ms after bounce, the interesting region may be around $r \sim 5 \times 10^6 - 10^7 cm$, but the thermal energy is the same order as the Fermi energy, and our equations of state used for both magnetized and non-magnetized electron gas are not appropriate.

Assuming that only up to the first Landau Level is occupied, the magnetized electron pressure for the components perpendicular and parallel to the magnetic field are calculated (Fig. 26.3). This figure shows that the magnetized electron pressure can be much lower than the non-magnetized counterpart, and that the perpendicular pressure is much lower than the parallel pressure. The region between $r \sim 5 \times 10^6 - 10^7$ cm is the region of interest, but because of the thermal effect, magnetized gas pressure cannot be calculated and was set equal to that of the non-magnetized gas.

26.4 Discussion

If super-strong magnetic fields are generated during core collapse, the thermodynamic properties of the electron gas may change significantly depending upon how many Landau levels are occupied. We calculated the pressure of magnetized degenerate electron gas in parallel and perpendicular directions, as well as partially degenerate non-magnetized electron gas. For the case shown here, the magnetized electron pressure is less than that of non-magnetized gas and is anisotropic. This pressure anisotropy is likely to be balanced by the $\mathbf{j} \times \mathbf{B}$ force of induced magnetization (Blandford & Hernquist 1982). The electron pressure will be reduced compared to calculations that ignore quantization, but it is not clear that will make a significant difference to the dynamics. In the extreme case where only the ground level is occupied, the gas becomes one dimensional and only moves along the field lines, implying that the magnetic force $\mathbf{j} \times \mathbf{B} = \mathbf{0}$. This may affect how dynamos operate in this high magnetic field regime.

Magnetic fields over the Q.E.D. limit may exist in the core collapse environment. Such high magnetic fields may be required by gamma-ray bursts (Kumar & Panaitescu 2003, Coburn & Boggs 2003), and magnetars are known to have super-strong magnetic fields (Duncan & Thompson 1992). We need to determine

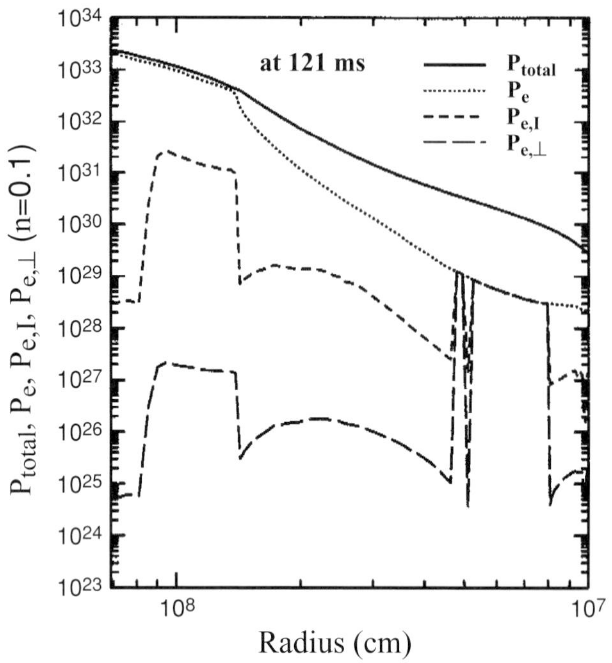

Fig. 26.3. The pressures of magnetized and non-magnetized electron gas are calculated and compared. The non-magnetized electron gas represents a large fraction of the total pressure because of the equation of state used. In general, magnetized gas pressure can be much lower than that of non-magnetized gas and is anisotropic in different directions.

if the quantum effects alter the classic behavior of charged particles and associated dynamics.

Acknowledgement

This work was supported in part by NSF AST-0098644 and by NASA NAG5-10766.

References

Akiyama, S., Wheeler, J. C., Meier, D. L., & Lichtenstadt, I. 2003, ApJ. 584, 954
Blandford, R. D., & Hernquist, L. 1982, J. Phys. C. 15, 6233
Canuto, V., & Ventura, J. 1977, Fundamentals of Cosmis Physics (Great Britain: Gordon and Breach Science Publishers Ltd.), 2, 203
Coburn, W., & Boggs, S. E. 2003, Nature, 423, 415
Duncan, R. C., & Thompson, C. 1992, ApJ. 392, L9
Fryer, C. L., & Heger, A. 2000, ApJ, 541, 1033
Fryer, C. L., & Warren, M. S. 2004, ApJ, 601, 391
Koide, S., Meier, D. L., Shibata, L., & Hudoh, T. 2000, ApJ, 536, 668
Kotake, K., Yamada, S., & Sato, K. 2003, ApJ, 595, 304
Kumar, P., & Panaitescu, A. 2003, MNRAS, 346, 905

LeBlanc, J. M. & Wilson, J. R. 1970, ApJ, 161, 541
Leonard, D. C., Fillippenko, A. V., Ardila, D. R. & Brotherton, M. S. 2000, ApJ, 533, 861
Leonard, D. C., Fillippenko, A. V., Barth, A. J., & Matheson, T. 2001, ApJ, 536, 239
Müller, E., & Hillebrandt, W. 1981, A&A, 103, 358
Möchmeyer, R., & Müller, E. 1989, in Timing Neutron Stars, ed. H. Ögelman & E. P. J. van den Heuvel (NATO ASI Ser. C, 262; Dordrecht: Kluwer), 549
Ott, C. D., Burrows, A., Livne, E., & Walder, R. 2004, ApJ, 600, 834
Pun, C. S. J., & The Supernova Intensive Studies (SINS) Collaboration. 2001, AAS Meeting, 199, 94.02
Shapiro, S. L., & Teukolsky, S. A. 1983, Black Holes, White Dwarfs, and Neutron Stars (New York: Wiley)
Symbalisty, E. M. D. 1984, ApJ, 285, 729
Wang, L., Wheeler, J. C., Li, Z. W., & Clocchiatti, A. 1996, ApJ, 467, 435
Wang, L., Howell, D. A., Höflich, P., & Wheeler, J. C. 2001, ApJ, 550, 1030
Wang, L., Wheeler, J. C., Höflich, P., Khokhlov, A., Baade, D., Branch, D., Challis, P., Filippenko, A. V., Fransson, C., Garnavich, P., Kirshner, R. P., Lundqvist, P., McCray, R., Panagia, N., Pun, C. S. J., Phillips, M. M., Sonneborn, G., Suntzeff, N. B. 2002, ApJ, 579, 671
Wheeler, J. C., & Hansen, C. J. 1971, Ap&SS, 11, 373
Wheeler, J. C., Yi, I., Höflich, P., & Wang, L. 2000, ApJ, 537, 810
Wheeler, J. C., Meier, D. L., & Wilson, J. R. 2002, AoJ, 568, 807
Yamada, S., & Sato, K. 1994, ApJ, 434, 268

27

Non-radial instability of stalled accretion shocks: advective-acoustic cycle

T. Foglizzo and P. Galletti

Service d'Astrophysique, CEA/DSM/DAPNIA, CE-Saclay, 91191 Gif-sur-Yvette, France

Abstract

The linear stability of stalled accretion shocks is investigated in the context of core collapse of type II supernovae. We focus on a particular instability mechanism based on the coupling of acoustic perturbations with advected ones (vorticity, entropy). This advective-acoustic cycle takes place between the shock and the nascent neutron star. Both adiabatic and non-adiabatic processes may contribute to this coupling, but only adiabatic ones are considered in this first approach. The growth time of the adiabatic instability scales like the advection time, and is dominated by low degree modes $l = 0,1,2$. Non radial modes ($l = 1,2$) found unstable by Blondin *et al.* (2003) can be related to this mechanism.

27.1 Introduction

Shocked accretion onto the surface of a compact star is known to be unstable in the context of magnetized white dwarfs, leading to shock oscillations (from Langer, Chanmugam & Shaviv 1981, hereafter LCS81, to Saxton & Wu 2001). Houck & Chevalier (1992, hereafter HC92) made a linear stability analysis of shocked accretion onto a neutron star, and found an instability reminiscent of the instability found by LCS81. HC92 showed specific cases where the cooling occurs mostly in a thin layer at the surface of the neutron star, while the flow is essentially adiabatic above it. The mechanism of the instability was described by LCS81 and subsequent authors as a kind of thermal instability: if the shock surface is moving outwards, the higher incident velocity in the frame of the shock produces a higher temperature blob, which pushes the shock further out if the increased cooling time exceeds the increased advection time. This cycle, however, resembles the unstable adiabatic cycle described by Foglizzo & Tagger (2000, hereafter FT00), and Foglizzo (2001, 2002, hereafter F01,F02) in the context of shocked spherical accretion onto a black hole. In this case the acoustic feedback is purely adiabatic and is due to the advection of vortical/entropic perturbations from the shock to the accretor. A similar coupling must also take place if the accretor is a neutron star. Is the instability found by HC92 due to a cooling process, in the spirit of LCS81, or is there a significant contribution of the adiabatic coupling between vortical/entropic and acoustic perturbations? The recent work of Blondin, Mezzacappa & DeMarino (2003, hereafter BMD03) seems to support this second hypothesis. However, is the acoustic feedback identified by BMD03 linear or due to turbulence? Understanding the physical mechanism underlying this instability could prove useful in order to evaluate its role when realistic non-adiabatic processes are taken into account.

27.2 The perturbed adiabatic flow viewed as a forced oscillator

In order to distinguish between adiabatic processes and non adiabatic ones, the flow structure between the shock $r_{\rm sh}$ and the surface of the nascent neutron star $r_\star$ is schematized as an adiabatic flow above a cooling layer $r_{\rm cool}$. The present study focuses on the stability of the adiabatic part of the flow $r_{\rm cool} < r < r_{\rm sh}$. The entropy and vorticity equations can be integrated explicitly as in F01. The Rankine-Hugoniot conditions at the shock impose that entropy and vorticity perturbations δS, δw are simply related through:

$$(\delta w_r, \delta w_\theta, \delta w_\varphi) = \left(0, -\frac{c^2}{\gamma r v \sin\theta}\frac{\partial \delta S}{\partial\varphi}, \frac{c^2}{\gamma r v}\frac{\partial \delta S}{\partial\theta}\right). \tag{27.1}$$

The differential system satisfied by perturbations is the same as in F01 (Eqs. (B18–B19)), only the functions $\mathcal{M}$, v, c describing the stationary flow are different. Pressure perturbations satisfy a differential equation with a source term due to entropy/vorticity perturbations (Eq. (4) of F01):

$$\left\{\frac{\partial^2}{\partial r^2} + a_1\frac{\partial}{\partial r} + a_0\right\}\frac{\delta p}{p} = \frac{-\Delta}{v^2(1-\mathcal{M}^2)}\frac{\delta_{ps}}{p}. \tag{27.2}$$

The source term δ_{ps} is:

$$\frac{\delta_{ps}}{p} \equiv -\frac{1}{c^2}\frac{v}{i\omega}\frac{\partial}{\partial r}\left\{\left(1-\frac{2v}{i\omega}\frac{\partial\log\mathcal{M}}{\partial r}\right)\frac{\omega^2 c^2}{\Delta}\right\}\delta S, \tag{27.3}$$

$$\Delta \equiv \omega^2 + l(l+1)\frac{v^2}{r^2} + 2i\omega\frac{\partial v}{\partial r},. \tag{27.4}$$

Eq. (27.3) thus describes the local production of pressure perturbations due to advection of entropy and vorticity perturbations in a inhomogeneous flow. It characterizes the "excitator," whereas the left hand side of Eq. (27.2) characterizes the "oscillator". According to Eq. (27.3), the local strength of the excitator depends on the value of the ratio $\omega r/v$.

$$\frac{\Delta}{v^2}\left(\frac{\delta_{ps}}{p}\right)_{\frac{\omega r}{v}\gg 1} \sim -\frac{1}{r^2}\frac{i\omega r}{v}\frac{\partial \log c^2}{\partial \log r}\delta S, \tag{27.5}$$

$$\frac{\Delta}{v^2}\left(\frac{\delta_{ps}}{p}\right)_{\frac{\omega r}{v}\ll 1} \sim \frac{2}{r^2}\frac{\partial \log c^2}{\partial \log r}\frac{\partial \log \mathcal{M}}{\partial \log r}\delta S \quad \text{if } l=0, \tag{27.6}$$

$$\sim -\frac{2}{r^2}\frac{\partial \log \mathcal{M}}{\partial \log r}\frac{\partial \log}{\partial \log r}\left(\frac{v}{c^2 r}\right)\delta S \quad \text{if } l\geq 1. \tag{27.7}$$

In view of Eqs. (27.5) to (27.7) and the advective-acoustic cycles described in F01, F02, two physical processes couple advected perturbations to the acoustic field:

(i) The gradient of temperature characteristic of the entropic-acoustic cycle is essential for spherical perturbations $l=0$ (Eq. (27.6) and FT00) and high frequency perturbations (Eq. (27.5) and F01).

(ii) Even in a isothermal flow (i.e. $\partial c^2/\partial r = 0$), non radial perturbations can excite acoustic waves in a vortical-acoustic cycle at low frequency (Eq. (27.7) and F02). The simple estimate in Eqs. (27.5) to (27.7) shows that the strength of the excitator is comparable for radial and non radial perturbations, and that its amplitude is highest near the lower boundary.

An efficient coupling between the excitator and the oscillator requires not only a strong amplitude of the excitator, but also a good matching of their spatial length-scales. The wavelength of the excitator ($\sim 2\pi v/\omega$) is approximately a factor $\mathcal{M}$ smaller than the wavelength of the oscillator ($\sim 2\pi(c \pm v)/\omega$). This contrasts with the simpler case of black hole accretion, in which case the excitator and oscillator have comparable wavelengths near the sonic radius ($\mathcal{M} \to 1$). The numerous oscillations of the excitator per acoustic wavelength should thus lead to a weak efficiency of the advective acoustic coupling in the inner regions where $\mathcal{M} \ll 1$. The inner regions, however, are precisely the place where the amplitude of the excitator is highest, because the adiabatic gradients are strongest there. What is the net effect? The choice of the lower boundary condition is crucial in answering this question.

27.3 Boundary condition at $r_{\rm cool}$

In order to separate the adiabatic effects from the non adiabatic ones, we choose to estimate the contribution of the adiabatic region by neglecting the acoustic feedback from the cooling layer as much as possible. The following assumptions are made at the lower boundary $r_{\rm cool}$:

(i) acoustic perturbations propagating downward are perfectly reflected out ($\omega \ll c_{\rm cool}/r_{\rm cool}$),
(ii) entropy and vorticity perturbations are freely advected below $r_{\rm cool}$ in the cooling region, where their coupling to acoustic waves is ignored.

Condition (ii) is equivalent to imposing that below $r_{\rm cool}$, entropy and vorticity perturbations cease to be source terms of the acoustic equation. The source term in Eq. (27.2) is thus artificially damped by multiplying it by a smooth transition function Φ_λ. The transition is assumed to take place over a length λ, comparable to the cooling length. This damping of the source term can either be viewed as an ad-hoc damping of the entropy perturbation itself, independent of its frequency, or as a smoothing of the flow gradient responsible for the coupling.

The equations corresponding to these assumptions are obtained by matching the pressure perturbation δp for $r \geq r_{\rm cool}$ with the homogeneous solution δp_0^0 associated to Eq. (27.2) for $r < r_{\rm cool}$.

27.4 Eigenmodes of shocked accretion

The Rankine-Hugoniot jump conditions are used to compute the perturbed quantities after the shock. These calculations are similar to those of Landau & Lifshitz (1987, Chap. 90) extended to the case of a non uniform flow, or Nakayama (1992) extended to non radial perturbations. For a perturbed shock velocity Δv in the strong shock limit,

$$\frac{\delta v_r}{v_{\rm sh}} = \frac{2}{\gamma+1}\left(\frac{i\omega r_{\rm sh}}{v_{\rm sh}} + \frac{\gamma}{2}\frac{5-3\gamma}{\gamma-1}\right)\frac{\Delta v}{i\omega r_{\rm sh}}, \tag{27.8}$$

$$\frac{\delta\rho}{\rho_{\rm sh}} = -\frac{\gamma}{\gamma+1}\frac{5-3\gamma}{\gamma-1}\frac{\Delta v}{i\omega r_{\rm sh}}, \tag{27.9}$$

$$\frac{\delta v_\Omega}{v_{\rm sh}} = -\frac{2}{\gamma-1}\frac{\partial}{\partial\Omega}\frac{\Delta v}{i\omega r_{\rm sh}}, \tag{27.10}$$

$$\frac{\delta p}{p_{\rm sh}} = -2\frac{\gamma-1}{\gamma+1}\left[\frac{i\omega r_{\rm sh}}{v_{\rm sh}} + \frac{5-3\gamma}{4}\frac{\gamma^2+1}{(\gamma-1)^2}\right]\frac{\Delta v}{i\omega r_{\rm sh}}. \tag{27.11}$$

The boundary value problem was solved numerically for $l = 0, 1$. A broad range of unstable modes grow on a timescale comparable to a fraction of the advection timescale. The radial mode is always the most unstable, closely followed by the non radial mode $l = 1$.

27.5 Evidence for the advective-acoustic cycle

Following the same method as F02, the discrete spectrum obtained in the boundary value problem is checked by computing, in two steps, the efficiency $Q_{\rm adv}$ of sound production by the advection of an entropy/vorticity perturbation (without a shock), and the efficiency $Q_{\rm sh}$ of entropy/vorticity production by an outgoing acoustic wave reaching a shock. $Q_{\rm adv}$ is defined by:

$$Q_{\rm adv} = \int_{r_{\rm cool}}^{r_{\rm sh}} \frac{1-\mathcal{M}^2}{2\gamma^2\mathcal{M}^2}\frac{\delta p_0^0}{p} {\rm e}^{i\omega\int_{r_{\rm sh}}^{r}\frac{1+\mathcal{M}^2}{1-\mathcal{M}^2}\frac{dr}{v}} \Phi_\lambda \frac{\partial}{\partial r}\left\{\left(1-\frac{2\eta v}{i\omega r}\right)\frac{\omega^2 c^2}{c_{\rm sh}^2\Delta}\right\} dr \qquad (27.12)$$

The efficiency $Q_{\rm sh}$ is obtained through a WKB approximation at high frequency:

$$|Q_{\rm sh}| \equiv \frac{1}{\mathcal{M}_{\rm sh}^{\frac{1}{2}}} \frac{2\gamma}{1+\frac{\gamma-1}{2\mathcal{M}_{\rm sh}}}. \qquad (27.13)$$

The global efficiency $Q \equiv |Q_{\rm adv} Q_{\rm sh}|$ of the advective-acoustic cycle leads to a first estimate of the growth rate at high frequency:

$$\omega_i = \frac{1}{\tau_{\rm tot}} \log Q, \qquad (27.14)$$

where $\tau_{\rm tot}$ is the total duration of the cycle (advection + acoustic). The exact resolution of the discrete eigenfrequencies was successfuly compared to the continuous WKB estimate (27.14).

27.6 Comparison with BMD03

Our stability analysis in an adiabatic flow should apply directly to the numerical simulations of BMD03. The authors recognized the existence of an advective-acoustic cycle similar to the one occuring in the vortical-acoustic instability of F02. Our linear study seems to agree qualitatively with their results concerning the mode $l = 1$. Nevertheless, the mode $l = 0$ is stable in their simulations, whereas it is the most unstable one according to our calculations, as well as in the work of HC92. This important difference may come from the "leaky" lower boundary condition of BMD03, which is different from ours even in the linear regime.

27.7 Conclusion

The temperature and velocity gradients in the subsonic flow between the shock and the accretor is responsible for an efficient *linear* coupling between entropy/vorticity and acoustic perturbations. Most of the sound comes from the region close to the lower boundary of the adiabatic region, despite the fact that the wave-length of advected perturbations is much smaller than acoustic ones. The acoustic waves reaching the shock produce new entropy/vorticity perturbations, in an unstable cycle. The growth time is comparable to the advection time. The most unstable modes correspond to $l = 0$, 1 perturbations. The identification of the advective-acoustic cycle as the mechanism responsible for the instability was checked numerically through the calculation of $Q_{\rm adv}$ and $Q_{\rm sh}$. Since the region contributing most efficiently to the instability is the vicinity of the lower boundary $r_{\rm cool}$, the role of cooling processes is at least crucial in determining the growth rate of the instability. The principle of a cycle between propagating and advected perturbations could be useful in interpreting the stability analysis of non adiabatic flows. Whether non adiabatic processes are partially stabilizing or even more destabilizing remains to be determined. The difficulties of 1-D numerical models in reaching an explosion (e.g. Buras *et al.* 2003) suggests that cooling processes are indeed important enough to significantly stabilize the entropic-acoustic cycle at work for radial perturbations.

Acknowledgements

The authors are grateful to T. Janka for his initial suggestion to study the effects of advective-acoustic cycles in the problem of core-collapse, and his permanent encouragements since then. Useful discussions with J. Blondin, A. Burrows and R. Chevalier are aknowledged.

References

Blondin, J., Mezzacappa, A. & DeMarino, C. 2003. *Astrophys. J.*, **584**, 971–980 (BMD03)
Buras, R., Rampp, M., Janka & H.-T., Kifonidis, K. 2003. astro-ph/0303171
Foglizzo, T. 2002. *Astron. Astrophys.*, **392**, 353–368 (F02)
Foglizzo, T. 2001. *Astron. Astrophys.*, **368**, 311–324 (F01)
Foglizzo, T. & Tagger, M. 2000. *Astron. Astrophys.*, **363**, 174–183 (FT00)
Houck, J. C. & Chevalier, R. A. 1992. *Astrophys. J.*, **395**, 592–603 (HC92)
Landau, L. & Lifshitz, E. 1987, Fluid Mechanics 6, Pergamon Press
Langer, S. H., Chanmugam, G. & Shaviv, G., 1981. *Astrophys. J.*, **245**, L23–L26 (LCS81)
Nakayama, K. 1992. *Mon. Not. R. astr. Soc.*, **259**, 259–264
Saxton, C. J. & Wu, K. 2001. *Mon. Not. R. astr. Soc.*, **324**, 659–684

28

Asymmetry effects in hypernovae

K. Maeda[1], K. Nomoto[1,2], J. Deng[1,2], P. A. Mazzali[2,3]

[1] *Department of Astronomy, School of Science, University of Tokyo, Hongo 7-3-1, Bunkyo-ku, Tokyo 113-0033, Japan*

[2] *Research Center for the Early Universe, School of Science, University of Tokyo, Hongo 7-3-1, Bunkyo-ku, Tokyo 113–0033, Japan*

[3] *INAF-Osservatorio Astronomico, Via Tiepolo, 11, 34131 Trieste, Italy*

Abstract

The basic explosion mechanisms of core-collapse supernovae (SNe) have not been clarified yet. The discovery of hypernovae with the isotropic kinetic energy (E) $E_{51} \equiv E \,/\, 10^{51}$ ergs $\gtrsim 5-10$ brought us a new light on this issue. Observational properties of hypernovae indicate that asymmetry may play an important role in the explosion. We discuss two classes of asymmetry effects related to hypernovae. (1) Effects of asymmetric ejecta on observed properties. Interpreting (late phase) optical light curves and spectra of hypernovae suggests that these objects are aspherical in nature. (2) Effects of asymmetric bipolar explosions on nucleosynthetic yields. An aspherical bipolar explosion provides high-velocity Fe-rich materials and low-velocity O-rich materials, which are in agreement with the observations. The unique yields of the bipolar explosions, e.g., enhanced (Zn, Co)/Fe and suppressed (Mn, Cr)/Fe, can account for the peculiar abundance patterns of extremely metal-poor stars, suggesting that they could have significantly contributed to the early Galactic chemical evolution.

28.1 Properties of hypernovae

Type Ic Hypernova SN 1998bw was probably linked to GRB 980425 (Galama *et al.* 1998), thus establishing for the first time a connection between gamma-ray bursts (GRBs) and core-collapse SNe. However, SN 1998bw was exceptional for a SN Ic: it was as luminous at peak as a SN Ia, indicating that it synthesized $\sim 0.5\ M_\odot$ of ^{56}Ni, and its isotropic E was estimated as $E_{51} \gtrsim 30$ (Iwamoto *et al.* 1998; Woosley, Eastman, & Schmidt 1999; see, however, Höflich, Wheeler, & Wang. 1999 for a different interpretation).

Subsequently, other "hypernovae" have been recognized, such as SN 1997ef (Iwamoto *et al.* 2000; Mazzali, Iwamoto, & Nomoto 2000), SN 1999as (Knop *et al.* 1999), and SN 2002ap (Mazzali *et al.* 2002). Although these SNe Ic did not appear to be associated with GRBs, the most recent "hypernova" SN 2003dh is clearly associated with GRB 030329 (Stanek *et al.* 2003; Hjorth *et al.* 2003; Kawabata *et al.* 2003; Matheson *et al.* 2003). Figure 28.1 (top) shows the near-maximum spectra of hypernovae. The hypernovae show broad spectral line, indicating a large amount of mass at high velocities, thus high explosion kinetic energies.

Figure 28.1 (bottom) shows (isotropic) E as a function of the main-sequence mass $M_{\rm MS}$ of the progenitor star obtained from fitting the optical light curves and spectra. These mass estimates place hypernovae at the high-mass end of SN progenitors. In contrast, SNe II 1997D and 1999br were very faint SNe with very low E (Turatto *et al.* 1998; Zampieri *et al.* 2003). Therefore, we propose that SNe from stars with $M_{\rm ms} \gtrsim 20\text{–}25\,M_\odot$. have different E and $M\,(^{56}\mathrm{Ni})$, with a bright, energetic "hypernova branch" at one extreme and a faint, low-energy SN branch at the other (Nomoto *et al.* 2003ab). For the faint SNe, the explosion energy was so small that most ^{56}Ni fell back onto the compact remnant. Thus the faint SN branch may become a "failed" SN branch at larger $M_{\rm ms}$. Between the two branches, there may be a variety of SNe (Hamuy 2003).

This trend might be interpreted as follows. Stars with $M_{\rm ms} \lesssim 20\text{–}25\,M_\odot$ form a neutron star, producing $\sim 0.08 \pm 0.03\,M_\odot$ ^{56}Ni as in SNe 1993J, 1994I, and 1987A. Stars with $M_{\rm ms} \gtrsim 20\text{–}25\,M_\odot$ form a black hole; whether they become hypernovae or faint SNe may depend on the angular momentum in the collapsing core, which in turn depends on the stellar winds, metallicity, magnetic fields, and binarity. Hypernovae might have rapidly rotating cores owing possibly to the spiraling-in of a companion star in a binary system.

28.2 Indications of asphericity in hypernovae

Modeling observations of core collapse supernovae/hypernovae, however, has been indicating that they are essentially aspherical. A direct HST image of SN 1987A (Wang *et al.* 2002) and large optical linear polarization in SNeIb/Ic (Wang *et al.* 2001) support this idea. In this section, we show some optical observations also indicate asphericity in hypernova explosions.

28.2.1 Light curves

The spherical hydrodynamical models used to fit the early phases of SNe 1998bw (Nakamura *et al.* 2001), 1997ef (Iwamoto *et al.* 2000; Mazzali *et al.* 2000), and 2002ap (Mazzali *et al.* 2002) do not reproduce well the later phase light curves (after ~50 days). Specifically, they predict late-phase optical luminosities much

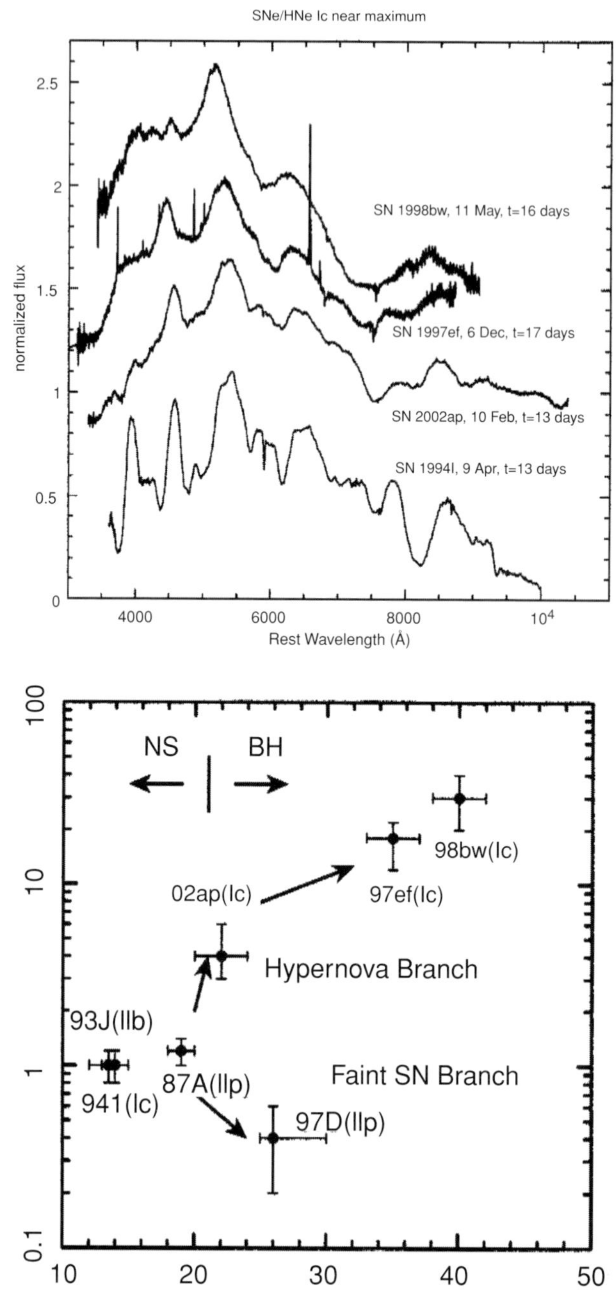

Fig. 28.1. Top: The near-maximum spectra of Type Ic SNe and hypernovae: SNe 1998bw, 1997ef, 2002ap, and 1994I. Bottom: The explosion energy as a function of the main sequence mass of the progenitors for several supernovae/hypernovae.

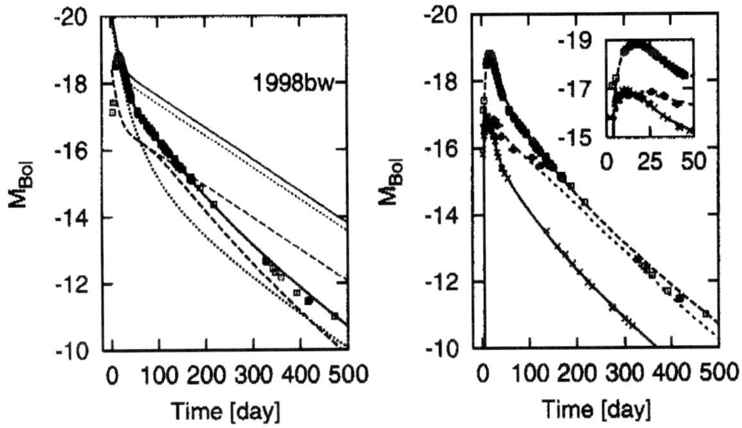

Fig. 28.2. Left: A γ-ray and e^+ deposition curve (thick-solid) as a result of a combination of an outer high-velocity large-E component ($E_{51} \sim 46$, $M(^{56}Ni) \sim 0.4\ M_\odot$; thick-dotted) and an inner low-velocity small-E component ($E_{51} \sim 0.6$, $M(^{56}Ni) \sim 0.1\ M_\odot$; thick-dashed). (Also shown is total luminosity emitted as γ-rays and e^+; thin lines.) Right: Light curves of SNe 1997ef (Mazzali *et al.* 2000; circles), 1998bw (Patat *et al.* 2001; squares), and 2002ap (Yoshii *et al.* 2003; crosses) as compared with synthetic light curves of two-component models computed with a Monte Carlo light-curve code (Maeda *et al.* 2003).

fainter than observed. (A similar problem may have been encountered for some SNe IIb/Ib/Ic. See Clocchiatti & Wheeler (1997), Clocchiatti *et al.* (1997).)

Figure 28.2 (left) shows the bolometric light curve of SN 1998bw (Patat *et al.* 2001) with analytical model curves (applicable to late epochs after $\gtrsim 50$ days, i.e., well after the peak). The large-E model (thick-dotted) minics the spherical hypernova model which is appropriate to explain the early phase observations. The model curve declines much faster than the observations between ~50 days and 200 days. After ~200 days, the observed curve declines almost on a parallel with the model curve at the rate predicted by the asymptotic behavior of the γ-ray deposition (Clocchiatti & Wheeler, 1997), but at brighter magnitudes.

This indicates that (1) the energy source is indeed radioactive decays of ^{56}Co, and any other contributions (e.g., background sources or energy inputs by the central remnant) are negligible as they will produce a different declining rate, and (2) the emission is dominated by a dense small-E component which is kept optically thick to γ-rays till ~100 days (Figure 28.2 (left); thick-dashed).

The entire evolution of the light curve of SN 1998bw is therefore well explained by a combination of the two components, i.e., an outer, large-E component and an inner, small-E component (Figure 28.2 (left); thick-solid: see also Chugai 2000). The model requires that a large fraction of ^{56}Ni should be in the outer component.

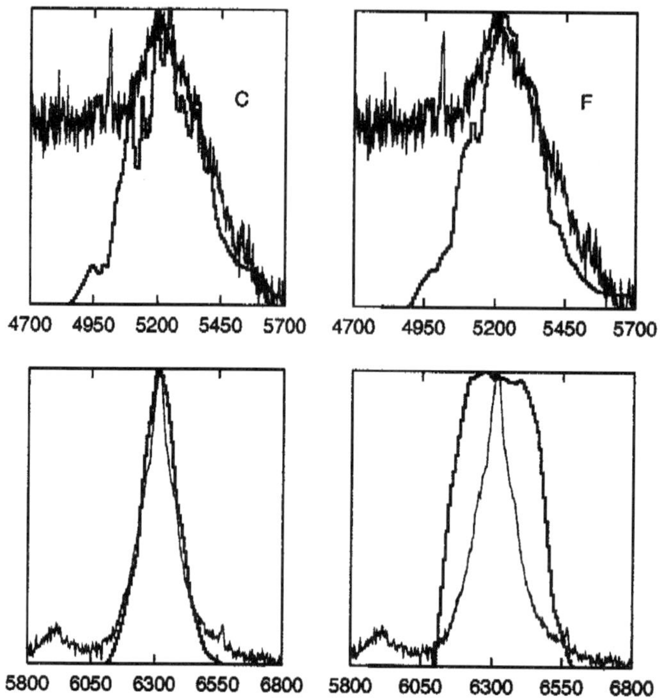

Fig. 28.3. Profiles of [Fe II] feature peaked at 5200 Å (upper panels) and of [O I] 6300, 6363 Å (lower panels) for an aspherical bipolar model (Maeda *et al.* 2002) viewed at 15° from the jet direction (left panels; thick lines) and for a spherical model (right panels). The observed lines at a SN rest-frame epoch of 216 days are also plotted for comparison (thin lines, Patat *et al.* 2001).

The same scenario is also applicable to SNe 1997ef and 2002ap (Figure 28.2 (right); Maeda *et al.* 2003).

Spherical explosion models so far do not predict a high density core near the center nor the presence of a large amount of ^{56}Ni at high velocities, but these features are required to fit the light curves.

28.2.2 Spectra

Mazzali *et al.* (2001) calculated synthetic nebular-phase spectra of SN 1998bw using a spherically symmetric NLTE nebular code based on the deposition of γ-rays from ^{56}Co decay in a nebula of uniform density and composition. They showed that the [O I] 6300 Å and the [Fe II] features peaked at 5200 Å can only be reproduced if different velocities are assumed for the two elements. A significant amount of slowly-moving O is necessary to explain the zero velocity peak of the [O I] line, and fast-moving Fe (faster then O) is needed to fit the broad-line Fe spectrum. This

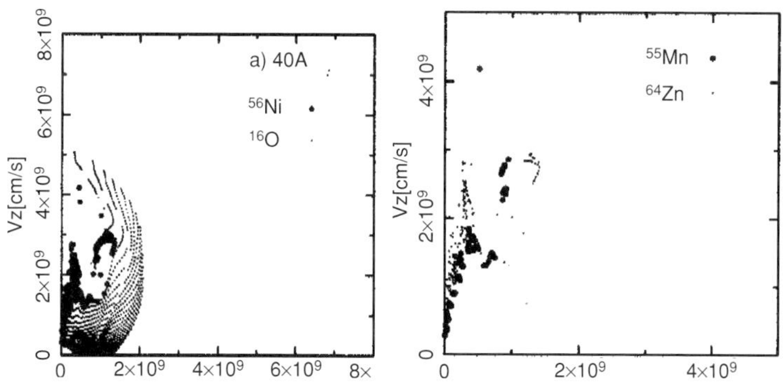

Fig. 28.4. Distributions of ^{56}Ni (which decays into ^{56}Fe) and ^{16}O (left panel) and of ^{55}Mn and ^{64}Zn (right panel) in the velocity space for the bipolar model.

suggests that an aspherical distribution of Fe and O may explain the observations, being consistent with the result of the light curve modeling in the previous section.

In order to verify the observable consequences of a bipolar explosion, Maeda *et al.* (2002) calculated the profiles of the [Fe II] blend and of [O I]. The bipolar model contains high velocity Fe along the jet axis, leaving a large amount of O along the equatorial direction (see the next section and Figure 28.4 for details).

The [Fe II] and [O I] profiles viewed at an angle of 15° from the jet direction and those for a spherical model are compared to the observed spectrum of 26 Nov 1998 in Figure 28.3. In a spherical explosion oxygen is located at higher velocities than iron, and the [O I] line is too broad. This is because of the deficiency of oxygen with small velocity along the line of sight. Therefore the observed line profiles are not explained with a spherical model.

In aspherical explosion models Fe is distributed preferentially along the jet direction, and so a larger ratio of the Fe and O line widths can be obtained. When the degree of asphericity is high and the explosion is viewed from near the jet direction, the component lines in the [Fe II] blend have double-peaked profiles, the blue- and red-shifted peaks corresponding to matter situated in the two opposite lobes of the jet, where Fe is mostly produced. Because of the high velocity of Fe, the peaks are widely separated, and the blend is wide (Fig. 28.3). In contrast, the [O I] line is narrower and has a sharper peak, because O is produced mostly in the r-direction, at lower velocities and with a less aspherical distribution.

28.2.3 SN 2003dh/GRB030329

Both late time light curves and spectra of hypernovae point to the need of high-velocity Fe-rich materials and low-velocity dense O-rich materials. This feature is contrary to what spherical explosion models predict. This characteristic is, however, qualitatively consistent with recent multi-dimensional models of jet-driven supernova explosions; they do predict such high density material (Khokhlov *et al.* 1999; Maeda *et al.* 2002) and a large amount of ^{56}Ni at high velocities (Maeda *et al.* 2002; Maeda & Nomoto 2003).

We note that SN 2003dh associated with GRB030329 also may share the same feature. SN 2003dh exhibited the optical spectra almost identical to SN 1998bw for a few weeks after the burst (Matheson *et al.* 2003). However, spectra at $\sim$1 month had changed somewhat, looking more like that of less energetic hypernova SN 1997ef (Kawabata *et al.* 2003). Mazzali *et al.* (2003) showed that the spectra and inferred light curve can be modeled adopting a density distribution similar to that used for SN 1998bw at $v > 25{,}000$ km s^{-1} (lerge-E component) but more like that of SN 1997ef at lower velocities (small-E component). Woosley & Heger (2003) also suggested a similar model containing two components for SN 2003dh.

28.3 Bipolar supernova/hypernova explosions

Motivated by previous hydrodynamic studies on jet-driven supernova explosions (Khoklov *et al.* 1999; MacFadyen, Woosley, & Heger. 2001), and by the observational indication in the previous section, we have investigated hydrodynamics and explosive nucleosynthesis in bipolar supernova/hypernova explosions (Maeda & Nomoto 2003). We have performed a series of 2D calculations, injecting a pair of jets at the interface between the initial compact remnant and the surrounding stellar envelopes. For the jet properties, we adopt a formalism similar to MacFadyen *et al.* (2001).

28.3.1 Nucleosynthesis

The strong outflow occurs along the z-axis (the jet direction), while materials accrete from the side (r-direction). The outcome is a highly aspherical explosion. As a result, materials along the r-axis with lower temperatures accrete onto the center, while materials undergoing higher temperatures are ejected along the z-axis. Thus, the bipolar explosions eject preferentially higher temperature materials. The consequences are as follows.

First, it results in highly aspherical distribution of nucleosynthetic products (Figure 28.4). The distribution of ^{56}Ni (which decays into ^{56}Fe) is elongated along the z-axis. O-rich materials occupy the central region whose density becomes very high via continuous accretion from the side. The result is, at least qualitatively, in agreement with the observational indications discussed in the previous section. For the same reason, ^{64}Zn is ejected with higher velocities than ^{55}Mn (Figure 28.4 (right)).

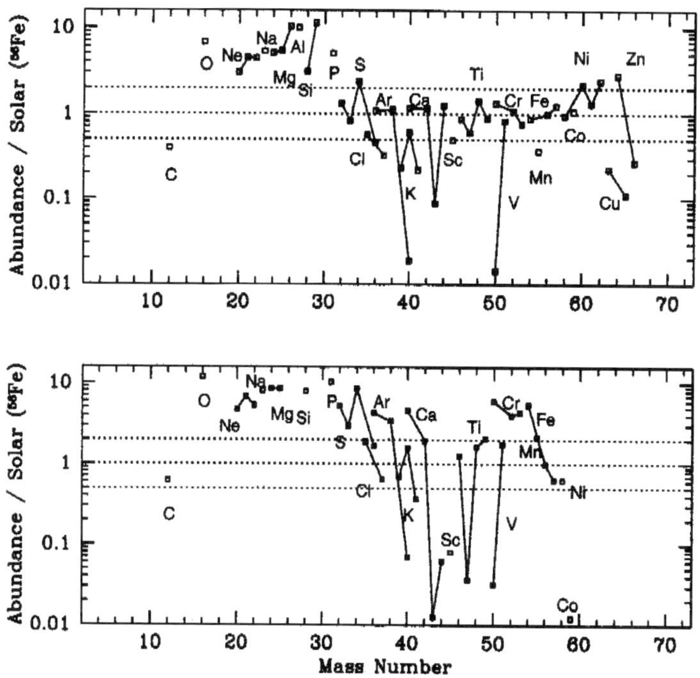

Fig. 28.5. Yields of the bipolar model (upper) and a spherical model (lower) with $E_{51} = 10$, $M(^{56}\mathrm{Ni}) = 0.1M_\odot$, and $M_{\mathrm{MS}} = 40M_\odot$.

In sum, the model leads to the velocity inversion as compared with conventional spherical models.

Second, the overall abundance patterns differ from those in spherical models as shown in Figure 28.5. For example, [Zn/Fe] and [Co/Fe] ($[X/\mathrm{Fe}] \equiv \log 10(X/\mathrm{Fe}) - \log 10(X/\mathrm{Fe})_\odot$) are enhanced because they are produced through strong α-rich freezeout, while [Mn/Fe] and [Cr/Fe] produced through incomplete Si-burning are suppressed. This is evident in Figure 28.4 (right), which shows that the velocity of ^{64}Zn is higher on average than that of ^{55}Mn, the latter more easily accreting onto the central remnant.

28.3.2 Effects on the galactic chemical evolution

The unique yields of our models may have important implications for early Galactic chemical evolution. Abundances in extremely metal-poor stars store the chemical information of an individual supernova, because those stars are likely to have been contaminated by only a single supernova at the early phase of the evolution of the Galaxy before it was chemically well-mixed.

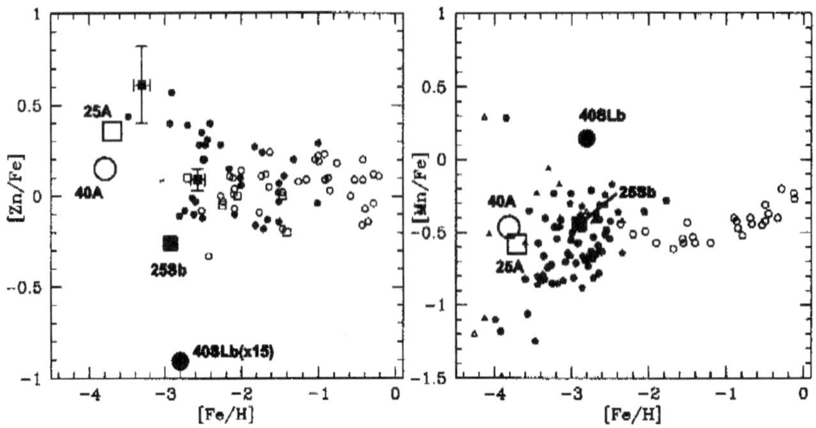

Fig. 28.6. Observed abundance ratios of [Zn, Co, Mn, Cr/Fe] (McWilliam *et al.* 1995; Ryan *et al.* 1996; Primas *et al.* 2000; Blake *et al.* 2001), and the theoretical abundance patterns for spherical supernovae (filled circles, $M_{\rm MS} = 40M_\odot, E_{51} = 1$; filled squares, $M_{\rm MS} = 25M_\odot$, $E_{51} = 1$) and the bipolar hypernova models (open circles, $M_{\rm MS} = 40M_\odot$, $E_{51} = 11$; open squares, $M_{\rm MS} = 25M_\odot$, $E_{51} = 7$).

Studies on metal-poor halo stars have uncovered that there exist interesting trends in the abundances of iron peak elements for [Fe/H] $\lesssim -2.5$ (Figure 1.6, Blake *et al.* 2001; McWilliam *et al.* 1995; Primas *et al.* 2000; Ryan, Norris, & Beers 1996). Both [Cr/Fe] and [Mn/Fe] decrease toward smaller [Fe/H], while [Co/Fe] and [Zn/Fe] increase to reach $\sim 0.3 - 0.5$ at [Fe/H] ~ -3. Also [O, Mg/Fe] are large (typically ~ 0.5) for [Fe/H] $\lesssim -2.5$, suggesting that the amount of Fe ejected in supernovae which are responsible for those stars is not so large.

We point out that the bipolar models presented here can naturally account for these features. Large [Zn, Co/Fe], small [Mn, Cr/Fe] and large [O, Mg/Fe] are simultaneously realized in the bipolar models (Figure 28.5). Figure 28.6 demonstrates how the trends in [Zn/Fe] and [Mn/Fe] are reproduced with our models. Given that the formation of metal-poor stars was driven by a supernova shock, [Fe/H] is determined by the ratio of the ejected mass of Fe to the explosion energy (Ryan *et al.* 1996; Shigeyama & Tsujimoto 1998). This places more energetic (bipolar) supernovae on lower [Fe/H]. The trends, therefore, are consistently reproduced.

28.4 Concluding remarks

Optical spectra and light curves indicate that at least hypernovae are intrinsically asymmetric (some supernovae may also be aspherical. See, e.g., Wang *et al.* 2001). Late-phase observations are especially useful to reveal asymmetry, and indicate the presence of high-velocity Fe-rich materials and low-velocity dense O-rich materials in hypernovae. A promising mechanism to create the above feature is a jet-driven, bipolar explosion. The unique nucleosynthetic yields of bipolar explosions can further account for the peculiar abundances of extremely metal-poor stars.

However, quantitative studies on the estimate of the real energy budget and the degree of asymmetry, which are related to the explosion mechanism of hypernovae, are still in their exploratory phase. In the future, a combination of more realistic explosion models (with effects of rotation, MHD, etc; See Burrows, Ott, & Meakin. 2003) and multi-dimensional radiation transport calculations (e.g., Höflich *et al.* 1999) will tell us the real nature of hypernovae and GRBs.

Acknowledgements

We would like to thank the organizers of this conference, J. C. Wheeler, P. Kumar, and P. Höflich, for their gracious and generous hospitality. This work was partly supported by the Grants-in-Aid for Scientific Research (14047206, 14540223, 15204010) of the Ministry of Education, Science, Culture, Sports, and Technology in Japan.

References

Blake L. A. J. *et al.* 2001, *Nucl. Phys. A.*, **688**, 502
Burrows, A., Ott, C. D., Meakin, C, in this volume
Chugai, N. N. 2000, *Astron. Lett.*, **26**, 797
Clocchiatti, A., & Wheeler, J. C. 1997, *Astrophys. J.*, **491**, 375
Clocchiatti, A., *et al.* 1997, *Astrophys. J.*, **483**, 675
Galama, T. J., *et al.* 1998, *Nature*, **395**, 670
Hamuy, M. 2003, *Astrophys. J.*, **582**, 905
Hjorth, J., *et al.* 2003, *Nature*, **423**, 847
Höflich, P., Wheeler, J. C., Wang, L. 1999, *Astrophys. J.*, **521**, 179
Iwamoto, K., *et al.* 1998, *Nature*, **395**, 672
Iwamoto, K., *et al.* 2000, *Astrophys. J.*, **534**, 660
Kawabata, K. S., *et al.* 2003, *Astrophys. J.*, **593**, L19
Khokhlov, A. M., *et al.* 1999, *Astrophys. J.*, **524**, L107
Knop, R., *et al.* 1999, *IAU Circ.*, 7128
MacFadyen, A. I., Woosley, S. E., & Heger, A. 2001, *Astrophys. J.*, **550**, 410
Maeda, K., *et al.* 2002, *Astrophys. J.*, **565**, 405
Maeda, K., *et al.* 2003, *Astrophys. J.*, **593**, 931
Maeda, K., & Nomoto, 2003, *Astrophys. J.*, in press (astro-ph/0304172)
Matheson, T., *et al.* 2003, *Astrophys. J.*, in press (astro-ph/0307435)
Mazzali, P. A., Iwamoto, K., Nomoto, K. 2000, *Astrophys. J.*, **545**, 407
Mazzali, P. A., Nomoto, K., Patat, F., Maeda, K. 2001, *Astrophys. J.*, **559**, 1047
Mazzali, P. A., *et al.* 2002, *Astrophys. J.*, **572**, L61
Mazzali, P. A., *et al.* 2003, *Astrophys. J.*, submitted (astro-ph/0309555)
McWilliam, A., *et al.* 1995, *Astron. J.*, **109**, 2757
Nakamura, T., *et al*, K. 2001, *Astrophys. J*, **550**, 991
Nomoto, K., *et al.* 2003a, in *IAU Symp 212, A massive Star Odyssey, from Main Sequence to Supernova*, eds Hucht, V. D., *et al.*, Astron. Soc. Pac., San Francisco, 395 (astro-ph/0209064)

Nomoto, K., *et al.* 2003b, in *Stellar Collapse*, ed Fryer, C. L., Astrophysics and Space Science, Kluwer, in press (astro-ph/0308136)
Patat, F. A., *et al.* 2001, *Astrophys. J.*, **555**, 900
Primas, F., *et al.* 2000, in *The First Stars*, eds Weiss, A., *et al.*, Springer, 51
Ryan, S. G., Norris, J. E. & Beers, T. C. 1996, *Astrophys. J.*, **471**, 254
Shigeyama, T., & Tsujimoto, T. 1998, *Astrophys. J.*, **507**, L135
Stanek, K. Z., *et al.* 2003, *Astrophys. J.*, **591**, L17
Turatto, M., *et al.*, 1998, *Astrophys. J.*, **498**, L129
Wang, L., Howell, D. A., Höflich, P., Wheeler, J. C. 2001, *Astrophys. J.*, **550**, 1030
Wang, L., *et al.*, 2002, *Astrophys. J.*, **579**, 671
Woosley, S. E., Eastman, R. G., Schmidt, B. P. 1999, *Astrophys. J.*, **516**, 788
Woosley, S. E., Heger, A. 2003, *Astrophys. J.*, submitted (astro-ph/0309165)
Yoshii, Y., *et al.* 2003, *Astrophys. J.*, **592**, 467
Zampieri, L., *et al.* 2003, *Mon. Not. R. astr. Soc.*, **338**, 711

29

Stellar abundances: the *r*-process and supernovae

J. J. Cowan

Department of Physics and Astronomy, University of Oklahoma
Norman, OK 73019, USA

C. Sneden

Dèpartment of Astronomy and McDonald Observatory, University of Texas
Austin, TX 78712, USA

Abstract

Stellar abundance observations are providing important clues about the relationship between supernovae (SNe) and the rapid neutron capture process (i.e., the *r*-process). Although the site for the *r*-process is still not identified, events in and around SNe have long been suspected. Abundances of heavy neutron-capture elements in a number of stars suggest a robust r-process operating over billions of years, constraining astrophysical and nuclear conditions in supernova models. Variations in lighter n-capture element abundances – observed only very recently in any stars – could be explained as a signature of certain supernova models, or might require multiple r-process sites with different mass ranges or frequencies of SNe. Recent observations of elemental abundance scatter in the early Galaxy are consistent with earlier suggestions of a restricted range of SNe responsible for the *r*-process.

29.1 Introduction

The elements heavier than iron are synthesized in neutron processes, either in the (s)-low or (r)-apid process. In the *s*-process the timescale for neutron capture (τ_n) is much longer than the electron (beta)-decay (τ_β) timescale. For the *r*-process, however, $\tau_n << \tau_\beta$ with many neutrons captured in a very short time period. As a result, neutron captures proceed into very neutron-rich regions far from the stable nuclei, where very little experimental nuclear data is available. This element synthesis is intimately connected to the late stages of stellar evolution, with the *s*-process occurring in the thermally pulsing helium shells of asymptotic giant branch (AGB) stars of low- and intermediate-mass (M $\sim$ 0.8–8 $M_\odot$) (see, e.g., the review by Busso, Gallino, & Wasserburg 1999). Supernovae have long been suspected as the site for the *r*-process (see recent reviews by Truran *et al.* 2002;

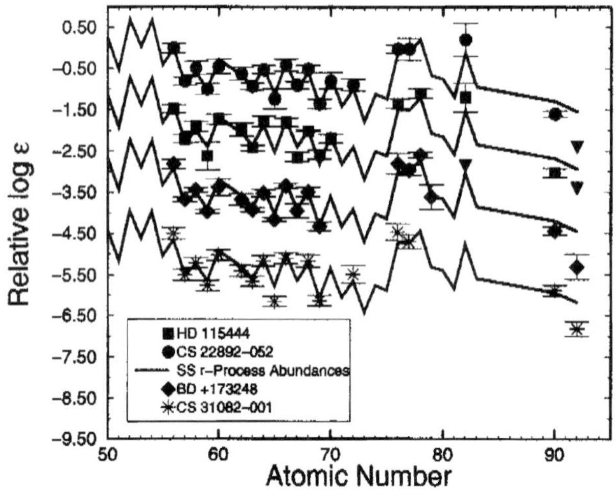

Fig. 29.1. A comparison of of the scaled solar system *r*-process abundances (solid line) with the *n*-capture element abundances in the the four Galactic halo, metalpoor stars CS 22892–052, HD 155444, BD +17°3248, and CS 31082–001 (see Westin *et al.* 2000; Cowan *et al.* 2002; Hill *et al.* 2002; Sneden *et al.* 2003). The abundances for CS 22892–052 are absolute, while the other abundances have been vertically shifted for display purposes. Upper limits are indicated by inverted triangles. ($\log \epsilon(A) \equiv \log_{10}(N_A/N_H) + 12.0$, for elements A and B.) (After Cowan & Sneden 2003.)

Sneden & Cowan 2003; Cowan & Sneden 2003). Recent abundance observations in metal-poor (i.e., low iron fractions) Galactic halo stars are providing new insights into these synthesis processes.

29.2 Abundance observations

A number of studies spanning decades have examined *n*-capture abundances in metal-poor ([Fe/H] < -1) Galactic halo stars (see *e.g.*, Spite & Spite 1978; Gilroy *et al.* 1988; McWilliam *et al.* 1995; Ryan, Norris, & Beers 1996; Sneden *et al.* 1996; Burris *et al.* 2000; Johnson & Bolte 2001; Hill *et al.* 2002). We show in Figure 29.1 the abundances of the heaviest *n*-capture elements (Ba and above, $Z \geq 56$) in four such stars: CS 22892–052 ([Fe/H] $= -3.1$, Sneden *et al.* 2003), HD 115444 ([Fe/H] $= -3.0$, Westin *et al.* 2000), BD +17°3248 ([Fe/H] $= -2.1$, Cowan *et al.* 2002) and CS 31082-001 ([Fe/H] $= -2.9$, Hill *et al.* 2002). The abundances for CS 22892-052 are absolute, while the others have been vertically displaced for display purposes. These stellar elemental abundances are compared with the solar system *r*-process abundances (solid lines in Figure 29.1) that have been scaled (*i.e.*, shifted downward to compensate for the difference in metallicities) to match the observed (*r*-process element) Eu. The *r*-process elemental abundances were determined from deconvolving the solar system abundances into their two components, and employed the "classical" model to obtain the individual *s*-process contributions (see Burris *et al.* 2000).

The comparisons indicate that there is an excellent overall agreement from Ba through the 3rd *r*-process peak between these metal-poor *n*-capture-rich stars and the scaled (or relative) solar system *r*-process abundances. These comparisons suggest further that early in the history of the Galaxy most (all?) of the *n*-capture elements observed in these stars were synthesized by the *r*-process, even those such as Ba that are formed in the *s*-process in solar system matter. These elements formed early in the Galaxy could not have been synthesized by low-mass stars, the primary sites for the *s*-process, with long evolutionary time scales. The detections of the *r*-process elements in these old halo stars, in fact, suggest that the *r*-process sites in the earliest stellar generations, the progenitors of the halo stars, were rapidly evolving. The agreement between the solar system abundances and the elemental abundances in the halo stars, for these *r*-process rich stars, also puts strong constraints on the conditions in, and suggests a robust nature of, the *r*-process.

29.3 The site or sites for the *r*-process

We can obtain additional clues about the nature of the *r*-process and the possible astrophysical sites by including the light *n*-capture elements in our abundance analyses. Only recently have there been detections of a significant number of these light elements in the metal-poor stars. We show in Figure 29.2 the most recent observational data for the star CS 22892-052, including detections and upper limits for 13 *n*-capture elements below Ba, from Sneden *et al.* (2003). The agreement between the solar system curve and the heavier elements ($Z > 56$) illustrated in Figure 29.1 does not seem to extend to the lighter elements. While Ru, Nb and perhaps Rh are consistent, other notable elements such as Ag, Pd and Mo fall noticeably below the same curve that matches the abundances of the heavier *n*-capture elements. We note that while this star has the most complete data set for these lighter elements, other metal-poor stars show a similar pattern.

The abundance comparisons shown in Figure 29.2, including the new lighter element data, could be explained in several manners. One possibility is that there are two sites for the *r*-process with SNe of different mass ranges and frequencies responsible for different (*i.e.,* the heavier and lighter element) ends of the abundance distribution (Qian & Wasserburg 2000). Some recent studies have examined the possibility of neutron-star (NS) binaries as sites for the *r*-process (Freiburghaus *et al.* 1999; Rosswog *et al.* 1999). If multiple sites are required to reproduce the heavy element abundance patterns observed in the metal-poor halo stars, then perhaps a combination of NS binaries and certain SNe might both contribute. Other studies

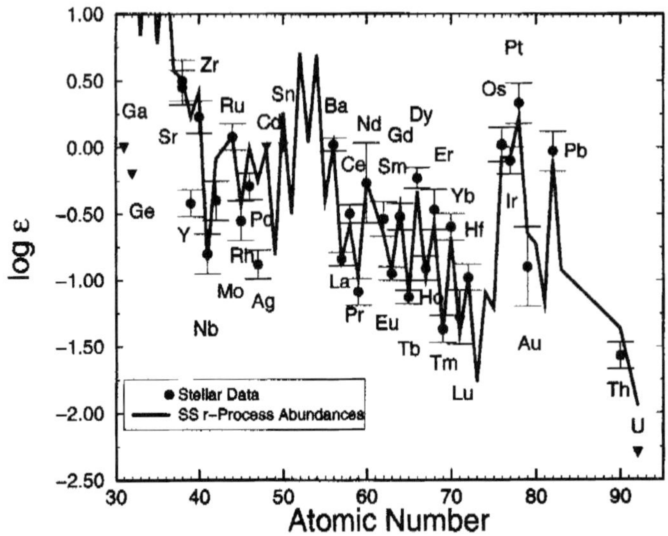

Fig. 29.2. Comparison of the observed *n*-capture abundances in CS 22892–052 from Sneden *et al.* (2003) and the solar system *r*-process abundance curve. Upper limits indicated by inverted triangles.

have suggested, in contrast, that only one site, a single core-collapse supernova, could synthesize the entire observed abundance distribution (Cameron 2001).

29.4 Abundance scatter in the early Galaxy

Observational studies have demonstrated the abundance scatter of *n*-capture elements in the early Galaxy (Gilroy *et al.* 1988; Burris *et al.* 2000). Specifically, it is observed that there is an increasing level of star-to-star scatter of [Eu/Fe] with decreasing metallicity (*i.e.*, below [Fe/H] ≈ -2.0), that is muted at higher metallicities. These results suggest an early, chemically unmixed and inhomogeneous Galaxy (Truran *et al.* 2002; Sneden & Cowan 2003). The observed abundance scatter also strongly suggests that the bulk of the *r*-process and the iron cannot be synthesized in the same stars (Fields, Truran, & Cowan 2002). Wheeler, Cowan, & Hillebrandt (1998) employed the (1) Galactic abundance scatter, (2) consistency of the abundance pattern for *r*-process elements in low-metallicity halo stars, (3) monotonic evolution of low-mass cores and (4) possibility of *n*-capture synthesis either in prompt or delayed explosions to suggest that the main *r*-process site was low-mass (8–10$M_\odot$), collapsing O/Ne/Mg cores. More recent studies seem to support that conclusion that low-mass SNe are promising *r*-process sites (Fields *et al.* 2002; Wanajo *et al.* 2003; Ishimaru *et al.* 2003). While it is not possible yet to exclude high-mass SNe, recent chemical evolution models, constrained by the observed abundance scatter, do seem to rule out NS binaries as the primary r-process site (Argast *et al.* 2003).

29.5 Summary and clues to the r-process

Abundance studies have identified a number of n-capture elements in the metal-poor Galactic halo stars. The presence of these elements in these old stars demonstrates that r-process nucleosynthesis occurred early in the Galaxy. It also suggests that the r-process sites, the earliest stellar generations and the progenitors of the halo stars, must have been rapidly evolving and points to massive stars and SNe. The heavy ($Z \geq 56$) n-capture element abundance pattern in the halo stars is consistent with the scaled solar system r-process distribution. This suggests a robust r-process operating over many Gyr and constrains the astrophysical conditions, and perhaps the supernova mass range, for the operation of the r-process. The abundance comparisons also demonstrate that the n-capture elements, including those such as Ba that are typically formed in the s-process, were predominantly synthesized by the r-process early in the history of the Galaxy. The lighter element n-capture abundances in the star CS 22892-052, in general, fall below the same solar system r-process abundance curve that matches the heavier ($Z \geq 56$) elemental abundances. This suggests either two separate r-process sites – different mass ranges and/or frequencies of supernovae or a combination of supernovae and NS-binaries – or two sets of conditions in the same supernova site. Star-to-star abundance scatter in the n-capture elements, such as Eu, is increasingly constraining the possible models for early Galactic nucleosynthesis suggesting that not all supernovae are responsible for synthesizing the r-process. Low-mass SNe appear to be a promising site for the r-process, although high-mass objects cannot be excluded at this time. NS binary models do not seem consistent with this early synthesis and suggest that they are not the primary r-process sites. Clearly, additional observational and theoretical studies will be required to help understand better the r-process and identify the astrophysical sites for this process.

Acknowledgments

This research has been supported in part by NSF grants AST-9986974 and AST-0307279 (JJC), AST-9987162 and AST-0307495 (CS), and by STScI grants GO-8111 and GO-08342.

References

Argast, D., Samland, M., Thielemann, F.-K., & Qian, Y.-Z., 2003. *Astron. Astrophys.*, in press.

Burris, D. L., Pilachowski, C. A., Armandroff, T. A., Sneden, C., Cowan, J. J., & Roe, H., 2000. *Astrophys. J.*, **544**, 302–319.
Busso, M., Gallino, R., & Wasserburg, G. J., 1999. *Ann. Rev. Astron. Astrophys.*, **37**, 239–309.
Cameron, A. G. W., 2001. *Astrophys. J.*, **562**, 456–469.
Cowan, J. J., *et al.*, 2002. *Astrophys. J.*, **572**, 861–879.
Cowan, J. J., & Sneden, C., 2003. To appear in *Carnegie Observatories Astrophysics Series, Vol. 4: Origin and Evolution of the Elements*, eds McWilliam, A. & Rauch, M. Cambridge, Cambridge Univ. Press.
Fields, B. D., Truran, J. W. & Cowan, J. J., 2002. *Astrophys. J.*, **575**, 845–854.
Freiburghaus, C., Rosswog, S. & Thielemann, F.-K., 1999. *Astrophys. J.*, **525**, L121–L124.
Gilroy, K. K., Sneden, C., Pilachowski, C. A. & Cowan, J. J., 1988. *Astrophys. J.*, **327**, 298–320.
Hill, V., *et al.*, 2002. *Astron. Astrophys.*, **387**, 560–579.
Ishimaru, Y., Wanjo, S., Aoki, W. & Ryan, S. G., 2003, *Astrophys. J.*, in press.
Johnson, J. A., & Bolte, M., 2001. *Astrophys. J.*, **554**, 888–902.
McWilliam, A., Preston, G. W., Sneden, C. & Searle, L., 1995. *Astron. J.*, **109**, 2757–2799.
Qian, Y.-Z., & Wasserburg, G. J. 2000, *Phys. Rep.*, **333–334**, 77–108.
Rosswog, S., Liebendorfer, M., Thielemann, F.-K., Davies, M. B., Benz,. W., & Piran, T., 1999. *Astron. Astrophys.*, **341**, 499–526.
Ryan, S. G., Norris, J. E., & Beers, T. C., 1996. *Astrophys. J.*, **471**, 254–278.
Sneden, C., *et al.*, 2003. *Astrophys. J.*, **591**, 936–953.
Sneden, C. & Cowan, J. J., 2003. *Science*, **299**, 70–75.
Sneden, C., McWilliam, A., Preston, G. W., Cowan, J. J., Burris, D. L. & Armosky, B. J., 1996. *Astrophys. J.*, **467**, 819–840.
Spite, M., & Spite, F., 1978. *Astron. Astrophys.*, **67**, 23–31.
Truran, J. W., Cowan, J. J., Pilachowski, C. A. & Sneden, C., 2002. *Pub. Astron. Soc. Pac.*, **114**, 1293–1308.
Wanajo, S., *et al.*, 2003. *Astrophys. J.*, **593**, 968–979.
Westin, J., Sneden, C., Gustaffson, B. & Cowan, J. J., 2000. *Astrophys. J.*, **530**, 783–799.
Wheeler, J. C., Cowan, J. J. & Hillebrandt, W., 1998. *Astrophys. J.*, **493**, L101–L104.

Part V

Magnetars, N-Stars, Pulsars

30

Supernova remnant and pulsar wind nebula interactions

R. A. Chevalier

Department of Astronomy, University of Virginia, P. O. Box 3818, Charlottesville, VA, USA

Abstract

I review several topics in the structure of supernova remnants. Hydro-dynamic instabilities in young remnants may give rise to the cellular structure that is sometimes observed, although structure in the ejecta might also play a role. The presence of ejecta close to the forward shock front of a young remnant can be the result of ejecta clumps or the dynamical effects of cosmic rays. Slower moving ejecta clumps can affect the outer shock structure of older remnants such as Vela. Young remnants typically show a circular structure, but often have a one-sided asymmetry; the likely reasons are an asymmetric circumstellar medium, or pulsar velocities in the case of pulsar wind nebulae. In older remnants, asymmetric pulsar wind nebulae can result from asymmetric reverse shock flows and/or pulsar velocities.

30.1 Introduction

Observations of supernova remnants frequently show complex structure that can have its origin in several ways: structure in the freely expanding ejecta, structure in the surrounding medium, and the growth of instabilities that result from the interaction of the supernova with its surroundings. If we are to infer properties of the initial explosion from the supernova remnant, consideration of these various influences is necessary. Pulsar wind nebulae (PWNe) provide an additional probe inside a supernova remnant and can lead to an asymmetry because of a pulsar velocity. Here, I review studies of these phenomena.

30.2 Instabilities in young remnants

The basic instability that results from the deceleration of the supernova ejecta by the surrounding medium is related to the Rayleigh-Taylor instability. The growth rates for the instability are such that it becomes saturated early in the evolution of the supernova remnant (Chevalier, Blondin, & Emmering 1992). The instability causes the growth and decay of Rayleigh-Taylor fingers that only persist a fraction of the region between the forward and reverse shock waves. Recent computations of the instability in 3-dimensions show that a cellular structure develops in which the cells continually grow and eventually split up (Blondin *et al.*, in preparation). An estimate of the appearance of the unstable structure by projecting the emission measure of the gas resembles the cellular structure observed in Cas A at radio (e.g., Fig. 22 of Rudnick 2002) and X-ray (Hughes *et al.* 2000) wavelengths. The model assumes that Cas A is interacting with a dense, freely expanding wind, which appears to be consistent with data on the shock positions and expansion rates in the remnant (Chevalier & Oishi 2003).

Tycho's remnant is believed to be the remnant of a Type Ia supernova and so is probably interacting with the constant density interstellar medium. A comparison of hydrodynamic models with Tycho's remnant shows that the observed density structure extends farther toward the forward shock than expected in the models. One way to have the structure extend farther toward the forward shock is to postulate clumps within the supernova ejecta (Wang & Chevalier 2001); there is direct evidence for clumps with specific chemical compositions in Tycho. Another way is to have cosmic ray pressure be an important component of the intershock pressure. A cosmic ray dominated shock has an adiabatic index approaching $4/3$, yielding a high shock compression; loss of energetic particles from the shock front can also lead to a larger compression ratio. Numerical studies of the instability in the case where cosmic ray pressure is important show a small distance between the forward and reverse shocks and the instability can grow to the vicinity of the forward shock (Blondin & Ellison 2001).

30.3 Clumpy/bubbly ejecta

Clumping of the ejecta can give rise to features that propagate through the shocked region to create protrusions on the forward shock wave (Kane, Drake, & Remington 1999). A remnant that clearly shows such knots is Cas A, which has fast moving knots that extend considerably beyond the forward shock front (Fesen 2001). The knots are observed by their optical emission which implies that the gas has radiatively cooled. Radiative cooling at the reverse shock gives rise to denser ejecta, but does not especially enhance the ability of the instable region to approach the forward shock front (Chevalier & Blondin 1995). The knots must thus be features created in the ejecta.

The origin of the knots is not known. One possibility is structure created by the expansion of ^{56}Ni that is synthesized in the explosion (the Ni bubble effect). One of the best pieces of evidence for this action is the finding in SN 1987A that the Fe occupies a large part of the inner volume, but only a small part of the mass (Li, McCray, & Sunyaev 1993). However, knots are observed at high velocities in Cas A (Fesen 2001), in a region where the Ni bubble effect is not expected to operate. It may occur at lower velocities, and Blondin, Borkowski, & Reynolds (2001) have modeled the hydrodynamics of bubble regions in ejecta interacting with a surrounding medium. The reverse shock front tends to move back rapidly through the low density bubbles. The result is added complexity to the structure that is already affected by instabilities.

The effects of ejecta clumps might also be present in older remnants if one considers lower velocity clumps ($\lesssim$3000 km s^{-1}). The Vela remnant, with an age of $\sim$10^4 years, shows evidence for protrusions at X-ray and radio wavelengths that may be ejecta clumps that are moving through the outer shock front (Aschenbach, Egger, & Trumper 1995; Strom *et al.* 1995). The evidence that these may be ejecta clumps include the pressure gradient associated with Knot A (Miyata *et al.* 2001) and the high pressure present in Knot D (Sankrit, Blair, & Raymond 2003). The clumps that are needed to bring about the protrusions involve a large density contrast and a relatively large mass, $\gtrsim 0.01 M_\odot$, and their origin is not clear (Wang & Chevalier 2002). The action of the Ni bubble effect is probably not sufficient to give the required compression. Another possibility that is applicable to Vela is the sweeping action in the supernova interior by a pulsar nebula, as can be seen in the Crab Nebula. The photoionized filaments in the Crab have a density that is $\sim$250 times the volume averaged density, and neutral gas, if present in such a situation, would be even denser. While the gas that is not affected by the pulsar nebula is decelerated by the surrounding medium, the pulsar nebula clumps can move out in the remnant. Although these clumps are dense, the question of whether they have the properties needed for the Vela protrusion remains.

Another aspect of the protrusions formed by clumps moving through the outer shock front is that ring features may be formed in the outer shock. Wang & Chevalier (2002) suggested that the X-ray feature RX J0852.0–4622, which has been interpreted as a separate supernova remnant projected on the Vela remnant (Aschenbach 1998), is actually a feature in the shell of the Vela remnant. This view would be negated by clear evidence that a central compact X-ray source in RX J0852.0 (Pavlov *et al.* 2001) is associated with the nebula. If the Vela interpretation is correct, other rings associated with protrusions might be expected. In fact, Carlin & Smith (2002) have found an optical emission line ring that appears to be associated with knot C. In this case, the shock front is radiatively cooling, so the ring appears in optical emission as opposed to X-rays.

The marked presence of protrusions in the outer shock of a remnant may occur at a particular phase of a supernova remnant when clumps at the inflection point in the supernova density profile are able to reach the forward shock front. In addition to the Vela remnant, protrusions at the outer shock boundary have been found in the remnant N63A in the Large Magellanic Cloud (Warren, Hughes, & Slane 2003).

30.4 Asymmetric young remnants

Among the young remnants, a surprising number show evidence for a one-sided asymmetry in velocity and/or position. In Cas A, the sphere occupied by the fast knots is redshifted by 770 km s^{-1} from the presumed expansion center at 0 velocity (Reed *et al.* 1995); the sphere is also displaced from the expansion center in the plane of the sky (Reed *et al.* 1995; Thorstensen *et al.* 2001). In Kepler's remnant, the velocities of slow knots show mean redshifted velocities and transverse velocities combining to a speed of 278 km s^{-1} (Bandiera & van den Bergh 1991) and the X-ray emission is stronger on the N side. In 0540–69 in the Large Magellanic Cloud, the velocities of optically emitting gas around the pulsar nebula are redshifted by +370 km s^{-1} from the local LMC rest velocity (Kirshner *et al.* 1989), and the pulsar nebula in G292.0 + 1.8 is displaced relative to the center of the surrounding remnant shell in the plane of the sky, corresponding to a transverse velocity ~770 km s^{-1} if it originated in the center of the shell (Hughes *et al.* 2001).

There are probably a variety of reasons for these asymmetries. In Cas A, Reed *et al.* (1995) attribute the asymmetry to a higher density in the surrounding medium on the near side. This view is supported by the facts that the quasi-stationary flocculi (shocked cloudlets) are predominantly blueshifted and the pressure appears to be higher on the near side. The hydrodynamic model of wind interaction (Chevalier & Oishi 2003) gives a specific scenario in which this hypothesis can be quantitatively tested. The model would require a large-scale asymmetry in the presupernova red supergiant wind. There are few detailed observations of such winds, but at least in the case of VY Canis Majoris, the wind shows a complex, asymmetric structure (Monnier *et al.* 1999); however, these observations refer to a scale of 0.04 pc, considerably smaller than the 2.5 pc radius of Cas A.

Although an explosion asymmetry is not indicated for Cas A, it is still interesting to check whether the offset of the recently discovered compact object in Cas A might be related to the one-sided asymmetry of the supernova remnant. The compact object is located to the S of the center of expansion (Thorstensen *et al.* 2001), but the center of the remnant ring is to the W of the center of expansion. Thus there is no direct evidence that the remnant asymmetry is related to an impulse given to the compact object.

In the case of Kepler, the asymmetric motions of the slow knots, which are presumably mass loss from the progenitor star, have been interpreted in terms of a space velocity of the progenitor (Bandiera 1987; Bandiera & van den Bergh 1991). Attractive features of this scenario are that it can explain the large distance of Kepler from the Galactic plane (400 pc for a supernova distance of 1.5 kpc) and the strong X-ray emission to the N can be explained as interaction with a bow shock structure (Bandiera 1987; Borkowski, Blondin, & Sarazin 1992). However, the proper motion of the X-ray emission shows a faster rate of expansion than would be expected in this picture (Hughes 1999). Many features of Kepler's remnant remain to be explained, but the basic picture of interaction with an asymmetric surrounding medium appears to apply.

The velocities inferred for the compact pulsar wind nebulae (PWNe) in 0540–69 and G292.0+1.8 can be attributed to the velocities given to neutron stars at birth. Initially, the PWN expands into the uniformly expanding supernova ejecta, similar to the Hubble flow. The inner regions of a supernova may not have a strong density gradient, so the surroundings of the pulsar are similar to what would be obtained if the pulsar were not moving.

Although there has been considerable interest in the possibility that core collapse supernovae have a bipolar structure, there has been little evidence for such structure in young supernova remnants. In Cas A, Fesen (2001) found evidence for possible bipolar structure in the fastest moving knots, along the NE–SW axis aligned with Minkowski's "jet" feature. Sulfur-rich knots preferentially appear along this axis. ¿From X-ray spectroscopy, Willingale *et al.* (2003) also found evidence for bipolar structure, but along a different axis, the N–S axis. Overall, the appearance of the bright X-ray and radio emission from Cas A is quite circular. One reason for this may be that the interaction region of a supernova with a circumstellar wind tends to show less asymmetry than the asymmetry present in the supernova ejecta. The pressure is higher in the lagging part of the interaction, which tends to reduce the asymmetry in the interaction region (Blondin, Lundqvist, & Chevalier 1996).

A recent study of the stability of standing accretion shocks showed evidence for a strong $l = 1$ instability which could influence core collapse supernovae (Blondin, Mezzacappa, & DeMarino 2003). Although a number of young remnants show evidence for an $l = 1$ asymmetry, there is no clear evidence for an asymmetry in the ejecta. More detailed modeling of the sources is needed to disentangle ejecta asymmetries from other asymmetry mechanisms.

30.5 Older pulsar wind nebulae

The interaction of pulsar nebulae with more evolved supernova remnants, in which the reverse shock front has moved back toward the center of the remnant, has been the subject of recent investigations. The PWN is crushed by the reverse shock front and eventually re-expands (Reynolds & Chevalier 1984; van der Swaluw *et al.* 2001). This process is strongly unstable if the PWN bubble can be assumed to behave as a hydrodynamic fluid (Blondin, Chevalier, & Frierson 2001). In the actual case, the magnetic field in the PWN may play a role, although the instability is still expected to operate across magnetic field lines. The instability might lead to the development of magnetic filaments, but further exploration of this topic is needed.

Once the PWN has started to interact with the reverse shocked material, there are two mechanisms that can lead to asymmetries in the supernova remnant. One involves interaction with an asymmetric surrounding medium, so that the existing PWN is pushed away from its central position in the supernova remnant (Blondin *et al.* 2001); the other is that the high space velocity of the pulsar carries it to one side of the remnant. The radio emitting electrons are typically long-lived compared to the age of the nebula, so the radio structure shows the positions of particles injected early in the development of the remnant. X-ray emitted electrons are typically shorter-lived because of synchrotron losses, so the X-ray emission is close to the present position of the pulsar.

There are a couple of ways that these mechanisms can be distinguished. The clearest is when the proper motion of the pulsar has been observed so it can be directly checked whether its velocity has carried it from the center of the remnant. This is possible for the Vela remnant whose pulsar has an optical counterpart. The proper motion vector does not take it back to the center of the Vela X radio PWN (Bock *et al.* 1998), so that an asymmetric interstellar interaction appears more likely for this case. Another argument against the pulsar velocity being the sole mechanism is if the trail left by the radio emission does not point back to the center of the remnant. This is the case for the remnant MSH 15–56 (Plucinsky 1998). Of course, both pulsar velocity and asymmetric interaction can be important in one remnant.

In summary, the structure of supernova remnants involves a complex set of phenomena which are now being elucidated by improved observational capability, such as that provided by the *Chandra* X-ray observatory. Detailed models are needed to sort out the effects of explosion asymmetries, asymmetries in the surrounding medium, and instabilities.

Acknowledgements

I am pleased to present this paper in honor of the 60th birthday of Craig Wheeler, who has been an inspiration to the supernova community for several decades. This work was supported in part by NASA grant NAGW5-13272 and by NSF grant AST-0307366.

References

Aschenbach, B. 1998. *Nature*, **396**, 141–142.
Aschenbach, B., Egger, R., & Trumper, J. 1995. *Nature*, **373**, 587–589.

Bandiera, R. 1987. *Astrophys. J.*, **319**, 885–892.
Bandiera, R., & van den Bergh, S. 1991. *Astrophys. J.*, **374**, 186–201.
Blondin, J. M., & Ellison, D. C. 2001. *Astrophys. J.*, **560**, 244–253.
Blondin, J. M., Borkowski, K. J., & Reynolds, S. P. 2001. *Astrophys. J.*, **557**, 782–791.
Blondin, J. M., Chevalier, R. A., & Frierson, D. M. 2001. *Astrophys. J.*, **563**, 806–815.
Blondin, J. M., Lundqvist, P., & Chevalier, R. A. 1996. *Astrophys. J.*, **472**, 257–266.
Blondin, J. M., Mezzacappa, A., & DeMarino, C. 2003. *Astrophys. J.*, **584**, 971–980.
Bock, D. C.-J., Turtle, A. J., & Green, A. J. 1998. *Astron. J.*, **116**, 1886–1896.
Borkowski, K. J., Blondin, J. M., & Sarazin, C. L. 1992. *Astrophys. J.*, **400**, 222–237.
Carlin, J. L. & Smith, R. C. 2002. *Bull. Amer. Astron. Soc.*, **34**, 1248–1248.
Chevalier, R. A., & Oishi, J., 2003. *Astrophys. J.*, **593**, L23–L26.
Chevalier, R. A., & Blondin, J. M., 1995. *Astrophys. J.*, **444**, 312–317.
Chevalier, R. A., Blondin, J. M., & Emmering, R. T., 1992. *Astrophys. J.*, **392**, 118–130.
Fesen, R. A. 2001. *Astrophys. J. Supp.*, **133**, 161–186.
Hughes, J. P. 1999. *Astrophys. J.*, **527**, 298–309.
Hughes, J. P., Rakowski, C. E., Burrows, D. N., & Slane, P. O. 2000. *Astrophys. J.*, **528**, L109–L113.
Hughes, J. P., Slane, P. O., Burrows, D. N., Garmire, G., Nousek, J. A., Olbert, C. M., & Keohane, J. W. 2001. *Astrophys. J.*, **559**, L153–L156.
Kane, J., Drake, R. P., & Remington, B. A. 1999. *Astrophys. J.*, **511**, 335–340.
Kirshner, R. P., Morse, J. A., Winkler, P. F., & Blair, W. P. 1989. *Astrophys. J.*, **342**, 260–271.
Li, H., McCray, R., & Sunyaev., R. A. 1993. *Astrophys. J.*, **419**, 824–836.
Miyata, E., Tsunemi, H., Aschenbach, B., & Mori, K. 2001. *Astrophys. J.*, **559**, L45–L48.
Monnier, J. D., Tuthill, P. G., Lopez, B., Cruzalebes, P., Danchi, W. C., & Haniff, C. A. 1999. *Astrophys. J.*, **512**, 351–361.
Pavlov, G. G., Sanwal, D., Kızıltan, B., & Garmire, G. P. 2001. *Astrophys. J.*, **559**, L131–L134.
Plucinsky, P. P. 1998. *Mem. Soc. Astron. Italiana*, **69**, 939–944.
Reed, J. E., Hester, J. J., Fabian, A. C., & Winkler, P. F. 1995. *Astrophys. J.*, **440**, 706–721.
Reynolds, S. P., & Chevalier, R. A. 1984. *Astrophys. J.*, **278**, 630–648.
Rudnick, L. 2002. *Pub. Astron. Soc. Pac.*, **114**, 427–449.
Sankrit, R., Blair, W. P., & Raymond, J. C. 2003. *Astrophys. J.*, **589**, 242–252.
Strom, R., Johnston, H. M., Verbunt, F., & Aschenbach, B. 1995. *Nature*, **373**, 590–591.
Thorstensen, J. R., Fesen, R. A., & van den Bergh, S. 2001. *Astron. J.*, **122**, 297–307.
van der Swaluw, E., Achterberg, A., Gallant, Y. A., & Tóth, G. 2001. *Astron. Astroph.*, **380**, 309–317.
Wang, C. & Chevalier, R. A. 2001. *Astrophys. J.*, **549**, 1119–1134.
Wang, C. & Chevalier, R. A. 2002. *Astrophys. J.*, **574**, 155–165.
Warren, J. S., Hughes, J. P., & Slane, P. O. 2003. *Astrophys. J.*, **583**, 260–266.
Willingale, R., Bleeker, J. A. M., van der Heyden, K. J., & Kaastra, J. S. 2003. *Astron. Astrophys.*, **398**, 1021–1028.

31

X-ray signatures of supernovae

D. A. Swartz

National Space Science and Technology Center
Marshall Space Flight Center, Huntsville, AL

Abstract

Combining sub-arcsec imaging with moderate spectral resolution and high throughput, the *Chandra* X-ray Observatory enables spectacular views of Galactic supernova remnants as well as X-ray studies of compact remnants, young extragalactic supernovae, and gamma-ray burst afterglows. In this contribution, I briefly review the capabilities of *Chandra* and then describe some recent observations of supernovae and supernova remnants made with *Chandra*.

31.1 The *Chandra* X-ray observatory – an overview

Chandra (see, e.g., Weisskopf *et al.* 2002) was launched from space shuttle Columbia 23 July 1999 and is now late into the fourth year of its ten year mission. The heart of the facility is the High-Resolution Mirror Assembly consisting of 4 nested mirror pairs with a 120 cm outer shell diameter. The mirrors provide about 800 cm^2 of collecting area at 1 keV and about 400 cm^2 at 5 keV. Most importantly, the mirror design results in less than $0.''5$ on-axis spatial resolution; an order-of-magnitude higher resolution than any other X-ray facility yet flown. This corresponds to a resolution of $\sim$100 AU at the distance of the Crab Nebula; making *Chandra* ideal for probing the fine structure of supernova remnants on spatial scales comparable to that achievable by some of the best ground-based optical telescopes. Equally important, high spatial resolution improves the sensitivity of X-ray measurements by concentrating source photons into a small area thereby minimizing the contribution from the underlying background. There are two types of focal plane instrument onboard *Chandra*. The one most utilized for supernova work is the Advanced CCD Imaging Spectrometer with a spectral resolution of about 150 eV at 1 keV. *Chandra* also has two retractable transmission gratings for high-resolution spectroscopy of bright, point-like, sources.

31.2 Spatially resolved spectroscopy of SNR

Perhaps the most recognizable *Chandra* image is that of the supernova remnant (SNR) Cassiopeia A. Cas A was the official first-light target for *Chandra*. Based on that single short exposure, the long-sought Cas A neutron star was at last discovered (Tannanbaum 1999). Since then, several other observations of Cas A have been made in several observing configurations.

Hughes *et al.* (2000) analyzed the X-ray spectra of small regions of Cas A at various locations within the remnant. These spectra show large differences in elemental abundance providing evidence for both explosive nucleosynthesis and mixing in the ejecta. Hughes's study is the first to find direct evidence of Fe-rich ejecta from explosive Si-burning in a SNR and shows that some of the Fe-rich ejecta lies outside the Si-rich material; an inversion resulting from rising bubbles in the neutrino-driven convection during the explosion.

Another early study of Cas A, by Una Hwang and collaborators (Hwang *et al.* 2000), mapped the spatial distribution of the elements by selecting narrow energy ranges around some of the prominent emission line features and then comparing the resulting narrow-band images. This study also shows clearly regions of intermediate-mass elements lying inside regions of Fe Kα emission.

Gotthelf *et al.* (2001) was the first to map the shock structure in Cas A. Their analysis of the Si XIII, 4–6 keV continuum, radio, and radio polarization across the northern portion of the remnant shows both forward and reverse shock structure. By taking radial profiles of emission in these bands, they find $\sim 3''$ wide tangential wisps that mark the location of the forward shock at a radius of $\sim 150''$. The radio data shows high surface brightness interior to this shock and a flat profile at larger radii while the synchrotron polarization angle shifts dramatically at the shock front. When the emissivity profile is deprojected; that is, account is made for the fact that the remnant is actually 3-dimensional; then the position of the reverse shock is also clearly seen as a steep rise in the X-ray emissivity at about $100''$ from the center of the remnant. There has also been sufficient monitoring of Cas A with *Chandra* to directly measure the shock expansion rate (DeLaney & Rudnick 2003).

Another study of the X-ray radial distribution, this time of the SNR 1E 0102.2-7219 using *Chandra*'s High-Energy Transmission Grating, resolves the individual X-ray spectral lines near the shock front. Gaetz *et al.* (2000) find the O VII emission peak is interior to the peak of O VIII emission; and conclude that the ionization state lags behind the observed electron temperature.

There are also measurements of non-equipartition between electron and ion temperatures in SNRs based on X-ray observations. As the best example (Burrows *et al.* 2000; Michael *et al.* 2002), models of a grating spectrum of SN 1987A gives an electron temperature $kT_e \sim 2.6$ keV which is far less than that inferred from the shock dynamics as measured from the widths of the spectral lines ($kT_s \sim 17$ keV). This shows that the electrons and ions have not yet thermalized in SN 1987A though partial to complete equilibration is deduced for older remnants, e.g., SNR DEM L71 in the LMC (Rakowski, Ghavamian, & Hughes 2003).

In summary, X-ray observations of SNRs are powerful tools for studying nucleosynthesis both in the hydrostatic pre-supernova evolution of the progenitor star and in the explosive stages that followed. Modern observations also reveal the complex dynamics occurring within SNRs. It is natural to take the next step and probe the very heart of SNRs in X-rays: the compact remnant and pulsar wind nebulae.

31.3 Compact objects and pulsar wind nebulae

High-resolution X-ray imaging is helping to locate neutron stars in SNRs, to probe neutron star cooling, and to map pulsar wind nebulae. Rivaling the Cas A image in public recognition, the complex environment of the Crab pulsar and nebula as revealed by *Chandra* (Figure 31.1) is the prototype of these studies. The basic pulsar-driven structure, consisting of an energetic synchrotron-emitting equatorial wind with particle jets along the pulsar spin axis, has now been observed in several nebulae. For example, the Crab-like nebula of G21.5–0.9 (Slane *et al.* 2000) has a power law spectrum that steepens with distance from the compact object as expected if the energetic particles loss energy through synchrotron emission as they move outward. Another intriguing object, 3C 58, has a complex morphology created by precessing jets and other outflows (Slane, *et al.* 2003). Several other objects show evidence of these structures prompting Gotthelf (2001) to suggest such X-ray signatures are perhaps common to all young rotation-powered pulsars.

Nevertheless, the old paradigm that core collapse SNe give rise to highly-magnetized, rapidly-rotating radio pulsars – such as the Crab – is not always the case. Many SNRs apparently lack the tell-tale signs. They are radio-quiet, non-plerionic, sometimes even lacking an apparent stellar remnant. As already mentioned above, the compact object in Cas A has just recently been discovered. Other remnants are also revealing tantalizing evidence for a compact object. But these are not pulsating. They show steep power law or blackbody spectra unlike the flat spectrum expected from acceleration in the magnetosphere of a classical young pulsar. In addition, the emission is not simply that from a cooling neutron star. The emission may be due instead to fallback accretion in a low magnetic field (if the field were stronger, then we would expect funneling onto a spot and pulsations). Temperatures of some of these compact objects, such as 3C 58, are well below what standard models for cooling neutron stars predict indicating that possibly some more rapid cooling process is present (Slane, *et al.* 2003).

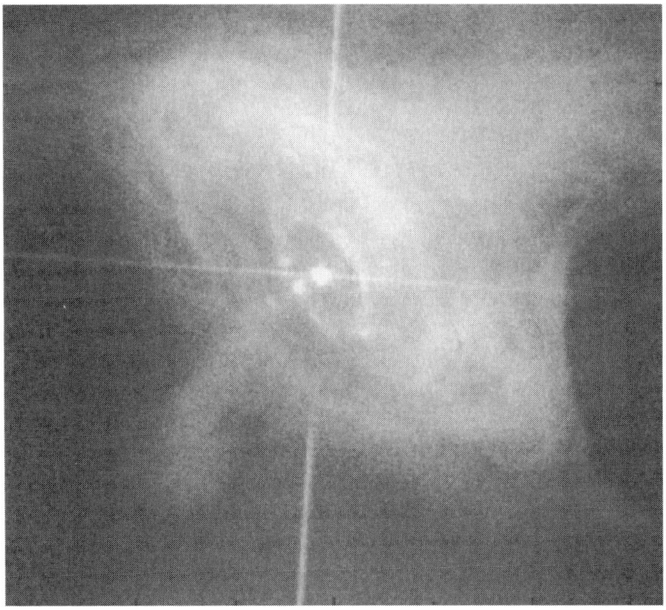

Fig. 31.1. *Chandra* X-ray Observatory image of the Crab pulsar and Nebula taken with the High-Energy Transmission Grating readout by the Advanced CCD Imaging Spectrometer. The dispersed spectrum of the pulsar produces the line of emission top-to-bottom in this figure while pileup of the zeroth-order pulsar image creates the streak or read-out trail across the image. Note the striking richness of X-ray structure including the jet extending towards the lower left and the circular patterns of equatorial emission. The image is approximately $150'' \times 150''$. Adapted from Weisskopf *et al.* 2000.

31.4 Young supernovae

Supernova ejecta, interacting with circumstellar debris, can be X-ray-bright for days and up to years after the explosion. This X-radiation is accompanied by radio emission and together can be used to probe the dynamics of the interaction region, examine the structure of the outer SN ejecta and the circumstellar material, and to study the late stages of stellar evolution.

Several young SNe have been observed with *Chandra* and other X-ray observatories. Immler & Lewin (2002) provide a thorough review. An example is our study of SN 1993J (Swartz *et al.* 2003). The *Chandra* X-ray spectrum of SN 1993J confirms the validity of the standard circumsteller interaction model (Chevalier 1982). We find multiple spectral components consistent with forward and reverse shock emission, and metal abundances that established SN 1993J is O-poor. The X-ray light curve (combining *Chandra* with earlier *ROSAT* and *ASCA* observations) is now declining more rapidly than in the past, but less rapidly than detailed numerical simulations predicted (Suzuki & Nomoto 1995).

At the time of the Workshop, I reported that the prospects for scientific return from X-ray observations of young supernovae was not promising. I based that statement on the paucity of accepted proposals for such observations in the past. I am happy to retract that statement now in light of the results of the most recent *Chandra* research announcement. There are, by my unofficial census, 4 approved SN targets-of-opportunity, deep exposures scheduled for SN 1986J, SN 1987A, SN 1988Z, and SN 1995N, and a very deep grating observation planned of SNR 1987A upcoming in the next year.

31.5 Gamma-ray burst X-ray afterglows

The first X-ray afterglow observed with *Chandra* is GRB 991216. Piro *et al.* (2000) report the detection of emission lines associated with this event. This result supports previous *Chandra* and *BeppoSAX* observations of gamma-ray bursts suggesting Fe line (and recombination continua) reprocessed from a hard power law spectrum. Piro and this team found evidence indicating that the burst had encountered an extremely dense gas. The properties of this gas suggest that it originated from a massive progenitor that exploded prior to the gamma-ray burst. The most impressive observation to date may be the *Chandra* grating observation of GRB 020813 (Butler *et al.* 2003) which shows evidence of intermediate-mass elements (Si and S) with line widths $\sim$0.1 c.

An important aspect of these observations is the study of the environments around GRBs. In combination with other missions, most notably HETE, future X-ray observations of GRBs should be very interesting.

31.6 Summary

Supernova remnants are providing some of the most spectacular *Chandra* X-ray images. Besides the pretty pictures, analyses of X-ray data are probing the complex physical processes at work on scales from remnant-wide abundance, ionization, and temperature gradients; to pressure-confined synchrotron-emitting jets and winds; and to the very surfaces of neutron stars (though the latter are not, of course, resolved). While X-ray observations of low-surface-brightness SNRs are confined primarily to Galactic and Magellanic Cloud sources, young supernovae out to the Virgo cluster and powerful GRBs at cosmological distances are now routinely targeted for X-ray spectral study. I have mentioned only a few of the highlights over the past few years but it is certain that the future of supernovae in X-ray light is bright indeed.

References

Burrows, D. N. *et al.* 2000, *ApJ*, **243**, L149.
Butler, N. R. *et al.* 2003, submitted to *ApJ*, astro-ph/0303539.
Chevalier, R. A. 1982, *ApJ*, **258**, 790.
DeLaney, T., Rudnick, L. 2003, *ApJ*, **589**, 818.
Gotthelf, E. V. 2001, in *20th Texas Symposium on Relativistic Astrophysics*, AIP conference proceedings, Vol. 586, ed. J. Craig Wheeler and Hugo Martel. p.513.
Gotthelf, E. V., Koralesky, B., Rudnick, L., Jones, T. W., Hwang, U., & Petre, R. 2001, *ApJ*, **552**, L39.
Gaetz, T. J., *et al.* 2000, *ApJ*, **534**, L47.
Hughes, J. P., Rakowski, C. E., Burrows, D. N., & Slane, P. O. 2000, *ApJ*, **528**, L107.
Hwang, U., Holt, S. S., & Petre, R. 2000, *ApJ*, **537**, L119.
Immler, S. & Lewin, W. H. G. 2002, in *Supernovae and Gamma-Ray Bursts*, ed. K. Weiler, astro-ph/0202231.
Michael, E. *et al.* 2002, *ApJ*, **574**, 166.
Piro, L. *et al.* 2000, *Science,* **5493**, 955.
Rakowski, C. E., Ghavamian, P., & Hughes, J. P. 2003, *ApJ*, **590**, 846.
Slane, P. O., *et al.* 2000, *ApJ*, **533**, L29.
Slane, P. O., Helfand, D. J., Gotthelf, E. V., Murray, S. S. 2003, in *Young Neutron Stars and their Environment*, IAU Symposium No. 218, p. 107.
Suzuki, T. & Nomoto, K. 1995, *ApJ*, **455**, 658.
Swartz, D. A. *et al.* 2003, *ApJS*, **144**, 213.
Tannanbaum, H. 1999, *IAUC*, 7246.
Weisskopf, M. C. *et al.* 2000, *ApJ*, **536**, L81.
Weisskopf, M. C., Brinkman, G., Canizares, C., Garmire, G., Murray, S., & Van Speybroeck, L. P. 2002, *PASP*, **114**, 1.

32

Neutron star kicks and supernova asymmetry

D. Lai

Center for Radiophysics and Space Research, Department of Astronomy Cornell University, Ithaca, NY 14853

Abstract

Observations over the last decade have shown that neutron stars receive a large kick velocity (of order a few hundred to a thousand km s^{-1}) at birth. The physical origin of the kicks and the related supernova asymmetry is one of the central unsolved mysteries of supernova research. We review the physics of different kick mechanisms, including hydrodynamically driven, neutrino – magnetic field driven, and electromagnetically driven kicks. The viabilities of the different kick mechanisms are directly related to the other key parameters characterizing nascent neutron stars, such as the initial magnetic field and the initial spin. Recent observational constraints on kick mechanisms are also discussed.

32.1 Evidence for neutron star kicks and supernova asymmetry

It has long been recognized that neutron stars (NSs) have space velocities much greater than their progenitors'. A natural explanation for such high velocities is that supernova (SN) explosions are asymmetric, and provide kicks to the nascent NSs. Evidence for NS kicks and NS asymmetry has recently become much stronger. The observations that support (or even require) NS kicks fall into three categories:

(1) *Large NS Velocities ($\gg$ the progenitors' velocities $\sim$30 km s^{-1}):*

- The study of pulsar proper motion give a mean birth velocity 200–500 km s^{-1} (Lorimer *et al.* 1997; Hansen & Phinney 1997; Cordes & Chernoff 1998; Arzoumanian *et al.* 2002), with possibly a significant population having $V \gtrsim 1000$ km s^{-1}. While velocity of $\sim$100 km s^{-1} may in principle come from binary breakup in a supernova (without kick), higher velocities would require exceedingly tight presupernova binary. Statistical analysis seems to favor a bimodal pulsar velocity distribution, with peaks around 100 km s^{-1} and 500 km s^{-1} (see Arzoumanian *et al.* 2002).

- Observations of bow shock from the Guitar nebula pulsar (B2224 + 65) implies $V \gtrsim 1000$ km s^{-1} (Cordes *et al.* 1993; Chatterjee & Cordes 2002).
- The studies of NS–SNR associations have, in some cases, implied large NS velocities, up to $\sim 10^3$ km s^{-1} (e.g., NS in Cas A SNR has $V > 330$ km s^{-1}; Thorstensen *et al.* 2001).

(2) *Characteristics of NS Binaries:* While large space velocities can in principle be accounted for by binary break-up (see Iben & Tutukov 1996), many observed characteristics of NS binaries demonstrate that binary break-up can not be solely responsible for pulsar velocities, and that kicks are required:

- The spin-orbit misalignment in PSR J0045-7319/B-star binary, as manifested by the orbital plane precession (Kaspi *et al.* 1996; Lai *et al.* 1995) and fast orbital decay (indicating retrograde rotation of the B star with respect to the orbit; Lai 1996; Kumar & Quataert 1997) require that the NS received a kick at birth. Similar precession of orbital plane has been observed in PSR J1740-3052 system (Stairs *et al.* 2003).
- The detection of geodetic precession in binary pulsar PSR 1913 + 16 implies that the pulsar's spin is misaligned with the orbital angular momentum; this can result from the aligned pulsar-He star progenitor only if the explosion of the He star gave a kick to the NS that misalign the orbit (Kramer 1998; Wex *et al.* 1999).
- The system radial velocity (430 km s^{-1}) of X-ray binary Circinus X-1 requires $V_{\rm kick} \gtrsim 500$ km s^{-1} (Tauris *et al.* 1999). Also, PSR J1141-6545 has $V_{\rm sys} \simeq 125$ km s^{-1}.
- High eccentricities of Be/X-ray binaries cannot be explained without kicks (van den Heuvel & van Paradijs 1997; but see Pfahl *et al.* 2002).
- Evolutionary studies of NS binary population (in particular the double NS systems) imply the existence of pulsar kicks (e.g., Fryer *et al.* 1998).

(3) *Observations of SNe and SNRs:* There are many direct observations, detailed in other contributions to this proceedings, of nearby supernovae (e.g., spectropolarimetry, Wang *et al.* 2003; X-ray and gamma-ray observations and emission line profiles of SN1987A) and supernova remnants which support the notion that supernova explosions are not spherically symmetric.

32.2 The problem of core-collapse supernovae and NS kicks

The current paradigm for core-collapse supernovae leading to NS formation is that these supernovae are neutrino-driven (see, e.g., Burrows & Thompson 2002; Janka *et al.* 2002 for a recent review): As the central core of a massive star collapses to nuclear density, it rebounds and sends off a shock wave, leaving behind a proto-NS. The shock stalls at several 100's km because of neutrino loss and nuclear dissociation in the shock. A fraction of the neutrinos emitted from the proto-NS get absorbed by nucleons behind the shock, thus reviving the shock, leading to an explosion on the timescale several 100's ms – This is the so-called "delayed mechanism."

It has been argued that neutrino-driven convection in the proto-NS and that in the shocked mantle are central to the explosion mechanism (e.g., Mezzacappa *et al.* 1998), although current 2D simulations with the state-of-the-art neutrino interaction and transport have not produced a successful explosion model (Buras *et al.* 2003; see Fryer & Warren 2003 for simulations in 3-D that use more approximate neutrino physics/transport). What is even more uncertain is the role of rotation and magnetic field on the explosion (see Rampp, Müller & Ruffert 1998; Fryer & Heger 2000; Ott *et al.* 2004 for simulations of collapse/explosion with rotation, and Wheeler *et al.* 2002 and Akiyama *et al.* 2003 and references therein for discussions of magnetic effects).

It is clear that our understanding of the physical mechanisms of core-collapse supernovae remains rather incomplete. The prevalence of neutron star kicks poses a significant mystery, and indicates that large-scale, global deviation from spherical symmetry is an important ingredient in our understanding of core-collapse supernovae. In the following sections, we review different classes of physical mechanisms for generating NS kicks, and then discuss possible observational constraints and astrophysical implications.

32.3 Kick mechanisms

32.3.1 Hydrodynamically driven kicks

(1) *Can convections lead to NS kicks?* The collapsed stellar core and its surrounding mantle are susceptible to a variety of hydrodynamical (convective) instabilities (e.g., Herant *et al.* 1994; Burrows *et al.* 1995; Janka & Müller 1996; Mezzacappa *et al.* 1998). It is natural to expect that the asymmetries in the density, temperature and velocity distributions associated with the instabilities can lead to asymmetric matter ejection and/or asymmetric neutrino emission. Most numerical simulations in the 1990s indicate that the local, post-collapse instabilities are not adequate to account for kick velocities $\gtrsim$100 km s^{-1}. Recently, Scheck *et al.* (2003) reported computer experiments in which they adjust the neutrino luminosity from the neutrinosphere (the inner boundary of the simulation domain) to obtain successful explosions. They found that for long-duration (more than a second) explosions, neutrino-driven convection behind the expanding shock can lead to global ($l = 1, 2$) asymmetries, accelerating the remnant NS to a range of velocities up to several hundreds of km s^{-1}. This result is encouraging. But note that, like most other SN simulations, the Scheck *et al.* simulations were done in 2D, the proton-NS was fixed on the grids (with the kick calculated by adding up the momentum flux across the inner boundary), and the explosions were obtained in an ad hoc manner. It is also not clear whether kick velocities of 500–1000 km s^{-1} can be easily obtained.

(2) *Asymmetries in pre-supernova cores:* It has been recognized that one way to produce large kicks is to have global asymmetric perturbations prior to core collapse (Goldreich *et al.* 1996; Burrows & Hayes 1996). One possible origin for the pre-SN asymmetry is

the overstable oscillations in the pre-SN core (Goldreich *et al.* 1996). The idea is the following. A few hours prior to core collapse, the central region of the progenitor star consists of a Fe core surrounded by Si-O burning shells and other layers of envelope. This configuration is overstable to nonspherical oscillation modes. It is simplest to see this by considering a $l = 1$ mode: If we perturb the core to the right, the right-hand-side of the shell will be compressed, resulting in an increase in temperature; since the shell nuclear burning rate depends sensitively on temperature (power-law index ~ 47 for Si burning and ~ 33 for O burning), the nuclear burning is greatly enhanced; this generates a large local pressure, pushing the core back to the left. The result is an oscillating g-mode with increasing amplitude. There are also damping mechanisms for these modes, the most important one being leakage of mode energy: The local (WKB) dispersion relation for nonradial waves is

$$k_r^2 = \left(\omega^2 c_s^2\right)^{-1} \left(\omega^2 - L_l^2\right) \left(\omega^2 - N^2\right), \tag{32.1}$$

where k_r is the radial wavenumber, $L_l = \sqrt{l(l+1)}c_s/r$ (c_s is the sound speed) and N are the acoustic cut-off (Lamb) frequency and the Brunt-Väisälä frequency, respectively. Since acoustic waves whose frequencies lie above the acoustic cutoff can propagate through convective regions, each core g-mode will couple to an outgoing acoustic wave, which drains energy from the core g-modes (see Fig. 32.1). In other words, the g-mode is not exactly trapped in the core. Our calculations (based on the $15M_\odot$ and $25M_\odot$ presupernova models of Weaver & Woosley) indicate that a large number of g-modes are overstable, although for low-order modes (small l and n) the results depend sensitively on the detailed structure and burning rates of the presupernova models (see Lai 2001). The typical mode periods are $\gtrsim 1$ s, the growth time $\sim$10–50 s, and the lifetime of the Si shell burning is $\sim$ hours. Thus there could be a lot of e-foldings for the nonspherical g-modes to grow. Our preliminary calculations based on the recent models of A. Heger and S. Woosley (Heger *et al.* 2001) give similar results (work in progress). Our tentative conclusion is that overstable g-modes can potentially grow to large amplitudes prior to core implosion, although several issues remain to be understood better. For example, the O-Si burning shell is highly convective, with convective speed reaching 1/4 of the sound speed, and hydrodynamical simulation may be needed to properly modeled such convection zones (see Asida & Arnett 2000).

We thus have a way of generating initial asymmetric perturbations before core collapse. During the collapse, asymmetries are amplified by a factor of 5–10 (Lai & Goldreich 2000; see also Lai 2000). How do we get the kick? The numerical simulations by Burrows & Hayes (1996) illustrate the effect. Suppose the right-hand-side of the collapsing core is denser than the left-hand side. As the shock wave comes out after bounce, it will see different densities in different directions, and it will move preferentially in the direction of lower density. We have an asymmetric shock propagation and mass ejection, a "mass rocket". The magnitude of kick velocity is proportional to the degree of initial asymmetry in the imploding core.

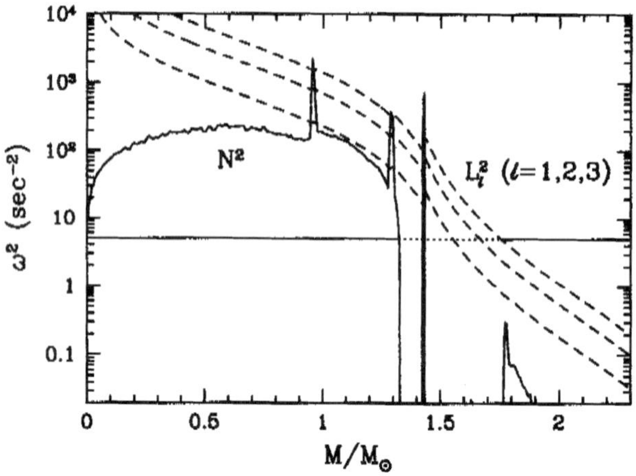

Fig. 32.1. Propagation diagram computed for a $15M_\odot$ presupernova model of Weaver and Woosley (1993). The solid curve shows N^2, where N is the Brunt-Väisälä frequency; the dashed curves show L_l^2, where L_l is the acoustic cutoff frequency, with $l = 1, 2, 3$. The spikes in N^2 result from discontinuities in entropy and composition. The iron core boundary is located at $1.3M_\odot$, the mass-cut at $1.42M_\odot$. Convective regions correspond to $N = 0$. Gravity modes (with mode frequency ω) propagate in regions where $\omega < N$ and $\omega < L_l$, while pressure modes propagate in regions where $\omega > N$ and $\omega > L_l$. Note that a g-mode trapped in the core can lose energy by penetrating the evanescent zones and turning into an outgoing acoustic wave (see the horizontal line). Also note that g-modes with higher n (the radial order) and l (the angular degree) are better trapped in the core than those with lower n and l.

32.3.2 Neutrino – magnetic field driven kicks

The second class of kick mechanisms rely on asymmetric neutrino emission induced by strong magnetic fields. Since 99% of the NS binding energy (a few times 10^{53} erg) is released in neutrinos, tapping the neutrino energy would appear to be an efficient means to kick the newly-formed NS. The fractional asymmetry α in the radiated neutrino energy required to generate a kick velocity $V_{\rm kick}$ is $\alpha = MV_{\rm kick}c/E_{\rm tot}$ ($= 0.028$ for $V_{\rm kick} = 1000$ km s^{-1}, NS mass $M = 1.4\,M_\odot$ and total neutrino energy radiated $E_{\rm tot} = 3 \times 10^{53}$ erg). There are several possible effects:

(1) *Parity violation:* Because weak interaction is parity violating, the neutrino opacities and emissivities in a magnetized nuclear medium depend asymmetrically on the directions of neutrino momenta with respect to the magnetic field, and this can give rise to asymmetric neutrino emission from the proto-NS. Calculations indicate that to generate interesting kicks with this effect requires the proto-NS to have a large-scale, ordered magnetic field of at least a few $\times 10^{15}$ G (see Arras & Lai 1999a,b and references therein).

(2) *Asymmetric field topology:* Another effect relies on the asymmetric magnetic field distribution in proto-NSs: Since the cross section for ν_e ($\bar{\nu}_e$) absorption on neutrons (protons) depends on the local magnetic field strength, the local neutrino fluxes from

different regions of the stellar surface are different. Calculations indicate that to generate a kick velocity of $\sim$300 km s^{-1} using this effect alone would require that the difference in the field strengths at the two opposite poles of the star be at least 10^{16} G (see Lai & Qian 1998). Note that only the magnitude of the field matters here.

(3) *Dynamical effect of magnetic fields:* A superstrong magnetic field may also play a dynamical role in the proto-NS. I has been suggested that a locally strong magnetic field can induce "dark spots" (where the neutrino flux is lower than average) on the stellar surface by suppressing neutrino-driven convection (Duncan & Thompson 1992). While it is difficult to quantify the kick velocity resulting from an asymmetric distribution of dark spots, order-of-magnitude estimate indicates that a local magnetic field of at least 10^{15} G is needed for this effect to be of importance.

(4) *Exotic neutrino physics:* There have also been several ideas of pulsar kicks which rely on nonstandard neutrino physics. For example, it was suggested (Kusenko & Segre 1996, 1998) that asymmetric ν_τ emission could result from the MSW flavor transformation between ν_τ and ν_e inside a magnetized proto-NS because a magnetic field changes the resonance condition for MSW effect. This mechanism requires neutrino mass of order 100 eV. A similar idea (Akhmedov *et al.* 1997; Grasso *et al.* 1998) relies on both the neutrino mass and the neutrino magnetic moment to facilitate the flavor transformation (resonant neutrino spin-flavor precession). Fuller *et al.* (2003) discussed the effect of sterile neutrinos. Analysis of neutrino transport (Janka & Raffelt 1998) indicates that even with favorable neutrino parameters, strong magnetic fields $B \gg 10^{15}$ G are required to obtain a 100 km s^{-1} kick.

32.3.3 Electromagnetically drievn kicks

Harrison & Tademaru (1975) showed that electromagnetic (EM) radiation from an off-centered rotating magnetic dipole imparts a kick to the pulsar along its spin axis. The kick is attained on the initial spindown timescale of the pulsar (i.e., this really is a gradual acceleration), and comes at the expense of the spin kinetic energy. A reexamination of this effect (Lai *et al.* 2001) showed that the force on the pulsar due to asymmetric EM radiation is larger than the original Harrison & Tademaru expression by a factor of four. Thus, the maximum possible velocity is $V_{\rm kick}^{(\rm max)} \simeq 1400\,(1{\rm kHz}/P_i)^2$ km s^{-1}. Nevertheless, to generate interesting kicks using this mechanism requires the initial spin period P_i of the NS to be less than 1–2 ms. Gravitational radiation may also affect the net velocity boost.

32.3.4 Other possibilities

(1) If rotation and magnetic fields play a dominate role in the explosion, bipolar jets may be produced. A slight asymmetry between the two jets will naturally lead to large kick (e.g., Khokhlov *et al.* 1999; Akiyama *et al.* 2003). While difficult to calculate, this is a serious possibility given the increasing observational evidence for bipolar explosions in many SNe (see other contributions to this proceedings).

(2) Colpi and Wasserman (2003) considered the formation of double proton-NS binary in a rapidly rotating core collapse; the lighter NS explodes after reaching its minimum mass limit (via mass transfer), giving the remaining NS a large kick ($\sim 10^3$ km s^{-1}). A related suggestion relies on the coalescence of proto-NS binary as providing the kick (Davies *et al.* 2002). The biggest uncertainty for such scenarios is that it is not clear core fragmentation can take place in the collapse (and numerical simulations seem to say no; see Fryer & Warren 2003).

32.4 Astrophysical constraints on kick mechanisms

The review in previous sections shows that NS kicks are not only a matter of curiosity, but intimately connected to other fundamental parameters of young NSs (initial spin and magnetic field). For example, the neutrino-magnetic field driven mechanisms are of relevance only for $B \gtrsim 10^{15}$ G. While recent observations have lent strong support that some neutron stars ("magnetars") are born with a super-strong magnetic field, it is not clear (perhaps unlikely) that ordinary radio pulsars (for which large velocities have been measured) had initial magnetic fields of such magnitude.

One of the reasons that it has been difficult to pin down the kick mechanisms is the lack of correlation between NS velocity and the other properties of NSs. The situation has changed with the recent X-ray observations of the compact X-ray nebulae of the Crab and Vela pulsars, which have a two sided asymmetric jet at a position angle coinciding with the position angle of the pulsar's proper motion (Pavlov *et al.* 2000; Helfand *et al.* 2001). The symmetric morphology of the nebula with respect to the jet direction strongly suggests that the jet is along the pulsar's spin axis. Analysis of the polarization angle of Vela's radio emission corroborates this interpretation (Lai *et al.* 2001). Thus, while statistical analysis of pulsar population neither support nor rule out any spin-kick correlation, at least for the Vela and Crab pulsars (and perhaps also for PSR J0538+2817, Romani & Ng 2003), the proper motion and the spin axis appear to be aligned.

The apparent alignment between the spin axis and proper motion raises an interesting question: Under what conditions is the spin-kick alignment expected for different kick mechanisms? Let us look at the three classes of mechanisms discussed before (Lai *et al.* 2001): (1) For the electromagnetically driven kicks, the spin-kick slignment is naturally produced. (Again, note that $P_i \sim 1$–2 ms is required to generate sufficiently large $V_{\rm kick}$). (2) For the neutrino–magnetic field driven kicks: The kick is imparted to the NS near the neutrinosphere (at 10's of km) on the neutrino diffusion time, $\tau_{\rm kick} \sim 10$ seconds. As long as the initial spin period P_i is much less than a few seconds, spin-kick alignment is naturally expected. (3) For the hydrodynamically driven kicks: because the kick is imparted at a large radius ($\gtrsim$100 km), to get effective rotational averaging, we require that the rotation period at ~100 km to be shorter than the kick timescale (~100 ms). This translates to $P_{\rm NS} \lesssim 1$ ms, which means that rotation must be dynamically important. On the otherhand, if rotation indeed plays a dynamically important role, the basic collapse and explosion may be qualitatively different (e.g., core bounce may occur at subnuclear density, the explosion is weaker and takes the form of two-sided jets (e.g. Khokhlov *et al.* 1999; Fryer & Heger 2000). The possibility of a kick in such systems has not been studied, but it is conceivable that an asymmetric dipolar perturbation may be coupled to rotation, thus producing spin-kick alignment.

Currently we do not know whether spin-kick alignment is a generic feature of all pulsars; if it is, then it can provide powerful constraint on the kick mechanisms and the SN explosion mechanisms in general.

Finally, it is worth noting that recent observations showed that black hole (BH) formation can be accompanied by SN explosion: The companion of the BH X-ray binary GRO J1655-40 (Nova Sco) and that of SAX J1819.3-2525 (V4641 SGR) have high abundance of α-elements (Israelian *et al.* 1999; Orosz *et al.* 2001), which can only be produced in a SN explosion (see Podsiadlowski *et al.* 2002). Apparently, the BH forms in an indirect process where a shock wave successfully makes an explosion and a NS forms temporarily followed by fall-back, or loss of angular momentum and thermal energy in the proto-NS which then collapses to a BH. This indirect process may explain the the relatively large space velocity of GRO J1655-40.

Acknowledgements

Support for this work is provided in part by NASA NAG 5-12034 and NSF AST 0307252. I thank my collaborators P. Arras, D. Chernoff, J. Cordes, P. Goldreich, A. Heger, Y.-Z. Qian and A. Shirakawa for their important contributions. I also thank the conference organizers for a stimulating meeting and travel support.

References

Akhmedov, E. K., Lanza, A., & Sciama, D. W. 1997, Phys. Rev. D, 56, 6117
Akiyama, S., *et al.* 2003, ApJ, 584, 954
Arras, P., & Lai, D. 1999a, ApJ, 519, 745
Arras, P., & Lai, D. 1999b, Phys. Rev. D60, 043001
Arzoumanian, Z., Chernoff, D. F., & Cordes, J. M. 2002, ApJ, 568, 289
Asida, S. M., & Arnett, D. 2000, ApJ, 545, 435
Buras, R., *et al.* 2003, Phys. Rev. Lett. 90, 241101
Burrows, A., Hayes, J., & Fryxell, B. A. 1995, ApJ, 450, 830
Burrows, A., & Hayes, J. 1996, Phys. Rev. Lett., 76, 352
Burrows, A., & Thompson, T. A. 2002, astro-ph/0210212
Chatterjee, S., & Cordes, J. M. 2002, ApJ, 575, 407
Colpi, M., & Wasserman, I. 2002, ApJ, 581, 1271.
Cordes, J. M., Romani, R. W., & Lundgren, S. C. 1993, Nature, 362, 133

Davies, M. B., *et al.* 2002, ApJ, 579, L63
Duncan, R. C., & Thompson, C. 1992, ApJ, 392, L9.
Fryer, C., Burrows, A., & Benz, W. 1998, ApJ, 498, 333
Fryer, C. L., & Heger, A. 2000, 541, 1033
Fryer, C. L., & Warren, M. S. 2003, ApJ, in press (astro-ph/0309539)
Fuller, G. M., *et al.* 2003, astro-ph/0307267
Goldreich, P., Lai, D., & Sahrling, M. 1996, in "Unsolved Problems in Astrophysics", ed. J. N. Bahcall and J. P. Ostriker (Princeton Univ. Press)
Grasso, D., Nunokawa, H., & Valle, J. W. F. 1998, Phys. Rev. Lett., 81, 2412
Hansen, B. M. S., & Phinney, E. S. 1997, MNRAS, 291, 569
Heger, A. *et al.* 2001, ApJ, 560, 307
Herant, M., *et al.* 1994, ApJ, 435, 339
Harrison, E. R., & Tademaru, E. 1975, ApJ, 201, 447
Iben, I., & Tutukov, A. V. 1996, ApJ, 456, 738
Israelian, G, *et al.* 1999, Nature, 401, 6749 GRO J1655–40
Janka, H.-T., & Müller, E. 1996, A&A, 306, 167
Janka, H.-T., & Raffelt, G. G. 1998, Phys. Rev. D59, 023005
Janka, H.-Th., *et al.* 2002, in "Core Collapse of Massive Stars" (astro-ph/0212316)
Kaspi, V. M., *et. al.* 1996, Nature, 381, 583
Khokhlov, A. M., *et al.* 1999, ApJ, 524, L107
Kramer, M. 1998, ApJ, 509, 856
Kumar, P., & Quataert, E. J. 1997, 479, L51
Kusenko, A., & Segré, G. 1996, Phys. Rev. Lett., 77, 4872
Lai, D. 1996, ApJ, 466, L35
Lai, D. 2000, ApJ, 540, 946
Lai, D., Bildsten, L., & Kaspi, V. M. 1995, ApJ, 452, 819
Lai, D., Chernoff, D. F., & Cordes, J. M. 2001, ApJ, 549, 1111
Lai, D., & Goldreich, P. 2000, ApJ, 535, 402
Lai, D., & Qian, Y.-Z. 1998, ApJ, 505, 844
Lorimer, D. R., Bailes, M., & Harrison, P. A. 1997, MNRAS, 289, 592
Mezzacappa, A., *et al.* 1998, ApJ, 495, 911.
Orosz, J. *et al.* 2001, ApJ, 555, 489
Ott, C. D., *et al.* 2004, ApJ, 600 (astro-ph/0307472)
Pavlov, G. G., *et al.* 2001, ApJ, 552, L129
Pfahl, E., *et al.* 2002, ApJ, 574, 364
Podsiadlowski, Ph., *et al.* 2002, ApJ, submitted (astro-ph/0109244)
Rampp, M., Müller, E., & Ruffert, M. 1998, A&A, 332, 969
Romani, R. W., & Ng, C.-Y. 2003, ApJ, 585, L41
Scheck, L. *et al.* 2003, PRL, submitted (astro-ph/0307352)
Tauris, T., *et al.* 1999, MNRAS, 310, 1165
Thorstensen, J. R., *et al.* 2001, AJ, 122, 297
van den Heuvel, E. P. J., & van Paradijs, J. 1997, ApJ, 483, 399.
Wang, L., Baade, D., Höflich, P., & Wheeler, J. C. 2003, ApJ, 592, 457
Weaver, T. A., & Woosley, S. E. 1993, Phys. Rep., 227, 65
Wex, N., Kalogera, V., & Kramer, M. 2000, ApJ, 528, 401
Wheeler, J. C., Meier, D. L., & Wilson, J. R. 2002, ApJ, 568, 807

33

Triggers of magnetar outbursts

R. C. Duncan

University of Texas at Austin TX USA

Abstract

Bright outbursts from Soft Gamma Repeaters (SGRs) and Anomalous X-ray Pulsars (AXPs) are believed to be caused by instabilities in ultramagnetized neutron stars, powered by a decaying magnetic field. It was originally thought that these outbursts were due to reconnection instabilities in the magnetosphere, reached via slow evolution of magnetic footpoints anchored in the crust. Later models considered sudden shifts in the crust's structure. Recent observations of magnetars give evidence that at least some outburst episodes involve rearrangements and/or energy releases within the star. We suggest that bursting episodes in magnetars are episodes of rapid plastic yielding in the crust, which trigger "swarms" of reconnection instabilities in the magnetosphere. Magnetic energy always dominates; elastic energy released within the crust does not generate strong enough Alfvén waves to power outbursts. We discuss the physics of SGR giant flares, and describe recent observations that give useful constraints and clues.

33.1 Introduction: a neutron star's crust

The crust of a neutron star has several components: (1) a Fermi sea of relativistic electrons, which provides most of the pressure in the outer layers; (2) another Fermi sea of neutrons in a pairing-superfluid state, present only at depths below the "neutron drip" level where the mass-density exceeds $\rho_{\rm drip} \approx 4.6 \times 10^{11}$ gm cm^{-3}; and (3) an array of positively-charged nuclei, arranged in a solid (but probably not regular crystalline) lattice-like structure throughout much of the crust. These nuclei become heavier and more neutron-bloated at increasing depths beneath the surface, until the swollen nuclei nearly "touch" and the quasi-body-centered-cubic nuclear array dissolves into rod-like and slab-like structures near the base of the crust: "nuclear spaghetti and lasagna" or "nuclear pasta" (e.g., Pethick & Ravenhall 1995).

In a magnetar, the crust is subject to strong, evolving magnetic stresses. Magnetic evolution within the crust occurs via Hall drift; while ambipolar diffusion and Hall drift of magnetic flux within the liquid interior strains the crust from below (Goldreich & Reisenegger 1992; Thompson & Duncan 1996, hereafter "TD96"). The crust and the magnetic field thus evolve together through a sequence of equilibrium states in which magnetic stresses are balanced by material restoring forces. Because a magnetar's field is so strong, this evolution inevitably involves crust-yielding episodes driven by the magnetic field on a variety of time-scales. Many complexities are likely to affect the crustal yielding threshold, and cause it to vary from place to place within a neutron star's crust. Moreover the very nature of neutron star crust-yield events are somewhat uncertain. In the deep crusts of magnetars, yields may resemble sudden and sporadic flow in an inhomogeneous liquid crystal, induced by evolving magnetic stresses.

Before discussing these complications, it is worthwhile to note how neutron star crust material, outside the pasta layers, differs from a terrestrial solid. A key difference (due to the high mass-density in a neutron star) is that the relativistic Fermi sea of electrons is only slightly perturbed by the Coulomb forces of the nuclei, and does not efficiently screen nuclear charges. With a nearly inert and uniform distribution of negative charge, pure neutron star crust-matter comes close to realizing the "ideal Coulomb crystal" or "Wigner crystal" (Wigner 1934). Such a body-centered cubic (bcc) lattice is expected to form in the low-temperature limit of a "one-component plasma" (Brush, Sahlin & Teller 1966; Ichimaru 1982; van Horn 1991). The crystallization temperature is $k_B T_c = \Gamma^{-1}(Ze)^2/a$, where Z is the ionic charge, a is the Wigner-Seitz radius satisfying $\frac{4}{3}\pi a^3 = n^{-1}$ with ion density n, and Γ is a numerical constant found in statistical mechanics to be $\Gamma \approx 170$. In the deep crust of a neutron star, $T_c \sim 10^{10}$ K.

The electrostatic structure of naturally-occurring terrestrial solids is more complex, with bound electrons and efficient screening. Only in the cores of old white dwarfs, which have cooled sufficiently to crystallize, does bulk material like the stuff of a neutron star's outer crust ($\rho < \rho_{drip}$) exist elsewhere in nature. Inner-crust matter is found nowhere outside of neutron stars.

However, nearly-ideal Coulomb crystals have recently been made in the laboratory, and the failure of these crystals under stress was studied (Mitchell *et al.* 2001). I will now describe one of these delicate and elegant experiments.

About 15,000 cold $^9Be^+$ ions were confined within a volume about half a millimeter in diameter in the laboratory at the National Institute of Standards in Boulder, Colorado. A uniform magnetic field confined the ions radially, while a static electric field with a quadratic potential trapped them axially (i.e., a Penning trap). The ions were cooled to millikelvin temperatures using lasers that were tuned to a frequency just below an ionic ground-state excitation level. The plasma crystallized into a disk, with a *bcc* lattice structure. (Note that trapping fields effectively provided the neutralizing background for this crystal.) Due to a weak radial component of the electric field, the charged crystal experienced $E \times B$ drift and rotated. The velocity of rotation was controlled and stabilized by the experimenters using a perturbing electric field. The ions, fluorescing in laser light and separated by 15 microns, were directly imaged and photographed. The crystal was then stressed by illuminating it with laser light from the side, and off-axis. Slips in the crystalline structure were detected. They were distributed, over at least three orders of magnitude, according to a power law with index between 1.8 and 1.2.

One cannot avoid mentioning here that this distribution of slip sizes resembles the energy-distribution of bursts from SGRs. Some would claim that this is a consequence of "universality" in self-organized critical systems. In any case it can have no deep implications about burst mechanisms, since the physics of SGR outbursts is much more complex than simple slips in a crystal.

These experiments verify some expectations about idealized, crystalline neutron star matter; but real neutron star crusts, and especially magnetar crusts, are likely to be complex and messy. As emphasized by Ruderman (1991), the circumference of a neutron star is about 10^{17} lattice-spacings, which is similar to the size of the Earth measured in the lattice-spacings of terrestrial rock.

According to the old, 20th century neutron star theory, the amount and distribution of crystalline imperfections in the crust is history-dependent. Two factors were thought to be involved: the rapidity of cooling when the solid originally formed, and the subsequent "working" of the solid by stresses and (in some cases) episodic reheating. More rapid initial cooling and solidification would generally produce more lattice imperfections and smaller lattice domains. Extremely rapid cooling, or "quenching" would produce an amorphous (glassy) solid rather than a crystal, which is really a long-lived metastable state: a super-cooled liquid. It was suggested that this occurs in neutron star crusts (Ichimaru *et al.* 1983); however, models of crust solidification in more realistic, neutrino-cooled neutron stars showed crystallization (de Blasio 1995). Subsequent strain-working of the crust would increase lattice imperfections, while episodes of (magnetic) re-heating followed by cooling would tend to anneal the solid.

That is the old picture. Jones (1999, 2001) recently turned the story on its head. He showed that, when the crust initially cools through the melting temperature, a substantial range of Z (i.e., nuclear proton-numbers) get frozen-in at every depth below neutron drip. This is due to thermal fluctuations in the nuclei, which are in equilibrium with the neutron bath. The energy separation of magic-number proton shells in the neutron-bloated nuclei is not large compared to the melt temperature. This is important because it means that the crust of a 21st-century neutron star is amorphous rather than crystalline. There exists some short-range crystalline order, but over distances greater than about $\sim 10\, a$ the variable Z's affect the inter-nuclear spacing enough to destroy all order. This has important implications for transport properties such as electrical conductivity, among other things.

Finally, there is the sticky issue of nuclear pasta (Pethick & Ravenhall 1995 and references therein).[1] Deep inside the crust, the neutron-bloated nuclei become elongated and join into "nuclear spaghetti": long cylindrical structures in a 2-D triangular array. As depth and density increase, these nuclear noodles join into slabs: "nuclear lasagna." At even higher densities this gives way (at least for some values of nuclear state parameters) to "inverse spaghetti": an array of cylindrical holes in the high-density fluid; followed by "inverse meatballs": a bcc lattice of spherical holes in otherwise continuous nuclear matter. Beneath that lies continuous nuclear fluid.

Because rod-like (or planar) structures can freely slide past each other along their length (and breadth) without affecting the Coulomb energy, nuclear pasta has an extremely anisotropic tensor of elasticity. Indeed, the elastic response of nuclear spaghetti resembles that of the *columnar phases of a liquid crystal*, and elastic nuclear lasagna resembles the *smectics A phase* (Pethick & Potekhin 1998). At sufficiently high temperatures, positionally-disordered (*nematic*) phases are also possible; but the threshold for this is $\sim 10^{10}$ K.

It has been suggested that up to half the mass of a neutron star's crust is in nuclear pasta (Pethick & Potekhin 1998). A detailed, realistic understanding of the response of a neutron star's crust to evolving stresses (especially stresses largely exerted from below by a magnetic field, as likely in a magnetar) may require understanding the size and coherence of nuclear pasta domains; their orientation relative to the vertical; the interactions of pasta with the magnetic field; and the yielding behavior of such liquid crystals, which plausibly depends upon instabilities in the pasta domain structure. These difficult issues have not begun to be addressed by astrophysicists.

Thus the range of complicating factors which could affect neutron star crust evolution is formidable. In the case of a magnetar, the crust is coupled to an evolving ultra-strong magnetic field and its generating currents, which penetrate the underlying core as well. Manifold uncertainties about field geometries and magnetic evolution compound the murkiness. Observations of SGRs and AXPs could provide the most sensitive probes of neutron star interiors available to astronomers, because magnetars are much less stable than other neutron stars. However, the intertwined complexities of neutron star magnetic activity must be unraveled.

[1] Nuclear pasta results from the competition between Coulomb and nuclear surface-energy terms when minimizing the energy. The pasta ground states exhibit spontaneously broken symmetries, although the underlying interactions between constituent nucleons are nearly rotationally-symmetric. This is different from the case with terrestrial liquid crystals, in which highly-anisotropic inter-atomic forces give rise to the broken global symmetries.

In this review we focus on one aspect of this problem: the triggering of bright outbursts. We will try to keep the discussion on a basic physical level, eschewing equations as much as possible. In Section 33.2 we review the physics and phenomenology of SGR outbursts. In Section 33.3 we compare rise-time observations with models for trigger mechanisms. Section 33.4 discusses other observations that offer clues. Section 33.5 discusses the general issue of crust-failure in magnetars. Section 33.6 gives conclusions.

33.2 Magnetar outbursts: a brief review

"Flares are triggered in magnetically-active main-sequence stars when convective motions displace the footpoints of the field sufficiently to create tangential discontinuities, which undergo catastrophic reconnection. Similar reconnection events probably occur in magnetars, where the footpoint motions are driven by a variety of diffusive processes."

– *Duncan & Thompson* (1992)

This quotation shows that magnetar outbursts were originally conceived as being triggered in the magnetosphere, as a consequence of the neutron star's slow, interior magnetic evolution. This was believed to apply to giant flares as well: "The field of a magnetar carries sufficient energy to power the 1979 March 5th event (5×10^{44} ergs at the distance of the LMC, assuming isotropic emission; Mazets *et al.* 1979)." (Duncan & Thompson 1992; DT92) Note that the March 5th event was the only giant flare that had been detected at that time, with energy >200 times greater than the second most-energetic SGR event.

Paczyński (1992), in work done soon after DT92, made this point more explicitly. He suggested that the March 5th event was "caused by a strong magnetic flare at low optical depth, which led to a thermalized fireball . . ."

By 1995, Thompson and I had realized that the evolving, strong field of a magnetar was capable of straining the crust more severely than it could bear. Thompson & Duncan (1995; TD95) thus discussed the relative merits of impulsive crustal shifts and pure magnetospheric instabilities as outburst triggers. TD95 favored scenarios in which both processes occurred, with the 1979 March 5th event involving profound exterior reconnection. TD95 and Thompson & Duncan 1996 (TD96) also discussed plastic deformation of magnetar crusts. Besides high plasticity at places where the temperature approaches $\sim 0.1 T_c$ (plausibly due to local magnetic heating), we noted that magnetic stresses dominate elastic stresses if $B > B_\mu = (4\pi\mu)^{1/2} = 4 \times 10^{15}\ \rho_{14}^{0.4}$ G, where μ is the shear modulus, and ρ_{14} is the mass-density in units of 10^{14} gm cm^{-3}. A magnetic field stronger than B_μ is thus like a 600-pound gorilla: "it does whatever it wants" in the crust. We noted possible implications of plastic deformation for glitches, X-ray light curve variations, and triggering catastrophic reconnection (TD95, TD96, Thompson *et al.* 2000; Thompson & Duncan 2001; TD01).

Observations made after 1996 have tended to fill in the "energy gap" between the 1979 March 5th event and other SGR outbursts. In particular, the 1998 August 27th giant flare was about 10^{-1} times as energetic as the March 5th event (Hurley *et al.* 1999a; Mazets *et al.* 1999a; Feroci *et al.* 2000); and two intermediate-energy events[2] have been observed: the 2001 April 18 flare from SGR 1900+14 (Kouveliotou *et al.* 2001; Feroci *et al.* 2002; Woods *et al* 2002), and the slow-rising, powerful 1998 June 18 flare from SGR 1627–41 (Mazets *et al.* 1999b) which had no long-duration soft tail. The emerging continuity in outburst energies makes it more plausible that giant flares and common SGR bursts differ in degree rather than in kind; while the profound differences between outbursts of comparable energies indicate that a wide variety of physical conditions and processes are involved.

Studies of SGR burst statistics since 1996 have also yielded important insights. Cheng, Epstein, Guyer & Young (1996) noted that the statistical distribution of SGR burst energies is a power law with index 1.6, resembling the Guttenberg-Richter law for earthquakes. Such a distribution can result from self-organized criticality (Katz 1986; Chen, Bak & Obukhov 1991). Cheng *et al.* found additional statistical resemblances between SGR bursts and earthquakes, as verified and further studied by Göğüş *et al.* (1999 and 2000) with a much larger sample of SGR events. AXP 2259+586's June 2002 active episode showed very similar burst statistics (Gavriil, Kaspi & Woods 2003). These results lend support to the hypothesis that SGR/AXP outbursts are powered by an intrinsic stellar energy source, which is plausibly magnetic. (Accretion-powered events, including Type I and Type II X-ray bursts, have much different statistics.) However, as Göğüş *et al.* pointed out, these burst statistics do not necessarily argue that SGR bursts are crustquakes. Similar statistical distributions have been found in solar flares (Crosby, Aschwanden & Dennis 1993; Lu *et al.* 1993).

In 2001, Thompson and I studied an idealized "toy model" for a giant flare. In this model, a circular patch of crust facilitates the release of magnetic energy by yielding along the circular fault and twisting.[3] Circular crust displacements are plausible because the crust is stably-stratified and strongly constrained in its motion, yet significant twisting movement could be driven by the magnetic field. Moreover, a magnetar's formation as a rapid rotator should significantly "wind up" the star's interior field (DT92); and twists of the exterior field, which would result from this kind of magnetic activity, could drive currents through the magnetosphere,

[2] Here, I call intermediate-energy events *flares* but not *giant flares*. Events releasing $\sim 10^{41}$ ergs or less are traditionally called *bursts*. I use *outburst* as a generic term for all magnetar events.

[3] The common, 0.1 s SGR bursts, on the other hand, "could be driven by a more localized and plastic deformation of the crust." (TD01)

contributing to the observed, quiescent X-ray emissions from magnetars (Thompson *et al.* 2000).

Note that the crust-yielding event in a giant flare, if it occurs, is not a brittle fracture. Neutron star crusts probably undergo plastic failure, at least outside the nuclear pasta (Jones 2003). The sudden yielding event could have been a widely-distributed plastic flow along circular flow-lines induced by rapidly-changing stresses exerted by the core field from below.

Alternatively, there may exist instabilities within the crust that cause rapid mechanical failure, triggering giant flares (and/or other events). The failure of nuclear pasta could involve sudden instabilities due to the interactions of domains with differently-oriented, strongly anisotropic elastic/liquid response. In the solid crust, a sufficiently long and localized plastic slip could drive melting along the fault, suppressing the normal elastic stress and mimicking a brittle fracture. Jones (2003) estimates that this could occur for slips longer than a few centimeters. The process might be facilitated, as a "mock-fracture" propagates, by the development of magnetic gradients within the fault plane, with localized magnetic heating.

In 2002, Thompson, Lyutikov & Kulkarni (hereafter "TLK") considered the possibility that giant flares are triggered in the magnetosphere, with no energetically-significant crust displacement on the time-scales of the flare. Section 5.6 of TLK discussed four pieces of observational evidence that bear on the question of which mechanism operates. TLK argued that three out of four favored crust-yielding. Finally, Lyutikov (2003) gave a new estimate of the rise-time for magnetospheric instabilities in magnetars. Since rise times are an important diagnostic we now discuss them in detail.

33.3 Outburst rise-times and durations

There are two rise-times of interest in SGR outbursts: the "growth time" τ_{grow} which is the e-folding time for the energy-flux growth during the initial, rapid brightening; and the "peak time" τ_{peak} which is the time from the initial onset of the event until the (highest) peak of the light curve.

A third time-scale of interest in the brightest SGR events is τ_{spike}, the duration of the initial, hard-spectrum, extremely bright phase of the event which we refer to as the "hard spike." In some flares, including both giant flares on record, this spike is followed by an intense "soft tail" of X-rays, modulated on the rotation period of the star. The soft tail is thought to be emitted by an optically-thick "trapped fireball" in the magnetosphere of a magnetar (TD95). The abrupt vanishing of soft tail emission at the end of the 1998 August 27th event seems to be due to fireball evaporation, corroborating this interpretation (Feroci *et al.* 2000; TD01).

The March 5th 1979 event reached its peak at $\tau_{peak} \approx 20$ ms, but the initial, fast rise through many orders of magnitude was unresolved by ISEE or the Pioneer Venus Orbiter, thus $\tau_{grow} < 0.2$ ms (Cline *et al.* 1980, Terrel *et al.* 1980; Cline 1982). The initial, hard-spectrum emission lasted for $\tau_{spike} \sim 0.15$ s (Mazets *et al.* 1979), during which time it showed variability on timescales of order $\sim$ 10–30 ms (Barat *et al.* 1983).

Thompson and I suggested interpretations of these time scales. TD95 noted (in eq. 16) that the Alfvén crossing time within the (fully-relativistic) magnetosphere of a magnetar is comparable to the light-crossing time of the star, roughly 30 microseconds. "Since reconnection typically occurs at a fraction of the Alfvén velocity, the growth time of the instability is estimated to be an order of magnitude larger . . . This is, indeed, comparable to the 0.2 msec rise time of the March 5 event." In other words, we suggested

$$\tau_{grow} \sim \frac{L}{0.1 V_A} \sim 0.3 \left(\frac{L}{10\,\mathrm{km}} \right) \mathrm{ms}, \tag{33.1}$$

where L is the scale of the reconnection-unstable zone, and $V_A \sim c$ is the (exterior) Alfvén velocity. TD95 further suggested that τ_{spike} is comparable to the *interior* Alfvén wave crossing time of the star, which applies if the event involves an interior magnetic rearrangement. This yields $\tau_{spike} \sim 0.1\ B_{15}^{-1}\, \rho_{15}\, (\Delta\ell/R_\star)$ s [TD95 eq. 17], in agreement with giant flare data: $\tau_{spike} = 0.15$ s [March 5th event] and 0.35 s [August 27th event; Mazets *et al.* 1999a].

Another physical time scale of possible relevance is the shear-wave crossing time of the active region of crust. (This is the elastic stress-equilibration timescale even when the generation of propagating shear waves is small.) The shear-wave velocity $V_\mu = (\mu/\rho)^{1/2}$ is insensitive to depth (or local density ρ) in the crust, at least in the zones outside the nuclear pasta: $V_\mu = 1.0 \times 10^3\ \rho_{14}^{-0.1}$ km s^{-1} [TD01, eq. 8]. For an active region of size ℓ, this gives a crossing time $\tau_\mu = \ell V_\mu = 3\ (\ell/3$ km) ms (TD95), thus $\tau_{grow} \le \tau_\mu \le \tau_{peak}$ for the March 5th flare.

The light curves of common, repeat bursts from SGRs were studied by Göğüş *et al.* (2001), using a data-base of more than 900 events measured by the Rossi X-ray Timing Explorer (RXTE). Göğüş *et al.* found that the distribution of burst durations (as measured by T_{90}, the time in which 90% of the burst counts accumulate) is lognormal, with a peak of order 100 ms. Most bursts rise faster than they decline, but many have roughly triangular light curves. In particular, about half of all bursts have $\tau_{peak} > 0.3T_{90}$. Thus the distribution of τ_{peak} peaks at $\sim$30 ms, and τ_{grow} peaks around $\sim$10 ms.

Lyutikov (2003) studied the growth of spontaneous reconnection in magnetar magnetospheres. Compared to better-understood conditions in the Solar chromosphere, radiative cyclotron decay times are extremely short, forcing currents to flow narrowly along field lines. The plasma is thus force-free and relativistic, being dominated by the magnetic field. Lyutikov suggested that a tearing-mode instability operates within current-sheets, involving the clustering of current filaments within the sheet and the formation of "magnetic islands." This has a rate $\tau_{rise} \sim \sqrt{\tau_A \tau_R}$, just as in the non-relativistic case, where $\tau_A \sim \ell/c$ is the Alfvén crossing time of the unstable zone, and τ_R is the resistive time-scale $\tau_R\ \ell^2/\eta$. Lyutikov conjectured that either Langmuir turbulence or ion sound turbulence provide the resistivity: $\eta \sim c^2/\omega_p$, where ω_p is the (electron or ion) plasma frequency. To evaluate this requires an estimate of the local particle density. Lyutikov adopted a value expected in the "globally-twisted magnetosphere" model of TLK, which yields

$$\tau_{grow} \sim 0.10 \left(\frac{L}{10\,\mathrm{km}}\right)^{3/2} \left(\frac{r}{100\,\mathrm{km}}\right)^{-7/8} \left(\frac{B_{pole}}{5 \times 10^{14}\mathrm{G}}\right)^{1/4} s, \tag{33.2}$$

for Langmuir turbulence, or smaller by a factor $(m_p/m_e)^{1/2} = 6.5$ for ion sound turbulence.[4] If reliable, this analysis represents an improvement over TD95's crude estimate [eq. (33.1) above]. But to match observed rise-times requires quite localized events, high in the magnetosphere, with fully-developed ion turbulence in the tearing layer. If the events happen closer to the stellar surface ($r \sim 10$ km) as likely,[5] then the rise time is closer to $\sim$1 s for electron turbulence and $\sim$0.1 s for ions. The problem is that this mechanism requires low particle densities to be efficient.

More realistic models of the magnetosphere than a simple global twist will greatly exacerbate the discrepancy, because reconnection occurs where the current density (and thus the charge density j/v where usually $v \sim c$) is especially large, in current sheets. That is, the magnetosphere can be locally as well as globally twisted; and the current density is determined by local magnetic shear.

Other mechanisms besides tearing modes coupled to ion sound turbulence probably operate in nonrelativistic astrophysical reconnection, and seem worthy of investigation in the magnetar context. One possibility is stochastic reconnection (Lazarian & Vishniac 1999), which requires some source of turbulence on scales that are larger than the current sheet width. In a magnetar this might be provided by crust-yielding motion which agitates the field near a developing magnetic discontinuity.

33.4 Other observational clues

There is evidence that at least some magnetar outbursts involve structural adjustments inside the star, with enhancements of magnetospheric currents:

[4] Note that the electron plasma frequency is given by $\omega_P \sim \sqrt{\omega_B c/r} \sim 3 \times 10^{11}$ rad/s, where the cyclotron frequency $\omega_B = (eB/mc)$ is evaluated at $r = 100$ km, outside a $R = 10$ km star with polar field $B_{pole} = 5 \times 10^4$ Gauss, assuming $B(r) = B_{pole}(r/R)^{-2-p}$ with $p = 1/2$ in a strongly twisted magnetosphere, $\Delta\phi \approx 2$ radians.

[5] The fraction of exterior magnetic energy lying beyond radius r, $f_B(> r)$ falls off substantially *faster* than the pure dipole contribution $f_B(> r) = (r/R)^{-1-2p}$, where $p = 1$ for no global twist. So the fraction of energy available for reconnection at $r > 100$ km is significantly less than 10^{-2}[$p = 1/2$; 2-radian twist] or 10^{-3} [untwisted].

- The 1998 June 18 event from SGR 1627–41 resembled a slow-rising giant flare with no soft tail (Mazets *et al.* 1999b). One plausible interpretation is that the star experienced a deep-crust adjustment that triggered little exterior reconnection or other energy dissipation on closed field-lines in the magnetosphere (relative to other powerful flares), thus no long-lasting trapped fireball. (Note that a patch of crust adjusting over a timescale $\tau_{peak} \sim 0.1$ s would produce little Alfvén wave emission on field lines shorter than $c \cdot \tau_{peak} \sim 3 \times 10^4$ km, helping to limit the outburst efficiency.) This is consistent with the results of Kouveliotou *et al.* (2003), who studied X-ray emissions from SGR 1627–41 following June 1998. For two years, the light curve was a 0.47-index power law, gradually leveling off to a "plateau"; and then, after 1000 days, dropping precipitously. Kouveliotou *et al.* found that the cooling crust of a 10^{15} Gauss neutron star could follow this pattern if the initial energy deposition (presumably on June 18th) extended deep into the crust, significantly below neutron drip. The integrated energy of the X-ray afterglow was comparable to the June 18 outburst energy; but the impulsive energy injection in the crust had to be much larger (by a factor $\sim 10^2$ in Kouveliotou *et al.*'s models) because of deep conduction and neutrino losses. This would then be a (relatively) crust-active, magnetosphere-quiet magnetar. One concern with this interpretation is that the observed afterglow spectrum was non-thermal and time-variable. It has not yet been shown that reprocessing by scattering in the magnetosphere can (fully) account for this. Other interpretations of SGR 1627–41 data might still be possible.
- The June 18 event peaked much more gradually than other flares: $\tau_{peak} \sim 0.1$ s. This could be consistent with slow crust failure, say at a rate $V \sim 0.1 V_\mu$, along a large fault-line or plastic shear-zone: $\tau_{peak} \sim 0.1(\ell/10 \text{ km})$ s (Mazets *et al.* 1999b).
- Timing studies of AXP 1E2259+586 revealed a glitch associated with a burst-active episode in June 2002, plausibly simultaneous with the onset of bursting (Kaspi *et al.* 2003; Woods *et al.* 2003). The star's rotation rate abruptly increased by 4.2 parts in 10^6, giving evidence for the redistribution of angular momentum between superfluid and non-superfluid components within the star. A sudden adjustment within the star is necessary, thus the bursting episode was probably not due to spontaneous magnetospheric instabilities.[6]
- No significant, persistent diminishment of $\dot{P}$ was detected in SGR 1900+14 following the 1998 August 27 giant flare (Woods *et al.* 1999). This puts constraints on large-scale rearrangements of the magnetosphere (Woods *et al.* 2001) in the context of the globally-twisted magnetosphere model (TLK). In this model, twists in the magnetosphere accelerate a magnetar's spindown rate. Such twists could be driven by a strong, "wound

[6] Note that the June 2002 glitch in AXP 2259 + 586 (Kaspi *et al.* 2003; Woods *et al.* 2003) was different from previous spindown irregularities in this star during the past 25 years. I say this because the star has been spinning down at a steady rate during the ~5 years since phase-coherent timing began (Kaspi, Chakrabarty & Steinberger 1999; Gavriil & Kaspi 2002), and the persistent (post-recovery) $\dot{P}$ changed by only ~2% during the glitch/bursting episode. If one extrapolates with this $\dot{P}$ back through the sparsely-sampled period history of the star, beginning with *Einstein* Observatory observations in 1979, one finds that the star must have experienced two episodes of *accelerated spindown*, or two spin *down* glitches, both with $(\Delta P/P) \sim +2 \times 10^{-6}$ The first occurred around 1985, between Tenma and EXOSAT observations. The second occurred after ASCA but before RXTE, during 1993–1996. One could alternatively fit the data with spin-*up* glitches, like the June 2002 glitch, as suggested by Usov (1994) and Heyl & Hernquist (1999), but this fit requires that the persistent value of $\dot{P}$ was larger in the past by ~25%. In either case, this star was behaving differently in the past.

up" interior field, dragging on the crust from below, which is a likely relic of magnetar formation (DT92). Exterior field-twists must be maintained in the force-free magnetosphere by currents flowing along field lines. As shown by TLK, global twists tend to shift field lines away from the star, enhancing the field strength at the light cylinder and hence the braking torque. If the August 27 flare involved a large-scale relief of global twists, analogous to the instabilities of Wolfson (1995) and Lynden-Bell & Boiley (1994), with significant diminishment of global currents (a possibility raised by TLK and Lyutikov 2003), then one would expect a diminishment in $\dot{P}$. In fact, there was no significant change in $\dot{P}$ immediately after the flare, but $\dot{P}$ significantly *increased* over the years which followed (Woods *et al.* 2002; Woods 2003a; Woods 2003b). This gives evidence that the net effect of the 1998 magnetic activity episode, including the giant flare, was to *increase* the global twist angle and global currents, in a way that did not immediately affect the near-open field lines, far from the star (C. Thompson, private communication). A complete discussion of SGR torque variations will be given elsewhere. Here I simply want to point out that models of giant flares that posit that the whole magnetosphere is significantly restructured, with largely dissipated currents, are not supported by SGR timing data.

This concludes my short list of new evidence. Thompson, Lyutikov & Kulkarni (2002; §5.6) gave three additional semi-empirical arguments for crustal shifts during the August 27 flare. They also gave one countervailing argument, based upon the softening of SGR 1900+14's spectrum after the giant flare. But later work (Lyubarski, Eichler & Thompson 2002) suggested that the post-burst emission was dominated by surface afterglow with a soft, thermal spectrum (which is presumably modified by scattering outside the star).

Lyutikov (2003) noted five pieces of evidence that favor magnetospheric instabilities. His first point was invoked by TLK to argue the other way, so I think that this evidence is ambiguous. Some of his other points might have alternative interpretations, as he himself notes. In particular, the mild statistical anti-correlation between burst fluence and hardness (Göğüş *et al* 2002) could be due to emission-physics effects, independent of trigger details, as suggested by Göğüş *et al.* Incidentally, this mild trend seems to go the other way in AXP bursts (Gavriil, Kaspi & Woods 2003).

33.5 Discussion: crust-yielding in magnetars

A magnetar's crust is a degenerate, inhomogeneous Coulomb solid in a regime of high pressure, magnetization, and stress which has no direct experimental analog. It lies atop a magnetized liquid crystal which is subject to unbearable Maxwell stresses from the core below. Its behavior is thus quite uncertain. The crust cannot fracture like a brittle terrestrial solid, which develops a propagating crack with a microscopic void (Jones 2003), but it may experience other instabilities, such as "mock fractures" (§33.2 above).

The magnetar model was developed in a series of papers by Thompson and myself that invoked magnetically-driven crust fractures. Most of this work will remain valid if magnetar crusts prove to yield only plastically with no instabilities. However, some of the outburst physics would return closer to the original conception of DT92 and Paczyński (1992), along with several other changes.

For example, TD96 considered "Hall fracturing" in the crusts of magnetars. This was a consequence of "Hall drift" (Jones 1988; Goldreich & Reisenegger 1992), whereby the Hall term in the induction equation drives helical wrinkles in the magnetic field, which stress the crust. In magnetars, but not in radio pulsars, the wrinkled field is strong enough to drive frequent, small-scale crust failure as ambipolar diffusion within the core forces flux across the crust from below. TD96 suggested that this crust-yielding occurs in small fractures which generate a quasi-steady flux of high-frequency Alvén waves, energizing the magnetosphere and driving a diffuse wind out from the star. There seemed to exist direct evidence for this: a bright, compact radio nebula that was believed to surround SGR 1806–20 (Kulkarni *et al.* 1994; Vasisht *et al.* 1995; Frail *et al.* 1997). However, the high power of this nebula required rather implausibly optimized physical parameters and efficiencies, which caused concern.[7] Then Hurley *et al.* (1999b) found evidence that SGR 1806–20 is not precisely coincident with the central peak of the radio nebula. *Chandra* measurements verified that the SGR is displaced 12″ from the radio core (Eikenberry *et al.* 2001; Kaplan *et al.* 2002). It now seems likely that many, or perhaps all, Hall-driven yields in a magnetar's crust are plastic, occurring via dislocation glide in the outer crust ($\rho < \rho_{drip}$) and microscopic shear-layers at depth. This probably dissipates magnetic energy locally as heat, rather than as Alfvén waves in a corona. Still the basic analysis of TD96 is valid with this reinterpretation.

I want to emphasize that crustquakes in magnetars cannot be ruled out.[8] Besides sudden yields of nuclear pasta and "mock fractures" involving fault-line

[7] Because the magnetar model predicted that SGR 1806–20 was rotating slowly (as later verified; Kouveliotou *et al.* 1998) this nebula could not have been rotation-powered. Fracture-driven Alfvén waves from an active, vibrating crust thus seemed necessary, beginning in October 1993 (when the radio nebula discovery was announced at the Huntsville GRB Workshop: Frail & Kulkarni 1994; Murakami *et al.* 1994) through most of the 1990's. The probability for a chance overlap of the radio plerion core with the ~ 1 arcmin ASCA X-ray box for the SGR (Murakhami *et al.* 1994) was initially estimated in the range $\lesssim 10^{-6}$. When a coincident, extremely reddened Luminous Blue Variable (LBV) star was discovered (Kulkarni *et al.* 1995; van Kerkwijk *et al.* 1995) smaller probabilities for chance coincidence were implied, so the LBV star was presumed to be a binary companion to the plerion-powering neutron star. It turns out that the LBV star may be the brightest star in the Galaxy, with luminosity $L \sim 5 \times 10^6 L_\odot$ (Eikenberry *et al.* 2002), thus it probably can power the radio nebula by itself (Gaensler *et al.* 2001; Corbel & Eikenberry 2003). The chance for this$\sim 200 M_\odot$ star to lie within 12″ of another nearly-unique galactic star, SGR 1806–20, is fantastically small. This seems to be a lesson in the dangers of *a posteriori* statistics, or the treachery of Nature, or both.

[8] There is evidence for quite localized crust shifts during some magnetar outbursts. The radiative area of the thermal afterglow of the 1998 August 29 event was only ~ 1 percent of the neutron star area (Ibrahim *et al.* 2001) consistent with an "aftershock" adjustment along a fault zone that was active in the August 27 flare. A similarly small radiative area was found following the 2001 April 28 burst which seemed to be an aftershock of the April 18 flare (Lenters *et al.* 2003). Observations of AXP 1E2259 + 586 during its June 2002 activity (Woods *et al.* 2003) showed an initial, hard-spectrum, declining X-ray transient with a very small emitting area during the first day following the glitch, while the emitting area of the slowly-declining thermal afterglow observed over the ensuing year was a sizable fraction of the star's surface. This suggests that there was a small region of the crust where the magnetic field was strongly sheared, perhaps along a fault; and a large area in which it experienced more distributed plastic failure.

liquification (§2 above), magnetic-mechanical instabilities in the outer layers of magnetars, where magnetic pressure is not insignificant compared to the material pressure, may be associated with the emergence of magnetic flux, as in solar activity (e.g., Solanski *et al.* 2003). Wherever crustal fields exceed $\sim 10^{16}$ Gauss, intrinsic magnetization instabilities may be possible (Kondratyev 2002). Finally, rapid stress-changes exerted on the crust by the evolving core field from below, or by a flaring corona from above, could drive catastrophic failure.

33.6 Conclusions

In conclusion, a magnetar is a sun with a crust. Both crustal and coronal instabilities are possible, as well as instabilities within the core, which is coupled to the crust from below by the diffusing magnetic field. Physical conditions are much more complicated than those that prevail on either the Earth or the Sun, and the available data is much more fragmentary, so the challenge of understanding these stars is great.

In this review, I have described evidence that rapid interior stellar adjustments occur during some magnetar outbursts and bursting episodes. Based on this evidence, it seems likely that plastic crust failure initiates bursting episodes in SGRs and AXPs, by triggering a sequence of reconnection instabilities in the magnetosphere that are observed as "ordinary" common SGR (and AXP) bursts. Ongoing, relatively rapid plastic motion of patches of crust during these burst-active episodes (compared to what occurs in the quiescent state) could explain why bursts come in "swarms" with the time between bursts much longer than the durations of the bursts themselves. No pure magnetospheric instability or cascade could show this disparity of timescales.

The "relaxation system" behavior found by Palmer (1999) may be due to the steady loading of magnetic free energy within the magnetosphere by the plastic motion of magnetic footpoints. Palmer's "energy reservoir" would then be the sheared (or twisted) components of the exterior magnetic field, steadily driven by plastic motion of the magnetic footpoints during active periods, and undergoing sporadic, catastrophic dissipation in bursts of reconnection.[9] If the reservoir is an arch of field lines with one footpoint anchored on a circular cap of crust of radius a that is slowly twisting at rate ϕ, then the loading rate is $\dot{E} \sim (1/4)a^3 B^2 \phi$, independent of the length of the arch in a first, crude estimate. For a cap diameter $\sim$ 1 km, comparable to the crust depth, this implies $\phi \sim 0.06\ (B_{14}/3)^{-2}\ (a/0.5\ \mathrm{km})^{-3}$ radians/day during Palmer's interval B, and 25 times slower during Palmer's interval A.

There is little doubt that profound exterior reconnection occurs in magnetar flares (TD95). There are two triggering possibilities:

[9] This differs from previous suggested explanations of the Palmer Effect, which involved energy reservoirs within the crust.

(1) A sudden, twisting crust-failure might occur at the flare onset, so that significant (magnetic) energy from within the star contributes to the flare, communicated outward by the crust motion. If the solid crust yields plastically, then this would require a sudden stress-change applied upon the solid from below; but crustal instabilities cannot be ruled out (§5).
(2) Flares might develop in the magnetosphere, with little energy communicated from below on the time-scale of the flare. This could be a spontaneous instability reached via incremental motion of the magnetic footpoints; but ongoing plastic failure of the crust seems more likely as a trigger.

Clearly, these are not fully-distinct possibilities. Let us set the dividing line at $\sim$0.1 of the energy coming from within. Then, at present, I favor mechanism (1) for the 1998 June 18 flare from SGR 1627–30; and mechanism (2) for the 1998 August 27 flare from SGR 1900 + 14.

The back-reaction of an exterior magnetic stress-change on the crust, which occurs when the exterior field-twist or shear is relieved by reconnection, could drive shallow crustal failure and heating that is consistent with the August 27 flare afterglow (Lyubarsky, Eichler & Thompson 2002). But this back-reaction could not account for the 1988 June 18 afterglow if interpreted as deep crust-heating (Kouveliotou *et al.* 2003). The slow-peaking, tail-free June 18 event (Mazets *et al.* 1999b) plausibly involved a deep-crust and/or core adjustment in a star with a relatively quiet, relaxed magnetosphere, far from the critical state, so that little exterior reconnection was induced (relative to the giant flares).

Note that even the August 27 event was probably not a spontaneous, pure magnetospheric instability. A soft-spectrum precursor-event detected 0.45 s before the onset of the 1998 August 27 event (Hurley *et al.* 1999a; Mazets *et al.* 1999a) suggests that the crust was experiencing an episode of accelerated plastic failure. Plastic creep probably continued during the first $\sim$40 seconds after the flare's hard spike, giving rise to the "smooth tail" part of the light curve (Feroci *et al.* 2001; §7 in TD01). Subsequent spindown measurements (§4) suggest that large-scale currents in the magnetosphere were enhanced rather than dissipated during the magneticallyactive, flaring episode.

Of course, much more work is needed to develop and test these hypotheses. Many mysteries persist, but it seems that the magnetar model has the physical richness needed to accommodate diverse observations of SGRs and AXPs. The path to full scientific understanding of these objects will no doubt be long and interesting.

Acknowledgements

I thank Chris Thompson, Malvin Ruderman and Ethan Vishniac for discussions.

References

Barat *et al.* 1983, A&A, 126, 400
Brush, S. G., Sahlin, H. L. & Teller, E. 1966, J. Chem Phys., 45, 2101
Chen, K., Bak, P. & Obukhov, S. P. 1991, Phys. Rev. A, 43, 625
Cheng, B., Epstein, R. I., Guyer, R. A. & Young, A. C. 1996, Nature, 382, 518
Cline, T. B. *et al.* 1980, ApJ, 237, L1
Cline, T. B. 1982, in *Gamma Ray Transients and Related Astrophysical Phenomena*, ed. R. E. Lingenfelter *et al.* (AIP: New York) p. 17
Corbel, S. & Eikenberry, S. 2003, A & A (in press) astro-ph/0311303
Crosby, N. B., Aschwanden, M. J. & Dennis, B. R. 1993, Sol. Phys., 143, 275
De Blasio, F. V. 1995, ApJ, 452, 359
Duncan, R. C. & Thompson, C. 1992, ApJ, 392, L9 (DT92)
Eikenberry, S. S. *et al.* 2001, ApJ, 563, L133
Feroci, M., Hurley, K., Duncan, R. C. & Thompson, C. 2001, ApJ, 549, 1021
Feroci, M., *et al.* 2003, ApJ, 596, 470
Frail, D. & Kulkarni, S. R. 1994, in *Gamma Ray Bursts: Second Huntsville Workshop*, eds. G. J. Fishman, J. J. Brainerd & K. Hurley (AIP: New York) p. 486
Frail, D., Vasisht, G. & Kulkarni, S. R. 1997, ApJ, 480, L129
Gaensler, B. M., Slane, P. O., Gotthelf, E. V. & Vasisht, G. 2001, ApJ, 559, 963
Gavriil, F. P., & Kaspi, V. M. 2002, ApJ, 567, 1067
Gavriil, F. P., Kaspi, V. M., & Woods, P. M. 2003, ApJ (in press) astro-ph/0310852
Goldreich, P. & Reisenegger, A. 1992, ApJ, 395, 250
Göğüş, E., Woods, P. M., Kouveliotou, C., van Paradijs, J., Briggs, M. S., Duncan, R. C., & Thompson, C. 1999, ApJ, 526, L93
Göğüş, E. *et al.* 2000, ApJ, 532, L121
Göğüş, E. *et al.* 2001, ApJ, 558, 228
Heyl, J. S. & Hernquist, L. 1999, MNRAS, 304, L37
Hurley, K. *et al.* 1999a, Nature, 397, 41
Hurley, K. *et al.* 1999b, ApJ, 523, L37
Ichimaru, S. 1982, Rev. Mod. Phys. 54, 1017
Ichimaru, S. *et al.*, 1983, ApJ, 265, L83
Ibrahim, A., Strohmayer, T. E., Woods, P. M., Kouveliotou, C., Thompson, C., Duncan, R. C., Dieters, S., van Paradijs, J. & Finger M. 2001, ApJ, 558, 237
Jones, P. B. 1988, MNRAS, 233, 875
Jones, P. B. 1999, Phys Rev Lett, 83, 3589
Jones, P. B. 2001, MNRAS, 321, 167
Jones, P. B. 2003, ApJ, 595, 342
Kaplan *et al.* 2002, ApJ, 564, 935
Katz, J. I. 1986, J. Geophys. Res., 91, 10,412
Kaspi, V. M., Chakrabarty, D. & Steinberger, J. 1999, ApJ, 525, L33
Kaspi, V. M. *et al.* 2003, ApJ, 588, L93
Kaspi, V. M. & Gavriil, F. P., 2003, ApJ, in press astro-ph/0307225
Kondratyev, V. N. 2002, Phys Rev Letters, 88, 221101
Kouveliotou, C., Dieters, S., Strohmayer, T., van Paradijs, J., Fishman, G. J., Meegan, C. A., Hurley, K., Kommers, J., Smith, I., Frail, D. & Murakhami, T. 1998, Nature, 393, 235
Kouveliotou, C., *et al.* 2001, ApJ. 558, L47
Kouveliotou, C., *et al.* 2003, ApJ, 596, L79
Kulkarni, S. R. *et al.* 1994, Nature, 368, 129
Kulkarni, S. R. *et al.* 1995, ApJ, 440, L61

Lazarian, A. & Vishniac, E. T. 1999, ApJ, 517, 700
Lenters, G. T. *et al.* 2003, ApJ, 587, 761
Lu, E. T. *et al.* 1993, ApJ, 412, 841
Lynden-Bell, D. & Boily, C. 1994, MNRAS, 267, 146
Lyutikov, M. 2003, MNRAS, 346, 540
Lyubarsky, E., Eichler, D., & Thompson, C. 2002, ApJ, 580, L69
Mazets *et al.* 1979, Nature, 282, 365
Mazets *et al.* 1999a, Astron. Lett., 25(10), 635
Mazets, E. P., Aptekar, R. L., Butterworth, P. S., Cline, T. L., Frederiks, D. D., Golenetskii, S. V., Hurley, K., & Il'inskii, V. N. 1999b, ApJ, 519, L151
Mitchell, T. B., Bollinger, J. J., Itano, W. M. & Dubin, D. H. E. 2001, Phys. Rev. Lett., 87, 183001
Murakami, T. *et al.* 1994, Nature, 368, 127
Murakami, T. *et al.* 1994, in *Gamma Ray Bursts: Second Huntsville Workshop*, eds. G. J. Fishman, J. J. Brainerd & K. Hurley (AIP: New York) p. 489
Paczyński, B. 1992, Acta Astron., 42, 145
Palmer, D. N. 1999, ApJ, 512, L113
Patel, S. K. *et al.* 2001, ApJ, 563, L45
Pethick, C. J. & Potekhin, A. Y. 1998, Phys. Lett. B, 427, 7
Pethick, C. J. & Ravenhall, D. G. 1995, Ann Rev Nuc Sci, 45, 429
Ruderman, M. 1991, ApJ, 382, 576
Solanski, S. K. *et al.* 2003, Nature, 425, 692
Terrell, J. *et al.* 1980, Nature, 285, 383
Thompson, C., & Duncan, R. C. 1995, MNRAS, 275, 255 (TD95)
Thompson, C., & Duncan, R. C. 1996, ApJ, 473, 322 (TD96)
Thompson, C., & Duncan, R. C. 2001, ApJ, 561, 980 (TD01)
Thompson, C., Duncan, R. C., Woods, P. M., Kouveliotou, C., Finger, M. H., & van Paradijs, J. 2000, ApJ, 543, 340
Thompson, C., Lyutikov, M., & Kulkarni, S. R. 2002, ApJ, 574, 332 (TLK)
Usov, V. V. 1994, ApJ, 427, 984
Van Horn, H. M. 1991, Science, 252, 384
Vasisht, G., Frail, D. A. & Kulkarni, S. R. 1995, ApJ, 440, L65
Wigner, E. 1934, Phys. Rev., 46, 1002
Wolfson, R. 1995, ApJ, 443, 810
Woods, P. M. 2003, in AIP Conf. Proc. 662 (AIP: N.Y.) p. 561 astro-ph/0204369
Woods, P. M. 2003, in *High Energy Studies of Supernova Remnants and Neutron Stars*, (COSPAR 2002) astro-ph/0304372
Woods, P. M., *et al.* 2000, ApJ, 535, L55
Woods, P. M., *et al.* 2001, ApJ, 552, 748
Woods, P. M., Kouveliotou, C., Göğüş, E., Finger, M. H., Swank, J., Markwardt, C. B., Hurley, K., van der Klis, M. 2002, ApJ, 576, 381
Woods, P. M. *et al.* 2003, ApJ (in press) astro-ph/0310575

34

Turbulent MHD jet collimation and thermal driving

P. T. Williams

Los Alamos National Laboratory, Los Alamos, NM, USA

Abstract

We have argued that MHD turbulence in an accretion disk naturally produces hoop-stresses, and that in a geometrically-thick flow these stresses could both drive and collimate an outflow. We based this argument on an analogy of turbulent MHD fluids to viscoelastic fluids, in which azimuthal shear flow creates hoop-stresses that cause a variety of flow phenomena, including the Weissenberg effect in which a fluid climbs a spinning rod.

One of the more important differences between the Weissenberg effect and astrophysical jets is the source of power. In our previous analysis, we only considered the power due to the spin-down torque on the central object, and thus found that we could only drive an outflow if the central object were maximally rotating. Here we take into account the energy that is liberated by the accreting matter, and describe a scenario in which this energy couples to the outflow to create a thermodynamic engine.

34.1 Introduction

We wish to discuss here in simple language some of our ideas regarding jet collimation and acceleration. In this paper, we will concentrate on the basic intuitive notions rather than the mathematics, which we have discussed in print elsewhere (see references below).

34.2 Review: turbulence models and jets

We have argued (Williams 2001; see also Ogilvie 2001) that the stress due to magnetohydrodynamic (MHD) turbulence in ionized accretion disks – such as, but not limited to, the turbulence driven by the magnetorotational instability (MRI) – behaves more like the stress in a viscoelastic fluid than the stress in a viscous fluid.

Viscoelastic fluids are a broad category of non-Newtonian fluids. Biological liquids with high nucleic acid or protein polymer concentrations, such as mucus, are one example. American popular culture provides the more pleasant example of Silly Putty® material, originally made by polymerizing silicone oil with boric acid.

These fluids are "goopy" and "stretchy"; the stress in them does not relax quickly, putting them somewhere between Newtonian fluids and elastic solids. The tangled polymers take time to rearrange themselves when the fluid suffers a distortion. Now let us replace these tangled polymers with a tangled magnetic field in a conducting fluid. Since we are taught that magnetic fields in ideal MHD are like elastic strings under tension, it seems natural to think of a tangled mess of magnetic fields as a goopy, stretchy fluid, as well. This is by no means a perfect analogy, but we argue that it is a better analogy than that provided by purely *viscous* fluids.

Deformation of a viscoelastic fluid will typically create different normal stresses in different directions. This is why it is possible to wrap Silly Putty® material around one's finger while temporarily maintaining the material under linear tension. Note that in an orthogonal coordinate system, normal stresses are the stress components along the diagonal of the stress tensor; shear stresses are off-diagonal. These fluids also have differing normal stresses in azimuthal shear flow, exhibiting tension in the azimuthal direction. This is a hoop-stress.

Such stress is responsible for various effects not seen in ordinary Newtonian fluids. One such effect we have discussed previously is that a spinning sphere immersed in a viscoelastic fluid may cause a meridional circulation – i.e., a poloidal flow – in which fluid is pulled in along the equator and pushed out at the poles. This is opposite to the meridional flow in a purely viscous fluid, and it is closely related to the Weissenberg effect, in which a fluid climbs a spinning rod.

By the analogy we have drawn between MHD turbulence and viscoelastic fluids, we expect that the turbulence in accretion disks naturally creates hoop-stresses as well. In fact, these hoop-stresses can be seen in simulations of the MRI (Hawley, Gammie & Balbus 1995), as pointed out by Williams (2003). Hawley *et al.* did not discuss these hoop-stresses, but one has only to look at their tables for the various components of the stress tensors to see them. In shearing-sheet simulations of disk turbulence, such as they performed, a hoop-stress appears as a stress in the direction of shear. These hoop-stresses are of the right order of magnitude to collimate jets. This is related to, but distinct from, the notion put forth (Akiyama & Wheeler 2002, Akiyama *et al.* 2003) that the MRI saturates to create an azimuthal field that collimates jets: The hoop-stresses in our picture are created entirely by a tangled field.

All of this has been discussed previously by us. Here we only briefly comment on the source of energy for an outflow. This is related in a simple way to the flow topology, which we discuss below.

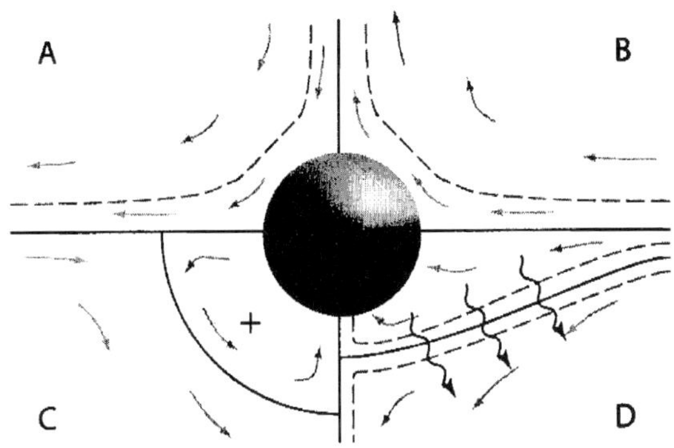

Fig. 34.1. A rough sketch of the meridional flow in three laboratory experiments (A–C), and the hypothetical flow in one astrophysical case (D). In the latter case, the sphere is somewhat fancifully representative of a generic central object. Flow is qualitatively indicated with curved arrows and fiducial streamlines (dashed lines). Separatrices are shown as solid lines. All flows are axisymmetric and vertically mirror-symmetric. Squiggles represent flow of thermal and mechanical energy; "+" denotes an o-point.

34.3 Flow topology and jet thermodynamics

Let us first sketch the basic arrangement of the streamlines of flow for the laboratory phenomenon mentioned above. Figure 34.1-A shows the meridional flow of a Newtonian fluid at low Reynolds number. The azimuthal Stokes flow is not shown, but it corresponds simply to corotation at the surface of the sphere, gradually dropping to zero at infinity. This Stokes flow is driven by the viscous torque or couple on the sphere and the outwards transport of angular momentum to the surrounding medium. This outward transport of energy is directly analogous to the outward "viscous" transport of energy in radiatively efficient thin-disk theory that is responsible for the infamous factor of three discrepancy between the local dissipation and the thermal emission (see Frank, King & Raine 2002, p. 86–87). The azimuthal flow drives the secondary meridional flow by the centrifugal ("inertial") forces.

The ratio of the stress relaxation time of the fluid to a representative timescale for the flow (such as the rotation timescale in this case) is variously called the Weissenberg number or the Deborah number. If we keep the Reynolds number low but gradually increase the Weissenberg-Deborah number, the secondary (i.e. meridional) flow reverses to the flow qualitatively sketched in Figure 34.1-B, as the elastic hoop-stresses begin to dominate the inertial forces.

The exact flow pattern depends not just on the Reynolds number and the Weissenberg number, but on the full rheological properties of the material. For example, the flow may become more topologically complex, with the introduction of a flow separatrix as shown in Figure 34.1-C. Broadly speaking, however, the viscoelastic fluid has stresses that tend to act opposite to the inertial stresses, as shown.

For the hypothetical analogous astrophysical phenomenon, we assume that the flow is steady and axisymmetric as it is in the laboratory case. Of course, if the astrophysical flow is turbulent, as we are assuming, then streamlines must be interpreted as some type of average over the turbulent fluctuations; we assume here that it is possible to define such an average.

The source of power for this flow is the torque of the sphere on the surrounding medium, so that for the analogous astrophysical problem, we found that, roughly, we could only power a jet strong enough to reach escape velocity if the central object were spinning near the centrifugal limit. This is because the natural scaling for the jet launch speed in this scenario is the surface rotation speed of the star. Of course, in the astrophysical case, there is also an inward advective transport of angular momentum, as well as gravity and compressibility, but we still feel this analogy may be informative.

The mass outflow in an astrophysical jet does not need to equal the mass inflow in the disk, unlike the laboratory phenomenon described above in which the central sphere acts as neither a source nor sink of material. Let us follow a fluid element in the accretion flow. This fluid element either ends up being accreted onto the central object or it does not. It makes sense then to introduce a streamline that separates the flow of material that ends up being accreted from the material that does not, *i.e.* a separatrix. Although in principle the topology of this separatrix could be quite complex, let us imagine it to be quite simple, as we have drawn it in Figure 34.1-D.

As has been long appreciated, a Keplerian thin disk accreting onto a stationary central sphere dissipates fully *half* of its energy in the boundary layer where the aziumthal velocity is slowed from Keplerian to zero. Even in the more general case where the flow is not Keplerian (nor geometrically thin), and the central object is rotating as we have assumed previously, a relatively very large amount of energy is released in the central few "stellar radii" of the accretion flow. If some of the energy released by material that ultimately accretes onto the central star could be tranmitted to the material that does not accrete, we might have a very effective way of powering a jet. This in itself is not a novel idea, but we discuss this notion within the context of our previous work. The common term "central engine" would be very appropriate for this driving mechanism; just as in the original steam engine, a working fluid (i.e. the material that ultimately ends up in the jet) would be compressed, heated by a furnace (which in this case is the central accretion region, powered by the release of gravitational binding energy of the matter that is ultimately accreted), and allowed to expand again. This would represent a jet driven in part by the reservoir of angular momentum in the central object, and in part by the energy of accretion.

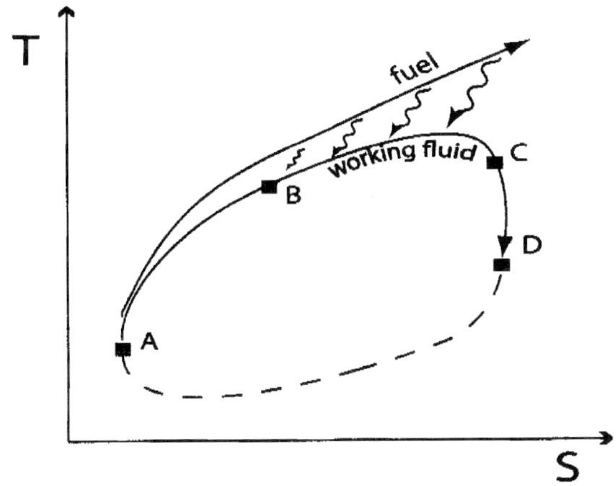

Fig. 34.2. The thermal cycle of the "central engine." The first leg ($A \rightarrow B$) is accretion in a relatively thin disk. Much more energy is transfered to the working fluid in the second leg, from the fuel that is accreted onto the central object. The third leg is expansion into an outflow. The fourth leg – reincorporation into the ISM followed by accretion again – is presented as a conceptual aid.

The contribution of *thermal* energy to the outflow may be visualized as a virtual thermodynamic engine cycle, as shown in Figure 34.2. This picture does not take into account the mechanical energy that we discussed previously, however, and the relative contribution of the two is uncertain.

34.4 Conclusion

We argue that independent of any *dynamo* process (except perhaps in the loosest sense of the word), collimating forces are easily created by the same MHD turbulence responsible for angular momentum transport in disks, so that angular momentum transport and outflow collimation may be different sides of the same coin. The argument may be made *a fortioriori* in the case of the MRI because it is an MHD rather than a hydro instability, but the turbulent MHD collimation will exist in either case. We have simply addressed here the consideration that the spindown stress is not the only source of power. Secondly, we wish to make quite explicit the notion of a central *engine* as a thermodynamic engine, in which there is a fuel and a working fluid. These materials are of the same substance, the difference between the two being whether the material is ultimately accreted or not.

Finally, we note here that simulations of nonradiative accretion flows seem to show that the MRI collimates jets (Hawley & Balbus, 2002). Note also though that this collimation was claimed to be due to magnetic *pressure*. The collimation we have described here and in our previous work, however, is due to the so-called curvature term – i.e, magnetic tension.

Acknowledgements

I am grateful to Paul Bradley, my advisor at Los Alamos, for giving me the time to present this work, and to Craig Wheeler for his encouragement.

References

Akiyama, S. & Wheeler, J. C., 2002. *astro-ph/***0211458**.

Akiyama, S., Wheeler, J. C., Meier, D. L. & Lichtenstadt, I., 2003. *Astrophys. J.*, **584**, 954–970.

Frank, J., King, A. & Raine, D. 2002. Accretion Power in Astrophysics, 3ed, (Cambridge: University Press).

Hawley, J. F. & Balbus, S. A., 2002. *Astrophys. J.*, **573**, 738–748.

Hawley, J. F., Gammie, C. F. & Balbus, S. A., 1995. *Astrophys. J.*, **440**, 742–763.

Ogilvie, G. I., 2001. *M. N. R. A. S.*, **325**, 231–248.

Williams, P. T., 2001. *astro-ph/***0111630**.

Williams, P. T., 2003. ASP Conf. Ser. Vol. 287, Galactic Star Formation Across the Stellar Mass Spectrum, ed. J. M. deBuizer & N. S. van der Bliek (San Francisco:ASP), 351–356. *astro-ph/***0206230**.

Williams, P. T., 2003. *astro-ph/***0212556**.

35

The interplay between nuclear electron capture and fluid dynamics in core collapse supernovae

W. R. Hix

Department of Physics and Astronomy, University of Tennessee, Knoxville, TN 37996 USA Physics Division, Oak Ridge National Laboratory, Oak Ridge, TN 37831 USA Joint Institute for Heavy Ion Research, Oak Ridge National Laboratory, Oak Ridge, TN 37831 USA

O. E. B. Messer

Department of Astronomy and Astrophysics, University of Chicago, Chicago, IL 60637 Center for Astrophysical Thermonuclear Flashes, University of Chicago, Chicago, IL 60637 Physics Division, Oak Ridge National Laboratory, Oak Ridge, TN 37831 USA

A. Mezzacappa

Physics Division, Oak Ridge National Laboratory, Oak Ridge, TN 37831 USA

Abstract

As we investigate the manifestly multi-dimensional nature of core collapse supernovae, the connection between microscopic physics and macroscopic fluid motion must not be forgotten. As an example, we discuss nuclear electron capture and its impact on the supernova shock. Though electron capture on nuclei with masses larger than 60 is the most important nuclear interaction to the dynamics of stellar core collapse, in prior simulations of core collapse it has been treated in a highly parameterized fashion, if not ignored. With a realistic treatment of electron capture on heavy nuclei come significant changes in the hydrodynamics of core collapse and bounce. We discuss these as well as their ramifications for the post-bounce evolution in core collapse supernovae.

35.1 Introduction

The many observations of asymmetries in core collapse supernovae, coupled with the failure of spherically symmetric simulations of the neutrino reheating paradigm to produce explosions, has persuaded the community that multidimensional effects like convection and other fluid instabilities must be vital elements of the supernova mechanism (Wilson & Mayle 1993, Herant *et al.* 1994, Burrows *et al.* 1995, Fryer & Warren 2002) though, even with these convective enhancements, explosions are not guaranteed (Janka & Müller 1996, Mezzacappa *et al.* 1998, Buras *et al.* 2003). This view has been reinforced in recent years by the failure of more accurate spherically symmetric multigroup Boltzmann simulations to produce explosions (Rampp & Janka 2000, Mezzacappa *et al.* 2001, Liebendörfer *et al.* 2001, Thompson *et al.* 2003). Proto-neutron star (PNS) instabilities, such as convection or *neutron fingers* (which are responsible for Wilson & Mayle's explosions) are driven by lepton and entropy gradients caused by the emission of neutrinos from the PNS. Convection behind the shock originates from gradients in entropy that result from the stalling of the shock and grow as the matter is heated by interaction with the neutrinos streaming from below. Therefore investigations of the multi-dimensional nature of supernovae must include study of the wide variety of nuclear and weak interaction physics that are also important to the supernova mechanism.

35.2 Electron capture in supernovae

Once the supernova shock forms, emission and absorption of electron neutrinos and antineutrinos on the dissociation-liberated free nucleons is the dominant process. However, during core collapse, electron capture on nuclei plays a dominant role by significantly altering the electron fraction and entropy, thereby determining the structure of the star and, in particular, the strength and location of the initial supernova shock. Therefore, improvements in the treatment of electron capture can alter the initial conditions for the entire postbounce evolution of the supernova, including the gradients responsible for driving fluid instabilities.

Calculation of the rate of electron capture on heavy nuclei in the collapsing core requires two components: the appropriate electron capture reaction rates and knowledge of the nuclear composition. The inclusion of electron capture within a multigroup neutrino transport simulation adds an additional requirement: information about the spectra of emitted neutrinos. Unlike stellar evolution and supernova nucleosynthesis simulations, wherein the nuclear composition is tracked in detail via a reaction network (see, e.g., Hix & Thielemann 1999 and references therein), in simulations of the supernova mechanism the composition in the iron core is calculated by the equation of state assuming nuclear statistical equilibrium. Typically, the information on the nuclear composition provided by the equation of state is limited to the mass fractions of free neutrons and protons, α-particles, and the sum of all heavy nuclei as well as the identity of an average heavy nucleus, calculated in the liquid drop framework (Lattimer & Swesty 1991). In most recent supernova simulations the treatment of nuclear electron capture introduced by Bruenn (1985) is employed. This prescription treats electron capture on heavy nuclei through a generic ${}^1f_{7/2} \rightarrow {}^1f_{5/2}$ Gamow-Teller resonance [Bethe *et al.* 1979] in the average heavy nucleus identified by the equation of state. Because this treatment does not include additional Gamow-Teller transitions, forbidden transitions, or thermal unblocking [Langanke *et al.* 2003], electron capture on heavy nuclei ceases when the neutron number of the average nucleus exceeds 40. As a result, electron capture on protons was seen to dominate the later phases of collapse in prior simulations.

As a major advance over the simple treatment on nuclear electron capture used previously in supernova simulations, we have developed a treatment based on the shell model electron capture rates from Langanke & Martínez-Pinedo (2000, LMP) for $45 \leq A \leq 65$ and 80 reaction rates from a hybrid shell model- random phase approximation (RPA) calculation (Langanke *et al.* 2001a, Langanke *et al.* 2003) for a sample of nuclei with $A = 66$–112. For the distribution of emitted neutrinos, we use the approximation described by Langanke *et al.* (2001b). To calculate the needed abundances of the heavy nuclei, a Saha-like nuclear statistical equilibrium (NSE) is used, including Coulomb corrections to the nuclear binding energy (Hix 1995, Bravo & García-Senz 1999), but neglecting the effects of degenerate nucleons (El Eid & Hillebrandt 1980). We use the combined set of LMP & hybrid model rates to calculate an average neutrino emissivity per nucleus. The full neutrino emissivity is then the product of this average with the number density of heavy nuclei calculated by the equation of state. With the limited coverage of rates for $A > 65$, this approach provides the most reasonable estimate of what the total electron capture would be if rates for all nuclei were available. A more detailed description of our method, including tests of some assumptions made, will be presented in a forthcoming article (Hix *et al.* 2003).

35.3 Effects on core collapse

Simulations of the collapse, bounce, and post-bounce evolution of a $15M_\odot$ model (Heger *et al.* 2001) were carried out using both the Bruenn prescription as well as our LMP + hybrid treatment in our spherically symmetric AGILE-BOLTZTRAN code (Mezzacappa & Messer 1998, Liebendörfer *et al.* 2003). In these simulations, AGILE-BOLTZTRAN is employed in the O(v = c) limit with 6-point Gaussian quadrature to discretize the neutrino angular distributions and 12 energy groups spanning the range from 3 to 300 MeV to discretize the neutrino spectra.

Our improved treatment of nuclear electron capture has two competing effects. In lower density regions, where the average nucleus is well below the $N = 40$ cutoff of electron capture on heavy nuclei, the Bruenn parameterization results in more electron capture than the LMP + hybrid case. The lesser electron capture in our improved simulations also results in a slower collapse of the outer layers, reinforcing the diminished electron capture rate by reducing the density. In denser regions, the dominance of electron capture on heavy nuclei over electron capture on protons results in more electron capture in the LMP + hybrid case. The results of these competing effects can be seen in the upper panel of Figure 35.1, which shows the distributions of Y_e throughout the core at bounce (maximum central density).

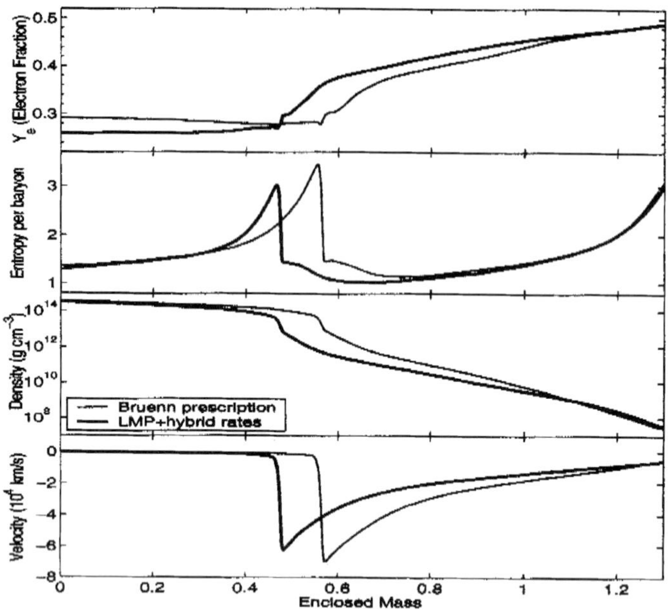

Fig. 35.1. The electron fraction, entropy and velocity as functions of the enclosed mass at the beginning of bounce for a 15 $M_\odot$ model. The thin line is a simulation using the Bruenn parameterization while the thick line is for a simulation using the LMP and hybrid reaction rate sets.

In addition to the marked reduction (12%) in the electron fraction in the interior of the PNS, this improved treatment also results in a nearly 20% reduction in the mass of the homologous core. At bounce this change in the homologous core manifests itself as a reduction in the mass interior to the formation of the shock from .67 $M_\odot$ to .57 $M_\odot$ in the LMP + hybrid case, as demonstrated in the lower panel of Figure 35.1. A shift of this size is significant dynamically because the dissociation of .1 $M_\odot$ of heavy nuclei by the shock costs 10^{51} erg, the equivalent of the explosion energy. In the LMP + hybrid case there is also an 11% reduction in the central density and a 7% reduction in the central entropy at bounce. The shock is also weaker with a 10% smaller velocity difference across the shock and 14% smaller entropy jump when the shock is fully developed.

35.4 Post-Bounce evolution

The launch of a weaker shock with more of the iron core overlying it, caused by the increase in deleptonization at high densities that results from our improvement in the treatment of electron capture on nuclei, makes a successful explosion more difficult. However, the reduced electron capture this treatment predicts in

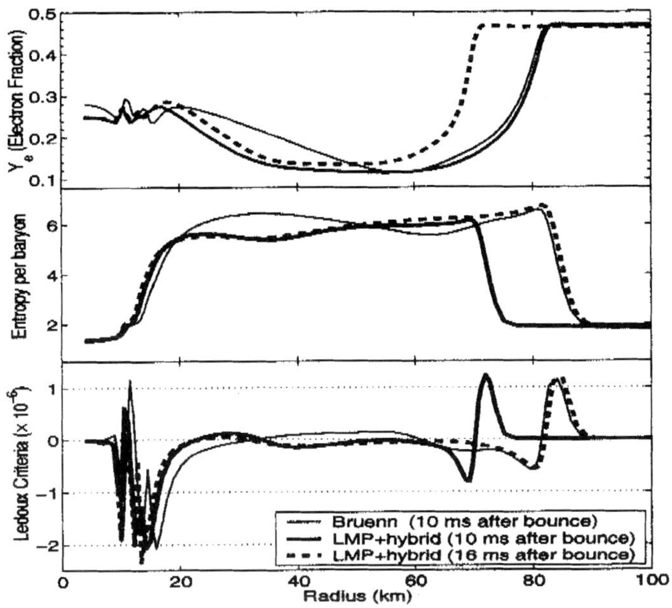

Fig. 35.2. The electron fraction, entropy and Ledoux criterion as functions of the radius soon after bounce. The thin line (B10) shows results a simulation using the Bruenn parameterization at 10 milliseconds after bounce while the thick lines are from the LMP + hybrid simulation set at 10 (solid, LMPH10) and 16 (dashed, LMPH16) milliseconds after bounce.

the lower density outer layers also plays an important role in the later behavior of the shock. Clearly evident in Figure 35.1 are reductions for the same mass element as large as a factor of 5 in density and 40% in velocity, greatly reducing the ram pressure at the shock from infalling matter. Though both models fail to produce explosions in the spherically symmetric limit, the slowed collapse of the outer layers allows the shock in LMP + hybrid case to reach 205 km, 9 km farther than in the Bruenn case. This maximum radial extent also occurs 30 ms ($\sim$25%) sooner in the LMP case than in the Bruenn case.

These changes also alter the gradients that drive the fluid instabilities in the PNS. Figure 35.2 shows the entropy and lepton gradients near the time of the onset of PNS convection. B10 and LMPH10 depict the results of the Bruenn and LMP + hybrid cases, both 10 ms after their respective maximum densities are reached. Clearly visible in the region between 20 km and 60 km are an inward displacement and steepening (flattening) of the negative lepton (entropy) gradient that results from our more accurate treatment of electron capture. LMPH16 presents the LMP + hybrid model at 16 ms after bounce, when its shock has reached a radius matching that of B10. Clearly the differences in the gradients described above are not transient, nor do they arise from the slower progress of the shock in the LMP + hybrid case. To illustrate the effects these changes in the entropy and lepton gradients could have on convection we have plotted the Ledoux criterion$(\partial \ln \rho/\partial \ln Y_e)(d \ln Y_e/dr) + (\partial \ln \rho/\partial \ln s)(d \ln s/dr)$ in the lower pane of Figure 35.2. Convection in the Ledoux approximation arises when this expression is positive. The broad spikes on the right that coincide with the shock are due to the numerical spreading of the shock and therefore should be ignored. In the model employing the Bruenn treatment there is an extended region of instability between 30 and 60 km, while in the LMP + hybrid case the unstable region is only 10 km across and displaced inward. This suggests that with accurate treatment of nuclear electron capture, PNS convection would be less extensive than previously thought and would occur deeper within the stellar core, providing an excellent example of the coupling of convective behavior to neutrino radiative processes.

35.5 Conclusion

We have demonstrated that supernova simulations with a modern treatment of electron capture differ significantly from previous models that employed more parameterized treatments. Though these improved models still fail to produce explosions in the spherically symmetric case, the differences in shock behavior are striking. The initial mass behind the shock when it is launched is reduced by 20%, with significantly ($\sim$ 10%) lower central densities, entropies, and electron fractions. In spite of an initially weaker shock, the maximum radius reached by the shock in these improved simulations is 5% larger and occurred 25% earlier. Furthermore the lepton and entropy gradients in the core differ significantly. Because these gradients will lead to changes in the convective behavior, we anticipate that PNS convection will be significantly less extensive than heretofore thought. These effects of improving nuclear electron capture serve as a concrete demonstration of the interplay between the microscopic physics and the macroscopic fluid motions and demonstrate the need for better understanding of physics at the smallest scales as well as at the largest scales.

Acknowledgements

This work would not have been possible without the efforts of our nuclear physics colleagues: K. Langanke, J. Sampaio, G. Martinez-Pinedo and D. J. Dean. The authors would also like to thank M. Liebendörfer, H.-T. Janka, S. W. Bruenn and F.-K. Thielemann for fruitful discussions. The work has been partly supported by the National Science Foundation under contracts PHY-0244783 and AST-9877130, by the Department of Energy, through the Scientic Discovery through Advanced Computing Programs, and by funds from the Joint Institute for Heavy Ion Research.

Oak Ridge National Laboratory is managed by UT-Battelle, LLC, for the U.S. Department of Energy under contract DE-AC05-00OR22725.

References

Bethe, H. A., Brown, G. E., Applegate, J., & Lattimer, J. M. 1979, Nucl. Phys. A, 324, 487
Bravo, E. & García-Senz, D. 1999, MNRAS, 307, 984
Bruenn, S. W. 1985, ApJS, 58, 771
Buras, R., Rampp, M., Janka, H.-T., & Kifonidis, K. 2003, Phys. Rev. Lett., 90, 241101
Burrows, A., Hayes, J., & Fryxell, B. A. 1995, ApJ, 450, 830
El Eid, M. F. & Hillebrandt, W. 1980, A&AS, 42, 215
Fryer, C. L. & Warren, M. S. 2002, ApJ, 574, L65
Heger, A., Langanke, K., Martínez-Pinedo, G., & Woosley, S. E. 2001, Phys. Rev. Lett., 86, 1678
Herant, M., Benz, W., Hix, W. R., Fryer, C. L., & Colgate, S. A. 1994, ApJ, 435, 339
Hix, W. R. 1995, PhD thesis, HARVARD UNIVERSITY.
Hix, W. R., Messer, O. E. B., Mezzacappa, A., Liebendoerfer, M., Sampaio, J. M., Langanke, K., Martinez-Pinedo, G., & Dean, D. J. 2003, Phys. Rev. Lett., submitted
Hix, W. R. & Thielemann, F.-K. 1999, ApJ, 511, 862
Janka, H.-T. & Müller, E. 1996, A&A, 306, 167
Langanke, K., Kolbe, E., & Dean, D. J. 2001a, Phys. Rev. C, 63, 32801
Langanke, K. & Martínez-Pinedo, G. 2000, Nucl. Phys. A, 673, 481
Langanke, K., Martínez-Pinedo, G., & Sampaio, J. M. 2001b, Phys. Rev. C, 64, 55801
Langanke, K., Martinez-Pinedo, G., Sampaio, J. M., Dean, D. J., Hix, W. R., Messer, O. E. B., Mezzacappa, A., Liebendoerfer, M., Janka, H.-T., & Rampp, M. 2003, Phys. Rev. Lett., 90, 241102
Lattimer, J. & Swesty, F. D. 1991, Nucl. Phys. A, 535, 331
Liebendörfer, M., Messer, O. E. B., Mezzacappa, A., Bruenn, S. W., Cardall, C. Y., & Thielemann, F.-K. 2003, ApJS, submitted
Liebendörfer, M., Mezzacappa, A., Thielemann, F.-K., Messer, O. E. B., Hix, W. R., & Bruenn, S. W. 2001, Phys. Rev. D, 63, 103004
Mezzacappa, A., Calder, A. C., Bruenn, S. W., Blondin, J. M., Guidry, M. W., Strayer, M. R., & Umar, A. S. 1998, ApJ, 493, 848 and 495, 911
Mezzacappa, A., Liebendörfer, M., Messer, O., Hix, W., Thielemann, F.-K., & Bruenn, S. 2001, Phys. Rev. Lett., 86, 1935
Mezzacappa, A. & Messer, O. E. B. 1998, J. Comp. Appl. Math, 109, 281
Thompson, T. A., Burrows, A., & Pinto, P. A. 2003, ApJ, 592, 434
Rampp, M. & Janka, H.-T. 2000, ApJ, 539, L33
Wilson, J. R. & Mayle, R. W. 1993, Phys. Rep., 227, 97

Part VI

Gamma-Ray Bursts

36

GRB 021004 and Gamma-Ray Burst distances

B. E. Schaefer

Louisiana State University, Baton Rouge Louisiana 70803 USA

Abstract

This article will cover two topics at the intersection of Gamma-Ray Bursts and supernovae that have been much studied by the Texas group with relevance for the 3-D structure of core collapse explosions. The first topic is the high-velocity and high-excitation absorption lines seen in GRB 021004 (and other more recent events). These lines must come from (likely clumpy) shells around the progenitor star, and hence can provide a unique means of knowing the nature of the exploding star. In particular, the lines imply that normal GRBs form from the core collapse of a massive star, and thus that GRBs are closely related to supernovae. The second topic is the four luminosity indicators for Gamma-Ray Bursts and their implications for cosmology. The validity of the luminosity (and hence distance) indicators is already well demonstrated, although the current accuracy of the distances is roughly a factor of three times worse than for Type Ia supernovae. With GRBs serving as standard candles visible out to redshifts of >12 or farther, they can be used for many of the same purposes in cosmology now reserved for supernovae at low redshifts. With the launch of Swift in 2004, hundreds of bursts can then be used to construct Hubble diagrams to $z \geq 5$, to measure the star formation rate to $z \sim 12$ or farther, and to serve as beacons for discovering the Gunn-Peterson effect.

36.1 GRB/SN connections

Gamma-Ray Bursts (GRBs) and supernovae (SNe) have long been connected. Before the discovery of GRBs, S. Colgate calculated that the shock breakout of a Type II SN should create a burst of gamma radiation (Colgate 1968; 1970; 1974). Based on Colgate's alert, Klebesadel, Strong, & Olson (1973) discovered the Gamma-Ray Burst phenomenon, although it was soon realized that the shock breakout from Type II SN would not occur at gamma ray photon energies. GRBs must necessarily pack large amounts of energy in a small volume, so attempts to link GRBs and SNe have persisted since 1973. From 1979 until the 1990's, a strong link (Felton 1982) was provided by the unique and bright burst seen on 5 March 1979 (Cline *et al.* 1980) coming from near the middle of a supernova remnant in the Large Magellanic Cloud (Evans *et al.* 1980). However, we now realize that this event is a separate subclass of bursts, called the Soft Gamma Repeaters, that apparently are magnetars and completely separate from the classical GRBs (Hurley 2000). With the discovery that GRBs are at cosmological distances (Metzger *et al.* 1997), some version of a supernova is apparently the only plausible means to get the required energy.

The first strong GRB/SN connection was made when GRB 980425 was found to have a coincidence in time and position to SN 1998bw (Galama *et al.* 1998). This GRB had a very smooth light curve (Fenimore & Ramirez-Ruiz 2000) plus a very long spectral lag (Norris, Marani, & Bonnell 2000) which implies that it must be extremely low in luminosity, while the SN was a highly unusual Type Ic SN with very high expansion velocities and a record breaking radio luminosity (Kulkarni *et al.* 1998). Despite this strong connection, both GRB and SN were so unusual that it was risky to generalize the connection to all events. With the motivation of the GRB980425/SN1998bw association, various groups started looking for other connections between SNe and GRBs. The first claim (Wang & Wheeler 1998) was that bright and well-observed Type Ib/c SNe are statistically correlated with GRB times and directions. This connection has been strongly rejected on statistical grounds (Deng 2001; Schaefer & Deng 2000) as well as through the use of better GRB error boxes (Kippen *et al.* 1998). In the last few years, there have been a variety of claims that poorly-observed small bumps in GRB afterglow light curves are due to an underlying SN (Bloom *et al.* 1999; Reichart 1999; Galama *et al.* 2000), but multiple bumps in the light curves occur in about half the well-observed afterglows on all time scales and at least four alternative physical mechanisms have been published to explain the bumps, so the claimed 'supernova bumps' have near zero utility. In all, by late 2002, the GRB community expected that normal GRBs are just some version of SNe, although there was essentially no evidence to support this bias.

In this setting, the bright GRB 021004 was well placed for spectroscopic observation with the Hobby-Eberly Telescope (HET) at McDonald Observatory, for which Craig Wheeler is the PI of a Target of Opportunity team. The HET spectra showed absorption lines from high excitation species which had a high velocity blue shift with respect to the host galaxy. As described in the next section, these lines could only have been formed in a massive shell around the GRB progenitor, and this implies that the progenitor was a very massive star at the stage of core collapse. As such, these lines suddenly became the first good evidence to connect normal GRBs with SNe.

Soon before the Wheeler Symposium, the HETE2 satellite discovered a relatively nearby ($z = 0.168$) burst (GRB 030329) which displayed an afterglow spectrum with a supernova component (like SN 1998bw) starting in the week after the burst (Stanek *et al.* 2003; Hjorth *et al.* 2003). With GRB 030329 being a normal event, a strong GRB/SN connection has been made and this can be extended to all classical GRBs. This is in agreement with the theoretical work of the Texas group that is identifying GRBs as coming from extreme cases of asymmetric core collapse.

SNe and GRBs are connected in a way other than the similarity of their origin; as recently GRBs have been realized to be standard candles. Here, 'standard candle' is used in the same sense as for Cepheids and Type Ia supernovae, in that a light curve property can be used to used to determine the sources' luminosity, and then its luminosity distance. The third section of the article will discuss the great potential of GRBs as tools of cosmology.

36.2 GRB 021004: a massive progenitor star surrounded by shells

GRB 021004 was discovered by the HETE2 satellite, and quickly discovered to have a bright optical afterglow. The HET was immediately used to record optical spectra (Schaefer *et al.* 2003). These spectra showed a series of low ionization lines with a redshift of 2.328, which represents the velocity of the host galaxy. Interestingly, there were also present deep absorption lines for CIV and SiIV which have multiple components with blue shifts between 560 and 3000 km s^{-1}. The possibility that these lines are caused by intervening galaxies is negligible due to the improbability of getting several chance galaxies along the line of sight with such large equivalent widths and so close to the host in redshift. (The clustering of galaxies does not help since the relative velocities are significantly greater than the escape velocities of clusters.) High-excitation lines (like CIV and SiIV) are also seen from AGN, but this possibility can be rejected as any AGN would be prominent and easily detected by many other means. The high velocities for the lines are much higher than can be obtained from any ordinary motions within the host galaxy. The only remaining possibility is that the absorbing gas is associated with a shell around the progenitor. This is not a surprising possibility, as the most massive stars around the time of core collapse all have massive stellar winds. The general picture is that the high-velocity and high-excitation absorption lines in GRB 021004 can only be from shells of gas surrounding a massive progenitor star, for example a Wolf-Rayet star that is suffering a core collapse.

Schaefer *et al.* (2003) offer a detailed physical analysis of the required conditions in the massive star wind around the GRB progenitor. One point is that the absorption line for N V is completely absent, and this demonstrates that the nitrogen in the circumstellar shells is underabundant (compared to the Sun) and hence that the progenitor cannot be a WN type Wolf-Rayet star. This likely points to the more massive WC type star as a progenitor, with original mass of $>30M_{\odot}$. A second point is that the high velocities (560 and 3000 km s^{-1}) could easily arise from either the original ejection velocities of the shells or from radiative accelerations of shell material placed roughly 0.2 and 0.5 parsecs from the star (or some combination of these two possibilities). A third point is that the ionization of the shell by the burst and afterglow radiation (see Lazzati & Perna 2002) can be easily avoided by invoking clumps in the shells with densities of 10^4 cm^{-3}. Such clumps are plausible since Wolf-Rayet winds show structure on all size scales and since the quasi-stationary flocculi (clumps of gas created in the Wolf-Rayet progenitor wind) in Cas A and the Kepler supernova remnants have even higher densities (Gerardy & Fesen 2001) after correcting by a factor of 4 for the shock jump conditions. Indeed, for GRB 021004, the presence of three bumps in the light curve might be due to high density clumps in the shells.

Do other bursts show high-excitation and high-velocity absorption lines? To answer this question, I have used the GRBlog web page to efficiently search all GCN notices (http://grad40.as.utexas.edu/grblog.php). This nice web site is designed and maintained by Robert Quimby and Erin McMahon (both graduate students under Craig Wheeler) along with Texas undergraduate Jeremy Murphy. With GRBlog and a small search time, I found that 15 GRBs have good early spectra of which 4 have high-velocity high-excitation absorption lines. This implies that the covering fraction for shells around GRB progenitors is of order 4/15. It could be that the shells around all GRBs are clumpy with 27% of the sky covered by clumps of adequate density, or it could imply that only 27% of GRB progenitors have shells surrounding them.

The GRB explosion destroys much of the evidence of what the progenitor was like, but now the high-excitation and high-velocity absorption lines offer a means to understand the stars that produce GRBs. In the future, I would hope and expect that very high resolution spectra might be able to resolve fine details (and possibly changes) in the velocity structure of the shells. Also, such spectra can provide the sensitivity to low abundance elements and resolve the lines so as to allow for real abundance analyses. With the better spectra, our community also needs better theoretical calculations that are time dependent (including recombination and the afterglow radiation), are three-dimensional (including clumping), and have more needed physics (like nonthermal knock-off electrons). The goal is to measure the abundances, densities, clumpiness, velocities, and radial distances of the shells around the progenitor. From this we can hope to understand the mass and evolutionary status of the progenitor in a unique direct manner.

36.3 Four GRB luminosity indicators

Type Ia supernovae have become a premier tool of modern cosmology. Yet as little as nine years ago, the literature had wide debate as to whether they were even standard candles while only two poorly observed events had ever been detected at distances farther than $z = 0.1$. The transformation from debatable-speculation to premier-tool has happened remarkably fast. Similarly, the main conclusion from the supernova cosmology projects (the 'Cosmological Constant') has been transformed from an ignored and maligned possibility to the mainstream paradigm within just a few years. The supernova success story demonstrates the power of standard candles in astrophysics.

The ground-based supernova cosmology projects could only get usable SNe out to $z = 0.94$. The Hubble Deep Field yielded one SN at $z \approx 1.7$ (Reiss *et al.* 2002), while further deep imaging with the Hubble Space Telescope might reveal a few additional events at comparable redshifts. The SNAP satellite (http://snap.lbl.gov) is hoped to be launched in 2008 and it will return exquisite light curves but only of SNe out to $z = 1.7$.

Lamb & Reichert (2000) point out that Swift can see GRBs out to redshifts >100 (should such exist). With GRBs coming from collapses of very massive stars with very short lifetimes, the GRB rate should be proportional to the star formation rate. So we expect GRBs to be seen by Swift out to $z \sim 20$ based on the WMAP results as well as on the best star formation models (Bromm & Loeb 2002), with ~25% of Swift bursts at $z > 5$. Also, variability has already shown BATSE bursts at $z \sim 12$ (Fenimore & Ramirez-Ruiz 2000).

This section will discuss two previously published GRB luminosity indicators plus two new luminosity indicators. With this, GRBs have become standard candles that are visible out to redshifts of ~12–20.

36.3.1 Log and variability

At the 5th Huntsville Gamma Ray Burst Symposium in October 1999, two GRB luminosity indicators were presented. When combined with the observed peak flux, the deduced luminosity can be turned into a luminosity distance and then (for some assumed cosmology) into a redshift. Both luminosity/distance indicators made use of gamma ray data alone, hence allowing distances for *all* GRBs at *low cost* with *no selection effects* other then from the gamma ray peak flux.

The first luminosity indicator is the spectral lag, which is like the time delay between the peak of light curve pulses as viewed with hard and soft photons (Norris, Marani, & Bonnell 2000). High luminosity bursts have the hard and soft peaks at nearly the same time while low luminosity bursts have the soft photons even up to seconds after the hard photons, with the luminosity going as the inverse of the lag. This relation is easily explained as a forced consequence of the empirical Liang-Kargatis relation, which itself is a simple consequence of the dominance of radiative cooling (Schaefer 2003a).

The second luminosity indicator is the variability (or spikiness or jaggedness) of the light curve (Fenimore & Ramirez-Ruiz 2000). The high luminosity bursts have very jagged light curves while the low luminosity events are very smooth. This relation is easily explained by both the peak luminosity and the variability being related as power laws to the bulk Lorentz factor of the jet expansion (Schaefer 2003b).

Both luminosity indicators have passed several critical tests by predictions: First, three new bursts have been added and these follow the original relations. Second, an independent set of 112 BATSE bursts displays the required lag/variability relation (Schaefer, Deng, & Band 2001). Third, Norris (2002) found that the longest lag bursts are concentrated towards the local supergalactic plane. In all, with these three successful predictions plus the good theoretical understanding, we can be confident in these two luminosity indicators to an accuracy of 33% in distance (Schaefer 2003c).

36.3.2 Minimum rise time

Theoretical work (Schaefer 2003b) to explain the variability/luminosity relation has both luminosity and variability as simple power laws of the bulk Lorentz factor of the jet (Γ). Luminosity goes as $\Gamma^{3.14\pm0.34}$ (Schaefer 2003b). From simulated light curves, the variability is found to be an imperfect measure of the pulse rise times, with the variability simply being inversely proportional to the rise times. The rise times are dictated by the geometry of the path length for light being emitted from a surface at the radial distance R as $R/(2c\Gamma^2)$. The values of R for individual pulses will depend on details of ejection times and velocities for the colliding shells. But R cannot be much less than a threshold value (R_{crit}) where the optical depth is of order unity ($\alpha M_{jet}^{0.5}/\Theta_{jet}$) as then the collision energy will be largely lost to adiabatic expansion and the pulse will be too faint to be seen. Nor is it likely to be greatly larger than R_{crit} because such collisions will have low probability and low luminosity. So the minimal rise time for pulses within a burst (τ_{min}) is likely to occur with $R \sim R_{crit}$. Panaitescu & Kumar (2001) have measured values from afterglows such that $M_{jet}^{0.5}/\Theta_{jet}$ is nearly a constant ($1-\sigma$ variations of 0.2 dex), so R_{crit} should be approximately constant from burst-to-burst. This implies that the minimum rise time in a GRB light curve should be roughly proportional to Γ^{-2}, and that the burst luminosity should scale approximately as $\tau_{min}^{1.57}$.

The previous paragraph presents a theoretical explanation of the variability/luminosity relation, and it makes a testable prediction of a specific

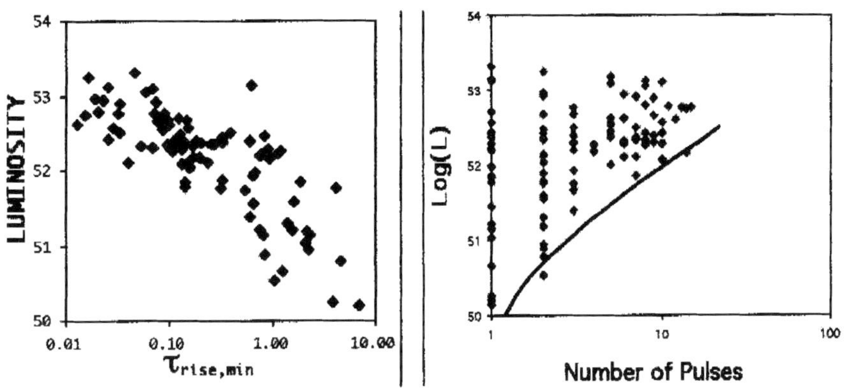

Fig. 36.1. The minimum rise time and the number of peaks are two new GRB luminosity indicators. The left panel shows the minimum rise time for 112 BATSE bursts as a function of their peak luminosity (Schaefer, Deng, & Band 2001). This shows the minimum rise time to be a luminosity indicator (with an accuracy of 0.17 dex in the distance) that is of moderate accuracy. There is good expectation that this measure can be substantially improved, much as the decline rate disproved the SNe 'standard candle' just eight years ago. The right panel shows the number of peaks in the gamma ray light curve as a function of the peak luminosity for the same 112 BATSE bursts. Theoretically, I predicted that low luminosity bursts must have few peaks, and this is demonstrated with this plot. Indeed, the number of peaks can be turned into a fourth luminosity indicator, where a *lower limit* on the luminosity is given.

τ_{min}/luminosity relation. The minimum rise time is shown to be more fundamental than the variability and indeed a new distance indicator is predicted. This predicted relation is found to be true for the BATSE bursts with measured redshifts (and hence luminosities) as well as for the 112 BATSE bursts (Schaefer, Deng, & Band 2001) with luminosities from the first two distance indicators (see Figure 36.1a). The model prediction is confirmed, so τ_{min} is a new GRB luminosity/distance indicator (as shown by the curve). The scatter in Figure 36.1 indicates a 1-σ uncertainty of 0.17 dex in the distance, while on theoretical grounds this can be substantially improved with corrections based on the measured spectral E_{peak}.

36.3.3 Number of peaks

For the general case where the pulse rise times (and hence durations; see Nemiroff 2000) are governed by geometrical delays, the pulse durations (D) will scale as $L^{2/3}$ (see previous section). For high luminosity bursts, all collisions will result in distinct pulses in the light curve, while low luminosity events will have many of the collisions resulting in overlapping broad pulses. Thus, a burst with only a few peaks could either be high luminosity (with few shell collisions) or low luminosity (with all the collisions producing merged peaks). A burst with many peaks can only be a high luminosity event as this is the only way to get narrow peaks that avoid merging together. Nevertheless, a high luminosity burst might have a small number of peaks due to the number of shell collisions being small. The maximum number of distinct peaks in a light curve with a duration of T_{90} is $1 + (T_{90}/D) * e^{-(1+D/T_{90})}$ for Poisson distribution of collisions. With $D \alpha L^{2/3}$, we can translate the observed number of peaks in a burst (N_{peak}) into a lower limit on the burst luminosity.

The previous paragraph outlines a theoretical prediction of how the N_{peak} can be used as a luminosity indicator. Even as a limit, it is useful, as for example a faint Swift burst with ten peaks must be at $z > 6.5$, and this can be determined within seconds of the end of the burst. This predicted limit has been tested against the BATSE bursts with redshifts as well as the 112 BATSE bursts with luminosities from the first two indicators. There is indeed a sharply demarcated region (towards low luminosity and many pulses) that is entirely devoid of bursts, with the line of demarcation following the predicted functional form (Figure 36.1b). This demonstrates that N_{peak} is a new GRB luminosity indicator, albeit as a limit on the luminosity.

36.3.4 The future

GRBs are now standard candles. And their distances can be measured from the gamma ray light curves alone (so as to avoid the grave difficulties and large unknown selection effects associated with optical spectroscopy). This opens up *all* GRBs as tools of cosmology. With the launch of Swift in 2004, we expect to get hundreds of bursts with independently measured redshifts, and this should allow for the accurate calibration of the four luminosity indicators. Just as with the introduction of SNe decline rate corrections, we should expect the accuracy of the luminosity indicators to be substantially improved both by empirical and theoretical corrections (for example, using the measured E_{peak} values). The prospect for the near future is a standard candle with an accuracy comparable to that of SNe for hundreds of Swift bursts plus thousands of BATSE bursts.

What big advances and discoveries can I predict for the next few years? (1) The Swift light curves can have their luminosity indicators calculated by the GCN computers and the derived redshift sent out to the world within seconds after the end of the burst. This will allow for 'precious' follow-up resources (like HST, SIRTF, VLA, Chandra, which can respond to only a very small fraction of Swift bursts) to respond to the very highest redshift bursts immediately. This will allow for large telescopes with infrared detectors to finally discover the Gunn-Peterson effect and measure the epoch of the reionization of the Universe. (2) The fast response of big telescopes with infrared detectors will also be able to position the GRB afterglow to sub-arc-second accuracy. These positions can then be examined very deeply with

HST, the Webb NGST, or with the Keck telescopes to get the first look at galaxies with $z > 7$. (3) Roughly 40% of Swift bursts will have independently measured redshifts, most with $z < 5$ but likely a few with redshifts up to ~12. From this, a GRB Hubble Diagram can be constructed (Schaefer 2003c). This will show the expansion history of the Universe from $1.0 < z < 1.7$ many years before SNAP. And it will show the various predicted quintessence (Weller & Albrecht 2001) and non-standard effects (Mannheim 2002) in the range $1.7 < z < 12$. The current paradigm of cosmology (the new inflation perhaps with quintessence) is so new and untested that surprises could easily await in the unknown regime of $z > 1$. (4) As GRBs come from the collapse of very massive stars with very short lifetimes, the GRB rate density should be proportional to the star formation rate of the Universe. By measuring the distances and luminosities to Swift bursts (Schaefer, Deng, & Band 2001; Fenimore & Ramirez-Ruiz 2000; Lloyd-Roning, Fryer, & Ramirez-Ruiz 2002), we will measure the star formation rate of our Universe from $0.2 < z < 20$.

GRBs have great potential as tools of cosmology. The reason is that they are the only reliable standard candles that can be seen with $z > 1.7$ and they are the only sources of any kind visible from redshifts 6.5 to ~20. This is a large fraction of the age of the Universe over which GRB are unique as being the only probe.

References

Bloom, J. S. *et al.* 1999. *Nature*, **401**, 453.
Bromm, V. & Loeb, A. 2002. *Astrophys. J.*, **575**, 111.
Cline, T. L. *et al.* 1980. *Astrophys. J.*, **237**, L1.
Colgate, S. A. 1968. *Canadian J. Phys.*, **46**, S476.
Colgate, S. A. 1970. *Acta Physica Academiae Scientiarum Hungaricae*, **29**, Suppl. 1, pp. 353–359.
Colgate, S. A. 1974. *Astrophys. J.*, **187**, 333.
Deng, M. 2001. *Ph. D. Thesis*, **Yale University**,.
Evans, W. D. *et al.* 1980. *Astrophys. J.*, **237**, L7.
Felten, J. E. 1982. *17th International Cosmic Ray Conference, Paris*, **9**, 52–55.
Fenimore, E. E. & Ramirez-Ruiz, E. 2000. *Astrophys. J.*, submitted (astro-ph/0004176).
Galama, T. *et al.* 1998. *Nature*, **395**, 670.
Galama, T. *et al.* 2000. *Astrophys. J.*, **536**, 185.
Gerardy, C. L. & Fesen, R. A. 2001. *Astron. J.*, **121**, 2781.
Hjorth, J. *et al.* 2003. *Nature*, **423**, 847.
Hurley, K. 2000. in *Gamma-Ray Bursts: 5th Huntsville Symposium*, ed. R. M. Kippen, R. S. Mallozzi, and G. J. Fishman (Melville NY, AIP Conf. Proc. 526), 763.
Kippen, R. M. *et al.* 1998. *Astrophys. J.*, **506**, L27.
Klebesadel, R. W., Strong, I. B., & Olson, R. A. 1973. *Astrophys. J.*, **182**, L85.
Kulkarni, S. *et al.* 1998. *Nature*, **395**, 663.
Lamb, D. Q. & Reichart, D. E. 2000. *Astrophys. J.*, **536**, 1.
Lazzati, D. & Perna, R. 2002. *MNRAS*, **330**, 383.
Lloyd-Ronning, N., Fryer, C., & Ramirez-Ruiz, E. 2002. *Astrophys. J.*, **574**, 554.
Mannheim, P. D. 2002. astro-ph/0204202.

Metzger, M. R, *et al.* 1997. *Nature*, **387**, 878.
Nemiroff, R. J. 2000. *Astrophys. J.*, **544**, 805.
Norris, J. P. 2002. *Astrophys. J.*, **579**, 386.
Norris, J. P., Marani, G., & Bonnell, J. 2000. *Astrophys. J.*, **534**, 248.
Panaitescu, A. & Kumar, P. 2001. *Astrophys. J.*, **560**, L49.
Reichart, D. E. 1999. *Astrophys. J.*, **521**, L111.
Riess, A. G. *et al.* 2001. *Astrophys. J.*, **560**, 49.
Schaefer, B. E. 2003a. *Astrophys. J.*, in press.
Schaefer, B. E. 2003b. *Astrophys. J.*, **583**, L71.
Schaefer, B. E., 2003c. *Astrophys. J.*, **583**, L67.
Schaefer, B. E. & Deng, M. 2000. in *Gamma-Ray Bursts: 5th Huntsville Symposium*, ed. R. M. Kippen, R. S. Mallozzi, and G. J. Fishman (Melville NY, AIP Conf. Proc. 526), 419–423.
Schaefer, B. E., Deng, M., & Band, D. L. 2001. *Astrophys. J.*, **563**, L123.
Schaefer, B. E. *et al.* 2003. *Astrophys. J.*, **591**, 387.
Stanek, K. Z. *et al.* 2003. *Astrophys. J.*, **591**, L17.
Wang, L. & Wheeler, J. C. 1998. *Astrophys. J.*, **504**, L87.
Weller, J. & Albrecht, A. 2001. *Phys. Rev. Lett.*, **86**, 1939.

37

Gamma-Ray Bursts as a laboratory for the study of Type Ic supernovae

D. Q. Lamb, T. Q. Donaghy and C. Graziani

Department of Astronomy & Astrophysics, University of Chicago, Chicago, IL 60637, USA

Abstract

HETE-2 has confirmed the connection between GRBs and Type Ic supernovae. Thus we now know that the progenitors of long GRBs are massive stars. HETE-2 has also provided strong evidence that the properties of X-Ray Flashes (XRFs) and GRBs form a continuum, and therefore that these two types of bursts are the same phenomenon. We show that both the structured jet and the uniform jet models can explain the observed properties of GRBs reasonably well. However, if one tries to account for the properties of both XRFs and GRBs in a unified picture, the uniform jet model works reasonably well while the structured jet model fails utterly. The uniform jet model of XRFs and GRBs implies that most GRBs have very small jet opening angles ($\sim$ half a degree). This suggests that magnetic fields play a crucial role in GRB jets. The model also implies that the energy radiated in gamma rays is $\sim$100 times smaller than has been thought. Most importantly, the model implies that there are $\sim 10^4$–10^5 more bursts with very small jet opening angles for every such burst we see. Thus the rate of GRBs could be comparable to the rate of Type Ic core collapse supernovae. Accurate, rapid localizations of many XRFs, leading to identification of their X-ray and optical afterglows and the determination of their redshifts, will be required in order to confirm or rule out these profound implications. HETE-2 is ideally suited to do this (it has localized 16 XRFs in $\sim$2 years), whereas *Swift* is not. The unique insights into the structure of GRB jets, the rate of GRBs, and the nature of Type Ic supernovae that XRFs may provide therefore constitute a compelling scientific case for continuing HETE-2 during the *Swift* mission.

37.1 Introduction

Gamma-ray bursts (GRBs) are the most brilliant events in the Universe. Long regarded as an exotic enigma, they have taken center stage in high-energy astrophysics by virtue of the spectacular discoveries of the past six years. It is now clear that they also have important applications in many other areas of astronomy: GRBs mark the moment of "first light" in the universe; they are tracers of the star formation, re-ionization, and metallicity histories of the universe; and they are laboratories for studying core-collapse supernovae. It is the last topic that we focus on here.

37.2 GRB–SN connection

There has been increasing circumstantial and tantalizing direct evidence in the last few years that GRBs are associated with core collapse supernovae [see, e.g. Lamb (2000)]. The detection and localization of GRB 030329 by HETE-2 (Vander-spek *et al.*, 2003a) led to a dramatic confirmation of the GRB – SN connection. GRB 030329 was among the brightest 1% of GRBs ever seen (see Figure 36.2). Its optical afterglow was $\sim 12^{\mathrm{th}}$ magnitude at 1.5 hours after the burst (Price *et al.*, 2003) – more than 3 magnitudes brighter than the famous optical afterglow of GRB 990123 at a similar time (Akerlof *et al.*, 1999). In addition, the burst source and its host galaxy lie very nearby, at a redshift $z = 0.167$ (Greiner *et al.*, 2003). Given that GRBs typically occur at $z = 1 - 2$, the probability that the source of an observed burst should be as close as GRB 030329 is one in several thousand. It is therefore very unlikely that HETE-2, or even *Swift*, will see another such event.

The fact that GRB 030329 was very bright spurred the astronomical community – both amateurs and professionals – to make an unprecedented number of observations of the optical afterglow of this event. Figure 37.1 (left panel) shows the light curve of the optical afterglow of GRB 030329 1–10 days after the burst. At least four dramatic "re-brightenings" of the afterglow are evident in the saw-toothed lightcurve. These may be due to repeated injections of energy into the GRB jet by the central engine at late times, or caused by the ultra-relativistic jet ramming into dense blobs or shells of material (Granot, Naka & Piran, 2003). If the former, it implies that the central engine continued to pour out energy long after the GRB was over; if the latter, it likely provides information about the last weeks and days of the progenitor star.

The fact that GRB 030329 was very nearby made its optical afterglow an ideal target for attempts to confirm the conjectured association between GRBs and core collapse SNe. Astronomers were not disappointed: about ten days after the burst, the spectral signature of an energetic Type Ic supernova emerged (Stanek *et al.*, 2003). The supernova has been designated SN 2003dh. Figure 36.1 (right panel) compares the discovery spectrum of SN 2003dh in the afterglow light curve of GRB 030329 and the spectrum of the Type Ic supernova SN 1998bw. The similarity is striking. The breadth and the shallowness of the absorption lines in the spectra of SN 2003dh imply expansion velocities of $\approx 36{,}000$ km s^{-1} – far higher than those seen in typical Type Ic supernovae, and higher even than those seen in SN 1998bw.

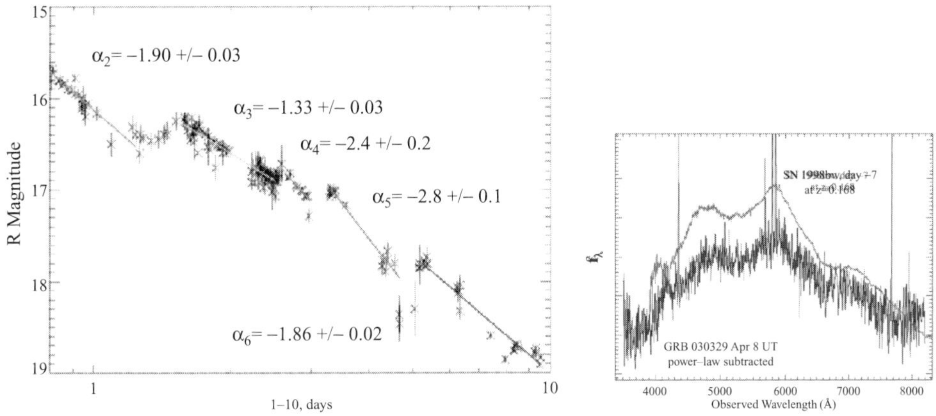

Fig. 37.1. Left panel: Successive rebrightenings of the optical afterglow of GRB 030329 during the 10 days following the burst. From (Fillipenko, 2003). Right panel: Comparison of the discovery spectrum of SN 2003dh seen in the afterglow of GRB 030329 at 8 days after the burst and the spectrum of the Type Ic supernova SN 1998bw. The similarity is striking. From Stanek *et al.* (2003).

It had been conjectured that GRB 980425 was associated with SN 1998bw [see, e.g., Galama *et al.* (1998)], but the fact that, if the association were true, the burst would have had to have been $\sim 10^4$ times fainter than any other GRB observed to date made the association suspect. The clear detection of SN 2003dh in the afterglow of GRB 030329 confirmed decisively the connection between GRBs and core collapse SNe.

The association between GRB 030329 and SN 2003dh makes it clear that we must understand Type Ic SNe in order to understand GRBs. The converse is also true: we must understand GRBs in order to fully understand Type Ic SNe. It is possible that the creation of a powerful ultra-relativistic jet as a result of the collapse of the core of a massive star to a black hole plays a direct role in Type Ic supernova explosions (MacFadyen, Woosley & Heger, 2001), but it is certain that the rapid rotation of the collapsing core implied by such jets must be an important factor in some – perhaps most – Type Ic supernovae. The result will often be a highly asymmetric explosion, whether the result of rapid rotation alone or of the creation of powerful magnetic fields as a result of the rapid rotation (Khokhlov *et al.*, 1999).

The large linear polarizations measured in several bright GRB afterglows, and especially the temporal variations in the linear polarization [see, e.g., Rol *et al.* (2003)], provide strong evidence that the Type Ic supernova explosions associated with GRBs are highly asymmetric. The recent dramatic discovery that GRB 021206 was strongly polarized (Coburn & Boggs, 2003) provides compelling evidence that GRB jets are in fact dominated by magnetic energy rather than hydrodynamic energy.

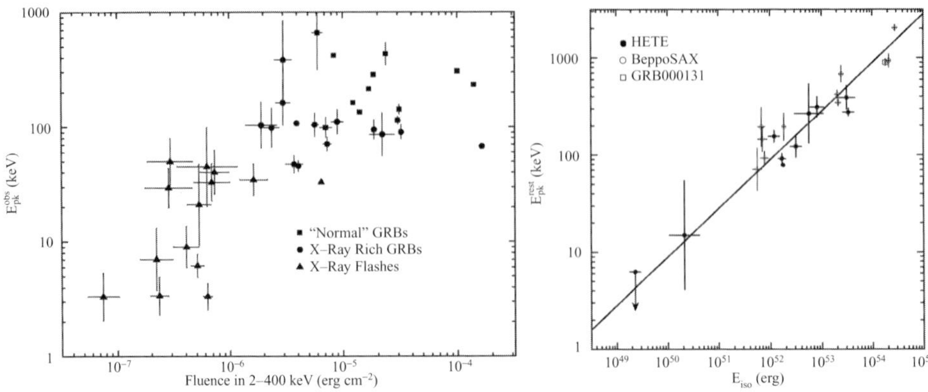

Fig. 37.2. Distribution of HETE-2 bursts in the [S(2–400 keV), $E_{\rm peak}^{\rm obs}$]-plane, showing XRFs (triangles), X-ray-rich GRBs (circles), and GRBs (squares) (left panel). From Sakamoto *et al.* (2003b). Distribution of HETE-2 and BeppoSAX bursts in the ($E_{\rm iso}$,$E_{\rm peak}$)-plane, where $E_{\rm iso}$ and $E_{\rm peak}$ are the isotropic-equivalent GRB energy and the peak of the GRB spectrum in the source frame (right panel). The HETE-2 bursts confirm the relation between $E_{\rm iso}$ and $E_{\rm peak}$ found by Amati *et al.* (2002), and extend it by a factor $\sim$300 in $E_{\rm iso}$. The bursts with the lowest and second-lowest values of $E_{\rm iso}$ are XRFs 020903 and 030723. From Lamb *et al.* (2003c).

In addition, the X-ray afterglows of several GRBs have provided tantalizing evidence of the presence of emission lines of α-particle nuclei (Reeves *et al.*, 2002; Butler *et al.*, 2003). These emission lines, if confirmed, provide severe constraints on models of GRBs and Type Ic supernovae [see, e.g., Lazzati, Ramirez-Ruiz & Rees (2002)]. They may also provide information on the abundances and properties of heavy elements that have been freshly minted in the supernova explosion.

It is therefore now clear that GRBs are a unique laboratory for studying, and are a powerful tool for understanding, Type Ic core collapse supernovae.

37.3 Nature of X-ray flashes and X-ray-rich GRBs

Two-thirds of all HETE-2–localized bursts are either "X-ray-rich" or X-Ray Flashes (XRFs); of these, one-third are XRFs* (Sakamoto *et al.*, 2003b). These events have received increasing attention in the past several years (Heise *et al.*, 2000; Kippen *et al.*, 2002), but their nature remains unknown.

Clarifying the nature of XRFs and X-ray-rich GRBs, and their connection to GRBs, could provide a breakthrough in our understanding of the prompt emission of GRBs. Analyzing 42 X-ray-rich GRBs and XRFs seen by FREGATE and/or the

* We define "X-ray-rich" GRBs and XRFs as those events for which $\log[S_X(2\text{–}30\ \text{kev})/S_\gamma(30\text{–}400\ \text{kev})] > -0.5$ and 0.0, respectively.

WXM instruments on HETE-2, Sakamoto *et al.* (2003b) find that the XRFs, the X-ray-rich GRBs, and GRBs form a continuum in the [S_γ(2–400 kev), $E^{\rm obs}_{\rm peak}$]-plane (see Figure 37.2, left-hand panel). This result strongly suggests that all of these events are the same phenomenon.

Furthermore, Lamb *et al.* (2003c) have placed 9 HETE-2 GRBs with known redshifts and 2 XRFs with known redshifts or strong redshift constraints in the ($E_{\rm iso}$, $E_{\rm peak}$)-plane (see Figure 37.2, right-hand panel). Here $E_{\rm iso}$ is the isotropic-equivalent burst energy and $E_{\rm peak}$ is the energy of the peak of the burst spectrum, measured in the source frame. The HETE-2 bursts confirm the relation between $E_{\rm iso}$ and $E_{\rm peak}$ found by Amati *et al.* (2002) for GRBs and extend it down in $E_{\rm iso}$ by a factor of 300. The fact that XRF 020903, one of the softest events localized by HETE-2 to date, and XRF 030723, the most recent XRF localized by HETE-2, lie squarely on this relation (Sakamoto *et al.*, 2003a; Lamb *et al.*, 2003c) provides strong evidence that XRFs and GRBs are the same phenomenon. However, additional redshift determinations are clearly needed for XRFs with 1 keV $< E_{\rm peak} <$ 30 keV energy in order to confirm these results.

37.4 XRFs as a probe of Type Ic supernovae

Frail *et al.* (2001; see also Bloom *et al.* 2003) have shown that most GRBs have a "standard" energy; i.e, if their isotropic equivalent energy is corrected for the jet opening angle inferred from the jet break time, most GRBs have the same radiated energy, $E_\gamma = 1.3 \times 10^{51}$ ergs, to within a factor of $\pm$2–3.

Two models of GRB jets have received widespread attention:

- The "structured jet" model (see the left-hand panel of Figure 37.3). In this model, all GRBs produce jets with the same structure (Rossi, Lazzati, & Rees, 2002; Woosley, Zhang, & Heger, 2003; Zhang & Mészáros, 2002; Mészáros, Ramirez-Ruiz, Rees, & Zhang, 2002). The isotropic-equivalent energy and luminosity is assumed to decrease as the viewing angle θ_v as measured from the jet axis increases. The wide range in values of $E_{\rm iso}$ is attributed to differences in the viewing angle θ_v. In order to recover the "standard energy" result (Frail *et al.*, 2001), $E_{\rm iso}(\theta_v) \sim \theta_v^{-2}$ is required (Zhang & Mészáros, 2002).
- The "uniform jet" model (see the right-hand panel of Figure 37.3). In this model GRBs produce jets with very different jet opening angles $\theta_{\rm jet}$. For $\theta < \theta_{\rm jet}$, $E_{\rm iso}(\theta_v) =$ constant while for $\theta > \theta_{\rm jet}$, $E_{\rm iso}(\theta_v) = 0$.

As we have seen, HETE-2 has provided strong evidence that the properties of XRFs, X-ray-rich GRBs, and GRBs form a continuum, and that these bursts are therefore the same phenomenon. If this is true, it immediately implies that the E_γ inferred by Frail *et al.* (2001) is too large by a factor of at least 100 (Lamb, Donaghy & Graziani, 2003). The reason is that the values of $E_{\rm iso}$ for XRF 020903 (Sakamoto

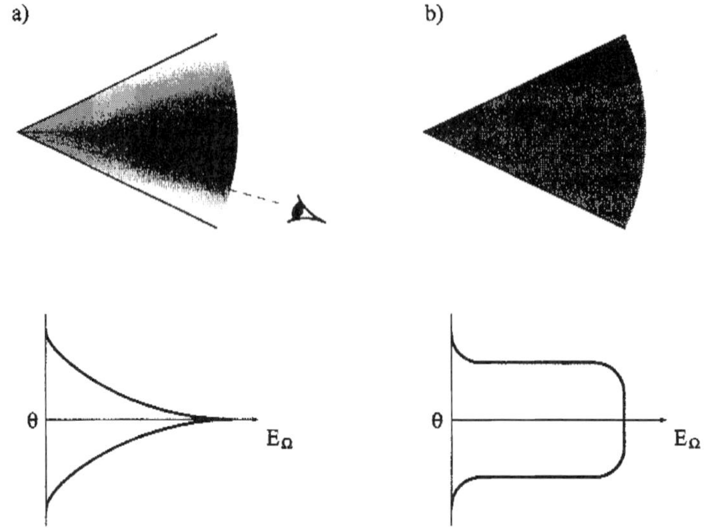

Fig. 37.3. Schematic diagrams of universal jet model and jet model of GRBs (Ramirez-Ruiz & Lloyd-Ronning, 2002). In the universal jet model, the isotropic-equivalent energy and luminosity is assumed to decrease as the viewing angle θ_v as measured from the jet axis increases. In order to recover the "standard energy" result (Frail *et al.*, 2001), $E_{\rm iso}(\theta_v) \sim \theta_v^{-2}$ is required. In the uniform jet model, GRBs produce jets with a large range of jet opening angles $\theta_{\rm jet}$. For $\theta < \theta_{\rm jet}$, $E_{\rm iso}(\theta_v) = $ constant while for $\theta > \theta_{\rm jet}$, $E_{\rm iso}(\theta_v) = 0$.

et al., 2003a) and XRF 030723 (Lamb *et al.*, 2003c) are ~100 times smaller than the value of E_γ inferred by Frail *et al.* – an impossibility.

HETE-2 has also provided strong evidence that, in going from XRFs to GRBs, $E_{\rm iso}$ changes by a factor $\sim 10^5$ (see Figure 37.2, right-hand panel). If one tries to explain only the range in $E_{\rm iso}$ corresponding to GRBs, both the uniform jet model and the structured jet model work reasonably well. However, if one tries to explain the range in $E_{\rm iso}$ of a factor $\sim 10^5$ that is required in order to accommodate both XRFs and GRBs in a unified description, the uniform jet works reasonably well while the structured jet model fails utterly.

The reason is the following: the observational implications of the structured jet model and the uniform jet model differ dramatically if they are required to explain XRFs and GRBs in a unified picture. In the structured jet model, most viewing angles θ_v are $\approx 90°$. This implies that the number of XRFs should exceed the number of GRBs by many orders of magnitude, something that HETE-2 does not observe (see Figures 37.2, 37.4, and 37.5). On the other hand, by choosing $N(\Omega_{\rm jet}) \sim \Omega_{\rm jet}^{-2}$, the uniform jet model predicts equal numbers of bursts per logarithmic decade in $E_{\rm iso}$ (and S_E), which is exactly what HETE-2 sees (again, see Figures 37.2, 37.4, and 37.5) (Lamb, Donaghy & Graziani, 2003).

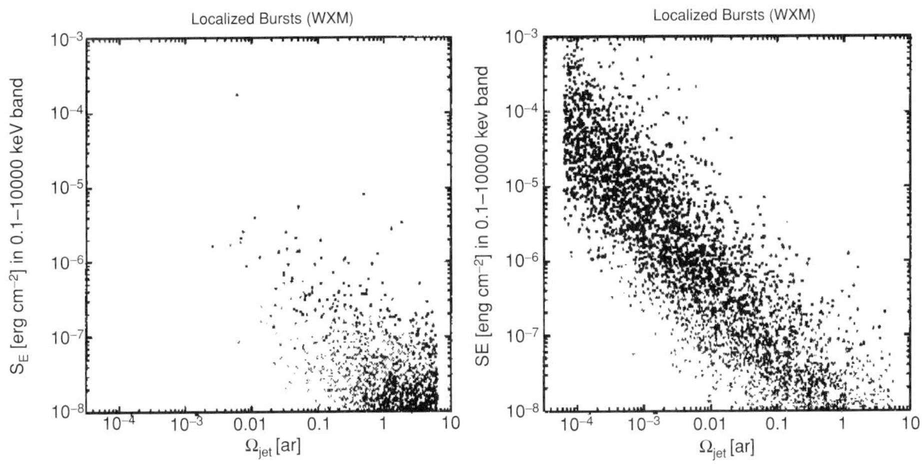

Fig. 37.4. Expected distribution of bursts in the ($\Omega_{\rm jet}$, S_E)-plane for the universal jet model (left panel) and uniform jet model (right panel), assuming that the Amati *et al.* (2002) relation holds for XRFs as well as for GRBs, as the HETE-2 results strongly suggest. From Lamb, Donaghy & Graziani (2003).

Thus, if $E_{\rm iso}$ spans a range $\sim 10^5$, as the HETE-2 results strongly suggest, the uniform jet model can provide a unified picture of both XRFs and GRBs, whereas the structured jet model cannot. This means that XRFs provide a powerful probe of GRB jet structure.

A range in $E_{\rm iso}$ of 10^5, which is what the HETE-2 results strongly suggest, requires a *minimum* range in $\Delta\Omega_{\rm jet}$ of 10^4–10^5 in the uniform jet model. Thus the unified picture of XRFs and GRBs in the uniform jet model implies that there are $\sim 10^4$–10^5 more bursts with very small $\Omega_{\rm jet}$'s for every such burst we see; i.e., the rate of GRBs may be ~ 100 times greater than has been thought.

In addition, since the observed ratio of the rate of Type Ic supernovae to the rate of GRBs in the observable universe is $R_{\rm Type\ Ic}/R_{\rm GRB} \sim 10^5$ (Lamb, 1999), a unified picture of XRFs and GRBs in the uniform jet model implies that roughly *all* Type Ic supernovae produce high-energy transients (Lamb, Donaghy & Graziani, 2003). More spherically symmetric jets yield XRFs and narrow jets produce GRBs. Thus XRFs and GRBs provide a combination of GRB/SN samples that would enable astronomers to study the relationship between the degree of jet-like behavior of the GRB and the properties of the supernova (brightness, polarization $\Leftrightarrow$ asphericity of the explosion, velocity of the explosion $\Leftrightarrow$ kinetic energy of the explosion, etc.). GRBs may therefore provide a unique laboratory for understanding Type Ic core collapse supernovae.

A unified picture of XRFs and GRBs in the uniform jet model also implies that most Type Ic supernovae produce narrow jets, which may suggest that the collapsing cores of most Type Ic supernovae are rapidly rotating. Finally, such a unified picture

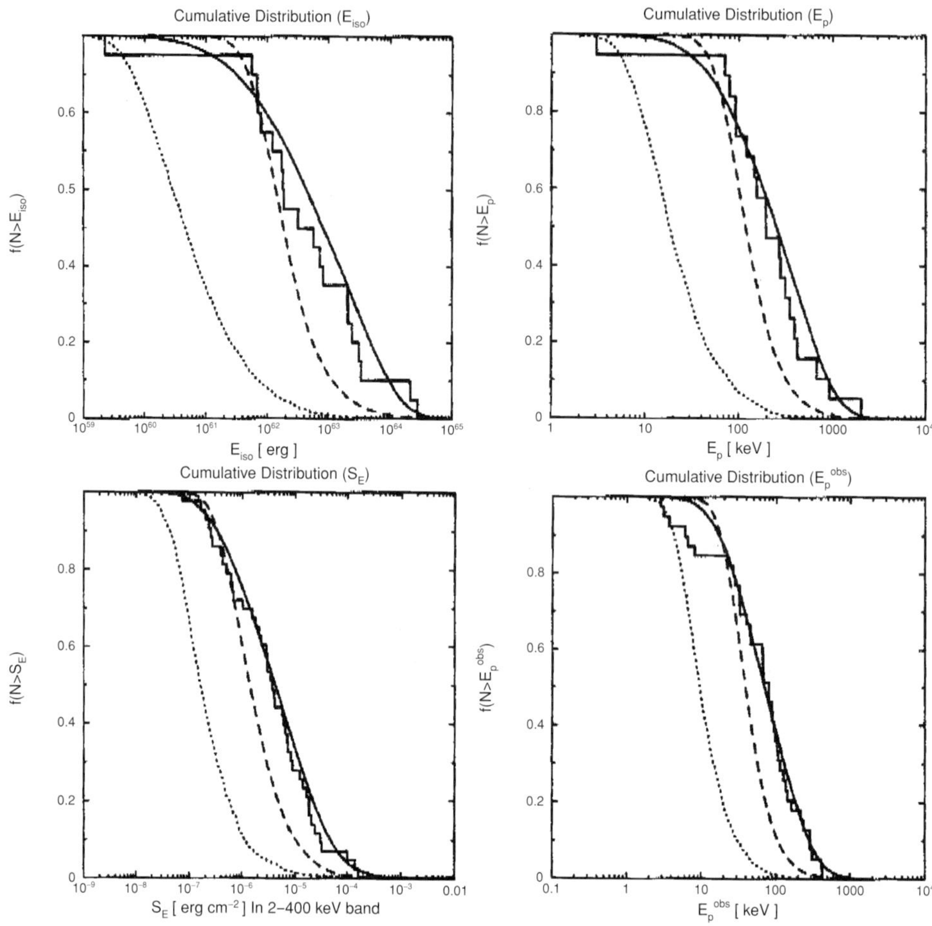

Fig. 37.5. Top row: cumulative distributions of S(2–400keV) (left panel) and $E_{\rm peak}^{\rm obs}$ (right panel) predicted by the structured and uniform jet models, compared to the observed cumulative distributions of these quantities. Bottom row: cumulative distributions of $E_{\rm iso}$ (left panel) and $E_{\rm peak}$ (right panel) predicted by the structured and uniform jet models, compared to the observed cumulative distributions of these quantities. The cumulative distributions corresponding to the best-fit structured jet model that explains XRFs and GRBs are shown as dotted lines, those corresponding to the best-fit structured jet model that explains GRBs alone are shown as dashed lines, and those corresponding to the best-fit uniform jet model are shown as solid lines. The structured jet model provides a reasonable fit to GRBs alone but cannot provide a unified picture of both XRFs and GRBs, whereas the uniform jet model can. From Lamb, Donaghy & Graziani (2003).

implies that the total radiated energy in gamma rays E_γ is ~100 times smaller than has been thought (Lamb, Donaghy & Graziani, 2003).

As we have seen, the HETE-2 results provide strong evidence that XRFs and GRBs are the same phenomenon. But the profound implications of these results in terms of the structure of GRB jets, the rate of GRBs, and the nature of Type Ic supernovae, require incontrovertible evidence.

Obtaining the incontrovertible evidence needed to sustain (or refute) these implications will require accurate, rapid localizations of XRFs, leading to identification of their X-ray and optical afterglows and the determination of their redshifts. Until very recently, only one XRF (XRF 020903; Soderberg *et al.* 2002) had even a probable optical afterglow and redshift. The reason why is that, as expected in the uniform jet picture, the X-ray (and therefore the optical) afterglows of XRFs are $\sim 10^3$–10^4 times fainter than those of GRBs (Lamb, Donaghy & Graziani, 2003). But this challenge can be met: the recent HETE-2–localization of XRF 030723 represents the first time that an XRF has been localized in real time (Prigozhin *et al.*, 2003); identification of its X-ray and optical afterglows rapidly followed (Fox *et al.*, 2003c). This event may well be the Rosetta stone for XRFs.

The exciting recent results involving XRF 030723 highlight the fact that HETE-2 is ideally suited to obtain the evidence about XRFs that is required to confirm or rule out the profound implications about the structure of GRB jets, the rate of GRBs, and the nature of Type Ic supernovae described above. HETE-2 will obtain this evidence, if the HETE-2 mission is extended, whereas *Swift* cannot. HETE-2's ability to accurately and rapidly localize XRFs – and study their spectra – therefore constitutes a compelling reason for continuing HETE-2 during the *Swift* mission.

References

Akerlof, C., *et al.* 1999, Nature, 398, 400
Amati, L., *et al.* 2002, A & A, 390, 81
Band, D. L. 2003, ApJ, in press (astro-ph/0212452)
Bloom, J., Frail, D. A. & Kulkarni, S. R. 2003, ApJ, 588, 945
Butler, N. R., *et al.* 2003, ApJ, in press
Coburn, W. & Boggs, S. E. 2003, Nature, 423, 415
Fillipenko, A. V. 2003, private communication
Fox, D. W., *et al.* 2003c, GCN Circular 2323
Frail, D. *et al.* 2001, ApJ, 562, L55
Galama, T., *et al.* 1998, Nature, 395, 670
Granot, J., Naka, E. & Piran, T. 2003, ApJ, in press (astro-ph/0304563)
Greiner, J., *et al.* 2003, GCN Circular 2020
Heise, J., in't Zand, J., Kippen, R. M., & Woods, P. M., in Proc. 2nd Rome Workshop: Gamma-Ray Bursts in the Afterglow Era, eds. E. Costa, F. Frontera, J. Hjorth (Berlin: Springer-Verlag), 16
Khokhlov, A., *et al.* 1999, ApJ, 524, L107

Kippen, R. M., Woods, P. M., Heise, J., in't Zand, J., Briggs, M. S., & Preece, R. D. 2002, in Gamma-Ray Burst and Afterglow Astronomy, AIP Conf. Proceedings 662, ed. G. R. Ricker & R. K. Vanderspek (New York: AIP), 244
Lamb, D. Q. 1999, A&A, 138, 607
Lamb, D. Q. 2000, Physics Reports, 333–334, 505
Lamb, D. Q., Donaghy, T. Q., & Graziani, C. 2003, ApJ, to be submitted
Lamb, D. Q., *et al.* 2003c, to be submitted to ApJ
Lazzati, D., Ramirez-Ruiz, E. & Rees, M. J. 2002, ApJ, 572, L57
Lloyd-Ronning, N., Fryer, C., & Ramirez-Ruiz, E. 2002, ApJ, 574, 554
MacFadyen, A. I., Woosley, S. E., & Heger, A. 2001, ApJ, 550, 410
Mészáros, P., Ramirez-Ruiz, E., Rees, M. J., & Zhang, B. 2002, ApJ, 578, 812
Price, P. A., *et al.* 2003, Nature, 423, 844
Prigozhin, G., *et al.* 2003, GCN Circular 2313
Ramirez-Ruiz, E. & Lloyd-Ronning, N. 2002, New Astronomy, 7, 197
Reeves, J. N., *et al.* 2002, Nature, 415, 512
Rol, E., *et al.* 2003, A&A, 405, L23
Rossi, E., Lazzati, D., & Rees, M. J. 2002, MNRAS, 332, 945
Sakamoto, T. *et al.* 2003a, ApJ, submitted
Sakamoto, T. *et al.* 2003b, ApJ, to be submitted
Soderberg, A. M., *et al.* 2002, GCN Circular 1554
Stanek, K. *et al.* 2003, ApJ, 591, L17
Vanderspek, R., *et al.* 2003, GCN Circular 1997
Woosley, S. E., Zhang, W. & Heger, A. 2003, ApJ, in press
Zhang, B. & Mészáros, P. 2002, ApJ, 571, 876

38

The diversity of cosmic explosions: Gamma-Ray Bursts and Type Ib/c supernovae

E. Berger

Division of Physics, Mathematics and Astronomy, California Institute of Technology, Pasadena CA 91125

Abstract

The death of massive stars and the processes which govern the formation of compact remnants are not fully understood. Observationally, this problem may be addressed by studying different classes of cosmic explosions and their energy sources. Here we discuss recent results on the energetics of γ-ray bursts (GRBs) and Type Ib/c Supernovae (SNe Ib/c). In particular, radio observations of GRB 030329, which allow us to undertake calorimetry of the explosion, reveal that some GRBs are dominated by mildly relativistic ejecta such that the total explosive yield of GRBs is nearly constant, while the ultra-relativistic output varies considerably. On the other hand, SNe Ib/c exhibit a wide diversity in the energy contained in fast ejecta, but none of those observed to date (with the exception of SN 1998bw) produced relativistic ejecta. We therefore place a firm limit of 3% on the fraction of SNe Ib/c that could have given rise to a GRB. Thus, there appears to be clear dichotomy between hydrodynamic (SNe) and engine-driven (GRBs) explosions.

38.1 The death of massive stars

The death of massive stars ($M \gtrsim 8M_{\odot}$) is a chapter of astronomy that is still being written. Recent advances in modeling suggests that a great diversity can be expected. Indeed, such diversity has been observed in the neutron star remnants: radio pulsars, AXPs, and SGRs. We know relatively little about the formation of black holes.

The compact objects form following the collapse of the progenitor core. The energy of the resulting explosion can be supplemented or even dominated by the energy released from the compact object (e.g. a rapidly rotating magnetar or an accreting black hole). Such "engines" can give rise to asymmetrical explosions (MacFadyen & Woosley 1999), but even in their absence the core collapse process appears to be mildly asymmetric (e.g. Wang *et al.* 2001). Regardless of the source of energy, a fraction, E_K, is coupled to the debris or ejecta (mass M_{ej}) and it is these two gross parameters which determine the appearance and evolution of the resulting explosion. Equivalently one may consider E_K and the mean initial speed of ejecta, v_0, or the Lorentz factor, $\Gamma_0 = [1 - \beta_0^2]^{-1/2}$, where $\beta_0 = v_0/c$.

Supernovae (SNe) and γ-ray bursts (GRBs), are distinguished by their ejecta velocities. In the former $v_0 \sim 10^4$ km s^{-1} as inferred from optical absorption features (e.g. Filippenko 1997), while for the latter $\Gamma_0 \gtrsim 100$, inferred from the non-thermal prompt emission (Goodman 1986; Paczynski 1986), respectively. The large difference in initial velocity arises from significantly different ejecta masses: $M_{\rm ej} \sim$ *few* $M_\odot$ in SNe compared to $\sim 10^{-5}\ M_\odot$ in GRBs.

In the conventional interpretation, $M_{\rm ej}$ for SNe is large because E_K is primarily derived from the (essentially) symmetrical collapse of the core and the energy thus couples to all the mass left after the formation of the compact object. GRB models, on the other hand, appeal to a stellar mass black hole remnant, which accretes matter on many dynamical timescales and powers relativistic jets (the so-called collapsar model; MacFadyen & Woosley 1999).

Still, as demonstrated by the association of the energetic Type Ic SN 1998bw ($d \approx 40$ Mpc) with GRB 980425 (Galama *et al.* 1998), as well as the association of SN 2003dh with GRB 030329 (e.g. Stanek *et al.* 2003), some overlap may exist. Here we take an observational approach to investigating the diversity of stellar explosions, their energetics, and the relation between them focusing in particular on GRBs and SNe Ib/c.

38.2 The energetics of γ-Ray Bursts

Recent studies revealed the surprising result that long-duration GRBs have a standard energy of $E_\gamma \approx 1.3 \times 10^{51}$ erg in ultra-relativistic ejecta when corrected for asymmetry ("jets"; Frail *et al.* 2001; Bloom, Frail & Kulkarni 2003). A similar result was found for the kinetic energies of GRB afterglow using the beaming-corrected X-ray luminosities as a proxy for the true kinetic energy (Figure 38.1; Berger, Kulkarni & Frail 2003). However, these studies have also highlighted a small group of subenergetic bursts, including the peculiar GRB 980425 associated (Galama *et al.* 1998) with SN 1998bw ($E_\gamma \approx 10^{48}$ erg). Until recently, the nature of these sources has remained unclear.

This question appears to now be resolved thanks to broad-band calorimetry of GRB 030329 (Berger *et al.* 2003a), the nearest cosmological burst detected to date (redshift, $z = 0.1685$). Early optical observations of the afterglow of GRB 030329 revealed a sharp break at $t = 0.55$ day (Figure 38.2; Price *et al.* 2003). The X-ray flux (Tiengo *et al.* 2003) tracks the optical afterglow for the first day, with a break consistent with that seen in the optical. Thus, the break at 0.55 day is not due to a change in the ambient density since for typical parameters (e.g. Kumar 2000) the

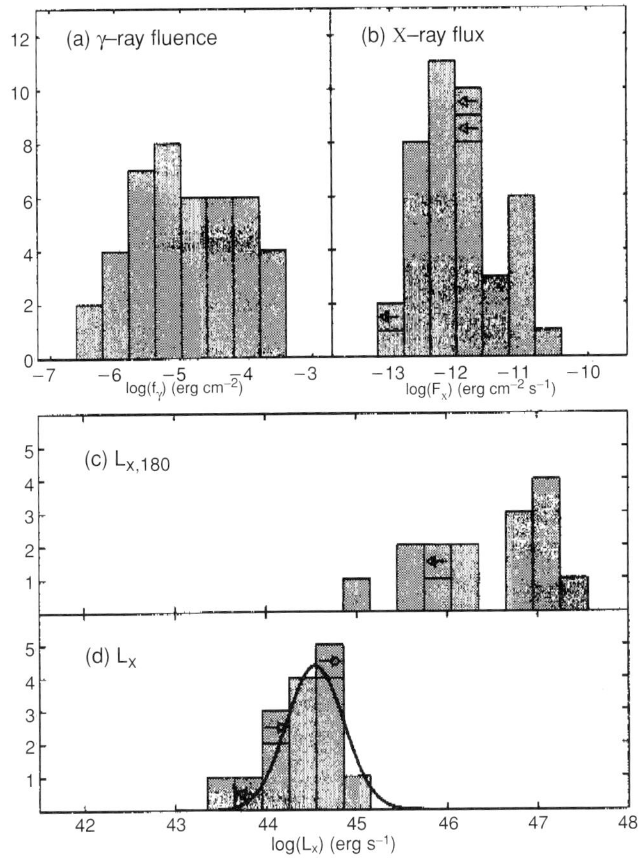

Fig. 38.1. (a) Distribution of γ-ray fluences; (b) Distribution of X-ray fluxes scaled to $t = 10$ hr after the burst; (c) Isotropic-equivalent X-ray luminosity plotted for the subset of X-ray afterglows with known jet opening angles and redshifts; (d) True X-ray luminosity corrected for beaming, a proxy for the afterglow kinetic energy.

X-ray emission is not sensitive to density. However, unlike the optical emission the X-ray flux at later time continues to decrease monotonically.

Given the characteristic $F_\nu \propto t^{-2}$ decay for both the X-ray and optical emission beyond 0.55 day, the break is reasonably modeled by a jet with an opening angle of 5°. The inferred beaming-corrected γ-ray energy is only $E_\gamma \approx 5 \times 10^{49}$ erg, significantly lower than the strong clustering around 1.3×10^{51} erg seen in most bursts. Similarly, the beaming-corrected X-ray luminosity at $t = 10$ hours is $L_{X,10} \approx 3 \times 10^{43}$ erg s^{-1}, a factor of ten below the tightly clustered values for "typical" bursts (Figure 38.1).

The radio afterglow of GRB 030329 (Berger *et al.* 2003a; Sheth *et al.* 2003) reveals a different picture. The increase in flux during the first 10 days, followed

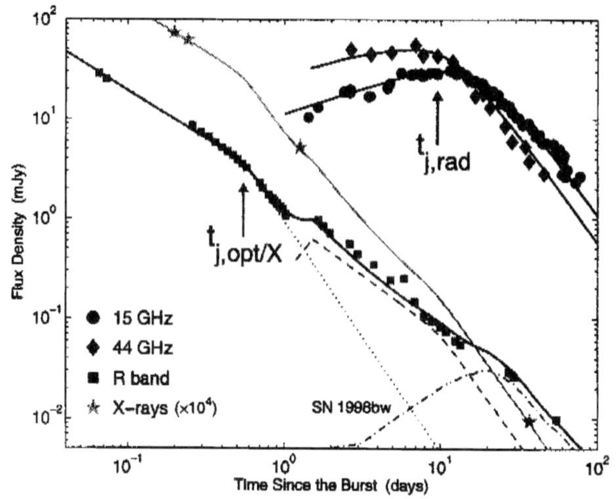

Fig. 38.2. Radio to X-ray lightcurves of the afterglow of GRB 030329, exhibiting the early jet break at 0.55 days in the optical and X-ray bands, as well as the subsequent rise in optical flux and the bright radio emission arising from a jet with an opening angle of 17°. Also plotted is the optical emission from SN 1998bw at the redshift of GRB 030329 as a proxy for SN 2003dh. The solid line is a combination of all the different emission components.

by a rapid decline, $F_\nu \propto t^{-1.9}$ at $t \gtrsim 10$ day and a decrease in peak flux at $\nu \lesssim$ 22.5 GHz (Figure 38.3) are indicative of a jet with an opening angle of 17°. The inferred beaming-corrected kinetic energy in the explosion is about 3×10^{50} erg, comparable to what is inferred from modeling of other afterglows (Panaitescu & Kumar 2002).

This result, combined with the resurgence in the optical emission at 1.5 days, is best explained in the context of a two-component explosion model. In this scenario the first component (a narrow jet, 5°) with initially a larger Lorentz factor is responsible for the γ-ray burst and the early optical and X-ray afterglow including the break at 0.55 day, while the second component (a wider jet, 17°) powers the radio afterglow and late optical emission (Figure 38.2; Berger *et al.* 2003a). The break at 10 days due to the second component has recently been inferred in the optical bands following a careful subtraction of the light from SN 2003dh which accompanied GRB 030329 (e.g. Matheson *et al.* 2003). Such a two-component jet finds a natural explanation in the collapsar model (Zhang, Woosley & Heger 2003).

The afterglow calorimetry of GRB 030329 has important ramifications for our understanding of GRB engines and the sub-energetic bursts. Namely, such bursts may have a total explosive yields similar to other GRBs (Figure 38.4), but their ultra-relativistic output varies considerably.

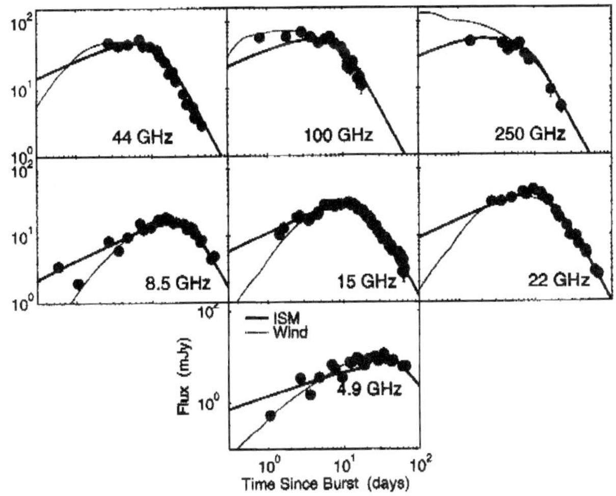

Fig. 38.3. Detailed radio lightcurves of the afterglow of GRB 030329. The solid lines are afterglow models of collimated ejecta expanding into circumburst media with a uniform and wind ($\rho \propto r^{-2}$) density profiles. The sharp turnover at about 10 days is indicative of a jet with an opening angle of 17°.

This leads to the following conclusions. First, radio calorimetry, which is sensitive to all ejecta with $\Gamma \gtrsim$ *few*, shows that the total energy yield of GRB 030329 is similar to those estimated for other bursts. Along these lines, the enigmatic GRB 980425 associated with the nearby supernova SN 1998bw also has negligible γ-ray emission, $E_{\gamma,\rm iso} \approx 8 \times 10^{47}$ erg; however, radio calorimetry (Li & Chevalier 1999) showed that even this extreme event had a similar explosive energy yield (Figure 38.4). The newly recognized class of cosmic explosions, the X-ray Flashes, exhibit little or no γ-ray emission but appear to have comparable X-ray and radio afterglows to those of GRBs (see § 38.4). Thus, the commonality of the total energy yield points to a common origin, but apparently the ultra-relativistic output is highly variable. Unraveling what physical parameter is responsible for this variation appears to be the next frontier in the field of cosmic explosions.

38.3 The incidence of engine in Type Ib/c supernovae

The inferences summarized in the previous section, coupled with the association of some GRBs with SNe Ib/c raises the question: is there a population of SNe that is powered by engines? Observationally there appear to be many distinctions (e.g. ejecta velocity and mass), but the association of the Type Ic SN 1998bw with GRB 980425 has indicated that some overlap exists. In particular, the radio emission from SN 1998bw revealed mildly relativistic ejecta with a complex structure indicative of a long-lived energy source. The expected fraction of similar events in the local

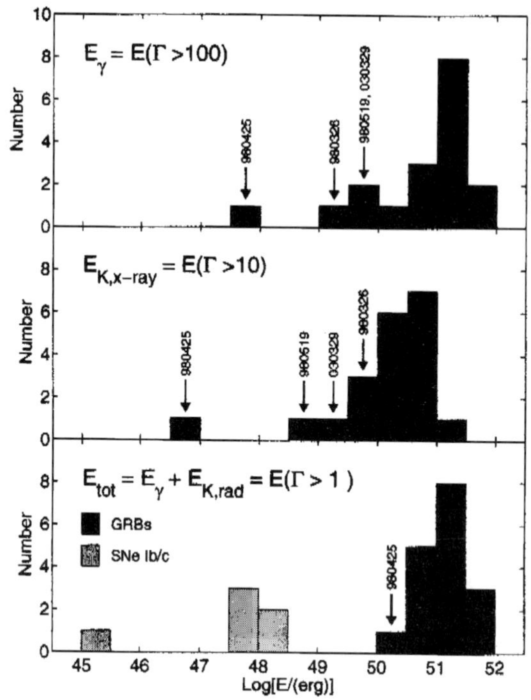

Fig. 38.4. The beaming-corrected energies of GRBs and SNe Ib/c in ejecta with a Lorentz factor ranging from $\gtrsim 100$ to order unity (*top:* the γ-ray energy, *middle:* the kinetic energy in the early afterglow; *bottom:* the total relativistic energy). The ultra-relativistic output of GRBs varies considerably despite a nearly standard total explosive yield. On the other hand, the significantly lower total energy in fast ejecta of SNe Ib/c points to a different energy source.

SN population, $\sim 0.5\%$ to 25%, depends on their origin: typical GRBs observed away from the jet axis versus an intermediate population of explosions.

To assess this fraction, and hence the origin of potential engine-driven SNe, directly, we have carried out since 1999 the most comprehensive radio survey of SNe Ib/c to date (Berger, Kulkarni & Chevalier 2002; Berger *et al.* 2003b). As was demonstrated in the case of SN 1998bw, such observations provide the best probe of relativistic ejecta (a proxy for an engine).

As seen in Figure 38.5 the luminosity function of SNe Ib/c is significantly broader than previously inferred, but none of the observed SNe approach the luminosity of typical GRB afterglows. We therefore place a limit of about 3% on the fraction of local SNe Ib/c that are powered by an engine or potentially gave rise to a GRB (Berger *et al.* 2003b).

In the majority of cases we find expansion velocities of $\lesssim 0.3c$ (Figure 38.6) as compared to $\Gamma \sim 2$ for SN 1998bw and $\Gamma \sim 5$ for GRB radio afterglows. Similarly, the energy carried by these ejecta can be accounted for in a hydrodynamic explosion

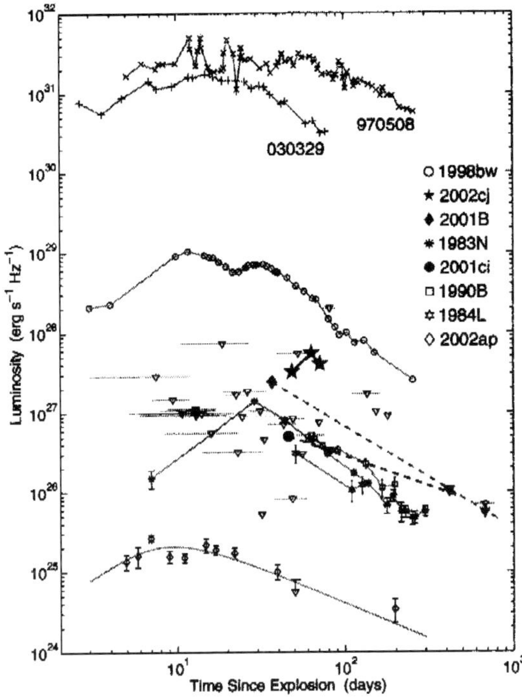

Fig. 38.5. Radio lightcurves of Type Ib/c SNe including the peculiar SN 1998bw/GRB 980425. A comparison to SN 1998bw and GRB afterglows reveals significantly less energy in high velocity ejecta, and thus constrains the fraction of SNe Ib/c that are powered by an engine to $<3\%$. There is therefore a dichotomy in the explosion mechanism of massive stars.

model. In fact, as seen in Figure 38.4 SNe Ib/c are significantly less energetic in terms of fast ejecta compared to GRBs. Thus, GRBs and the vast majority of SNe Ib/c do not share a common energy source. We can therefore rule out models of GRBs or SNe which suggest a significant overlap (e.g. Lamb *et al.* 2003).

38.4 Future directions

The recent recognition of a new class of cosmic explosion, the X-ray flashes (XRFs), provides an opportunity to extend the analysis presented above. These transients are defined as those with $\log[S_X(2\text{–}30\ \text{keV})/S_\gamma(30\text{–}400\ \text{keV})] > 0$, where S_X and S_γ are the fluences in the X-ray and γ-ray bands, respectively; i.e. the peak in the νF_ν spectrum lies in the X-ray band. With the exception of a significantly lower peak energy, XRFs share similar properties (e.g. duration, fluence) with GRBs.

Recent detections of XRF afterglows indicate that they likely arise at cosmological distances: they exhibit interstellar scintillation effects similar to those observed

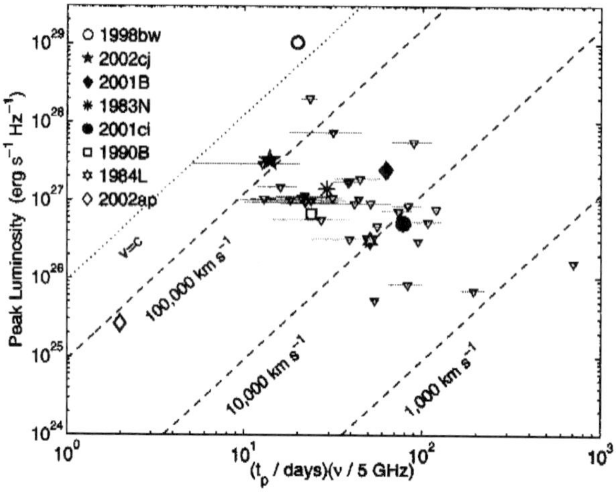

Fig. 38.6. Peak radio luminosity plotted versus the time of peak luminosity for Type Ib/c SNe. The diagonal lines are contours of constant average expansion velocity based on the assumption that the peak of the radio luminosity occurs at the synchrotron self-absorption frequency.

in GRB radio afterglows, they are associated with faint compact galaxies similar to the general population at $z \sim 1$, and in one case (XRF 020903), a redshift of 0.25 has been measured.

Using the measured redshift, the prompt energy release of XRF 020903 is only $\sim 2 \times 10^{49}$ erg, significantly lower than GRBs. The difference is even more pronounced when we consider that the spectrum peaked at ~5 keV. Thus, it is possible that XRF 020903 (and perhaps all XRFs) produce a negligible amount of highly relativistic ejecta, maybe as a result of higher baryon contamination in the ejecta. However, preliminary results indicate that the total relativistic output of XRF 020903, measured from the radio afterglow in the usual manner, is similar to that of GRBs (Figure 38.4; Soderberg *et al.* 2003). If so, XRFs may in fact share a common origin with GRBs.

To assess this possibility it is crucial to obtain a large sample of XRFs with measured redshifts. This, along with continued monitoring of GRBs and their afterglows (especially in the Swift era) and continued radio observations of SNe Ib/c, will allow us to determine more accurately the true diversity of cosmic explosions and the fraction of stellar deaths in each channel.

Acknowledgments

I would like to thank the conference organizers, C. Wheeler, P. Kumar, and P. Hoflich, and my many collaborators, in particular S. Kulkarni and D. Frail.

References

Berger, E., Kulkarni, S. R., and Frail, D. A. 2003, ApJ, 590, 379.
Berger, E., *et al.* 2003a, Nature in press; astro-ph/0308187
Berger, E., Kulkarni, S. R., and Chevalier, R. A. 2002, ApJ, 577, L5.
Berger, E., *et al.* 2003b, ApJ in press; astro-ph/0307228.
Bloom, J. S., Frail, D. A., and Kulkarni, S. R. 2003, astro-ph/0302210.
Filippenko, A. V. 1997, ARA&A, 35, 309.
Frail, D. A., *et al.* 2001, ApJ, 562, L55.
Galama, T. J., *et al.* 1998, Nature, 395, 670.
Goodman, J. 1986, ApJ, 308, L47.
Kumar, P. 2003, ApJ, 538, L125.
Li, Z. and Chevalier, R. A. 1999, ApJ, 526, 716.
MacFadyen, A. L. & Woosley, S. E. 1999, ApJ, 524, 262.
Paczynski, B. 2001, Acta Astronomica, 51, 1.
Panaitescu, A. and Kumar, P. 2002, ApJ, 571, 779.
Price, P. A., *et al.* 2003, Nature, 423, 844.
Sheth, K., *et al.* 2003, ApJ, 595, L33.
Soderberg, A. M., *et al.* 2003, in prep.
Stanek, K. Z., *et al.* 2003, ApJ, 591, L17.
Tiengo, A., *et al.* 2003, astro-ph/0305564.
Wang, L., *et al.* 2001, ApJ, 550, 1030.
Zhang, W., Woosley, S. E., & Heger, A. 2003; astro-ph/0308389.

39

A GRB simulation using 3-D relativistic hydrodynamics

J. K. Cannizzo

NASA/GSFC/Lab. for High Energy Astrophysics/Code 661, Greenbelt, MD 20771; also University of Maryland Baltimore County

N. Gehrels

NASA/GSFC/Lab. for High Energy Astrophysics/Code 661, Greenbelt, MD 20771;

E. T. Vishniac

Department of Physics and Astronomy, Johns Hopkins University, 3400 N. Charles Street, Baltimore, MD 21210

Abstract

We present the first unrestricted, three-dimensional relativistic hydrodynamical calculations of the blob of gas associated with the jet producing a gamma-ray burst as applied to the time when afterglow radiation is produced. Our main findings are that (i) gas ahead of the advancing blob does not accrete onto and merge with the blob material but rather flows around the blob, (ii) the decay light curve steepens at a time corresponding roughly to $\gamma^{-1} \approx \theta$ (in accord with earlier studies), and (iii) the rate of decrease of the forward component of momentum in the blob is well-fit by a simple model in which the gas in front of the blob exerts a drag force on the blob, and the cross sectional area of the blob increases quadratically with laboratory time.

39.1 Introduction

Gamma-ray bursts are the most powerful explosions in the Universe. If GRBs were isotropic, then the measured redshifts would imply total explosion energies of $\sim 10^{52}$–10^{54} ergs (Frail *et al.* 2001). Theoretical work on relativistic jet expansion, however, shows that one expects a steepening in the decay light curve if one is looking down the axis of a jet as the flow decelerates from a bulk Lorentz factor $\gamma^{-1} < \theta$ to $\gamma^{-1} > \theta$, where θ is the jet beaming angle (Rhoads 1999). The concept of a "break" corresponding to $\gamma^{-1} \simeq \theta$ has been used to infer the presence of strong beaming in GRBs (Frail *et al.* 2001; Panaitescu & Kumar 2001ab, 2002; Panaitescu, Mészáros, & Rees 1998; Piran, Kumar, Panaitescu, & Piro 2001). Frail *et al.* (2001) utilize the theoretical framework of Sari, Piran, & Halpern (1999) which takes the jet evolution to be spherical adiabatic expansion to show that, after correcting the "isotropic" energies to account for the specific beaming factor for each burst, the total burst energy reduces to a narrow range centered on $\sim 5 \times 10^{50}$ ergs. Frail *et al.* (2001) give a tabulation of 17 afterglows for which redshifts had been measured, up to 2001 January. The redshifts range from 0.433 to 4.5. Of these, 10 also have break times known to within $\sim$30%, ranging from 1 d to 25 d. (In addition, three GRBs have listed lower limits on the break time, and two have listed upper limits.) The combination of redshifts, fluences, and break times lead to estimates of the jet angles θ ranging from 1° to 25°, with an aggregation near 4°. (The number of GRB's with redshift determinations is currently $\sim$40. For an update of the work described in Frail *et al.* 2001, see Berger, Kulkarni, & Frail 2003.)

In the context of GRBs, work has been done using 2D and 3-D relativistic hydro codes to consider the evolution of the GRB jet as it propagates through the envelope of the progenitor star, up to the point where it breaks out of the stellar surface and produces the prompt GRB emission (Zhang, Woosley, & MacFadyen 2003, Zhang, Woosley, & Heger 2003). In this work we consider the evolution covering the afterglow time (i.e, after the period considered by Zhang *et al.* 2003ab). We utilize a three dimensional relativistic hydrodynamical code to study the propagation of an initially ultrarelativistic blob into a dense circumstellar medium (CSM). We study the spatial spreading of the blob both along the direction of propagation and orthogonal to it, as well as the evolution of γ in space and time. A full description of this study can be found in Cannizzo *et al.* (2004).

39.2 Results

The model used is that of Del Zanna & Bucciantini (2002). We performed runs in which the blob initially is either a small cone, sphere, or plate, with symmetry about the propagation vector. An initial spread is imparted to each fluid element by setting the y and z components of velocity such that $\theta = (v_y^2 + v_z^2)^{1/2}/v_x = 0.035$. The motivation for taking $\theta = 0.035$ is the study of Frail *et al.* We also perform one trial with $\theta = 0$, i.e., $v_y = v_z = 0$ at all points. This is the "null hypothesis" run.

Figure 39.1 shows values of total rest-mass energy within selected high γ cuts through the computational domain for the initially conical run, and the evolution of the x–component of momentum $\langle \gamma v_x \rangle$ for five different runs – the initial blob as a plate ("P1" and "P2"), sphere ("S"), cone ("C"), and cone with zero spreading ("$v_T(0) = 0$"). The weighting function used in evaluating $\langle \gamma v_x \rangle$ is $W(\gamma) = (\gamma^2 - 1)\rho\gamma$. The run P2 uses fewer grid points (500×50^2) than the other runs, thus material shunted aside by the bow shock reaches the edges of the computational domain earlier. The drag force on the blob increases as the effective area presented by the blob, which for a roughly constant lateral spreading goes as x^2. Therefore the integral of the relativistic impulse equation $d(M_{\rm blob}\gamma_{\rm blob}v_x)/dt = F_{\rm drag}$ provides

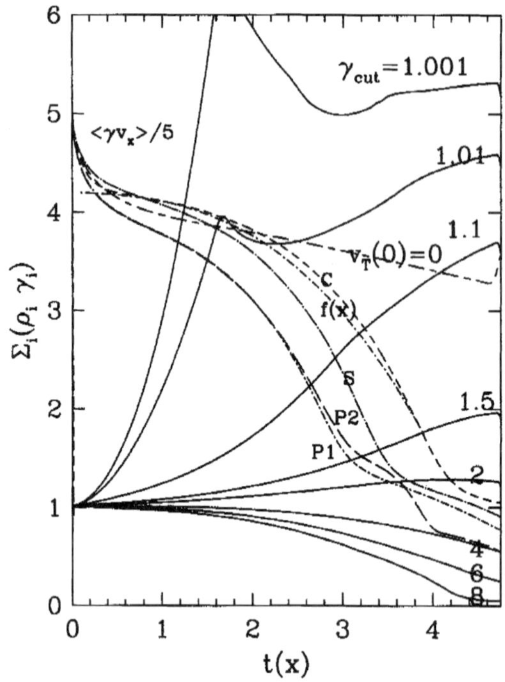

Fig. 39.1. The evolution of rest mass energy contained within various mass cuts, for the conical run (solid lines), and the evolution of the weighted x–component of the momentum $<\gamma v_x>$, divided by 5 so as to be on a common scale with the rest mass energy curves (dashed lines). The numbers associated with the solid lines indicate the γ value used in each rest mass energy cut. We show the $\langle\gamma v_x\rangle$ evolution for the conical run (C), the spherical run (S) and the two plate runs (P1: 500×100^2, P2: 500×50^2). For these four runs the initial spread in velocities is such that $\theta = 4°$, whereas for the run labelled $v_T(0) = 0$ there is no tangential component initially to the velocities. The function $f(x)$ represents a fitting to the conical run given by $21 - 0.15(x + 0.45)^3$.

a solution of the form $\langle\gamma v_x\rangle \simeq a - b(x + c)^3$, where our fitting for the conical run ($a = 21, b = 0.15$, and $c = 0.45$) is shown by the curve labeled "$f(x)$".

The curves in Figure 39.1 showing the variation of the rest-mass energy contained within different γ cuts follow the values $\gamma_{\rm cut} = 1.001$, 1.01, 1.1, 1.5, 2, 4, 6, and 8. The higher γ cuts $\gamma_{\rm cut} \gtrsim 2$ reveal that a negligible fraction of CSM matter gets accelerated to significant values. The curved bow shock shunts material laterally in front of the advancing blob, rather than accelerating it up to a significant fraction of the blob's bulk Lorentz factor ~10–20. The initially imposed tangential velocity v_T leads to the lateral expansion that effectively increases the cross section for interaction of the blob. The negligible decrease in $\langle\gamma v_x\rangle$ for the $v_T(0) = 0$ run demonstrates the importance of this effect.

Figure 39.2 shows the evolution of our canonical "luminosity" measure, namely $\Sigma_i p_i(1 + \beta_i \cos\phi)/(1 - \beta_i \cos\phi)^2$, for various viewing angles ϕ. The light curves

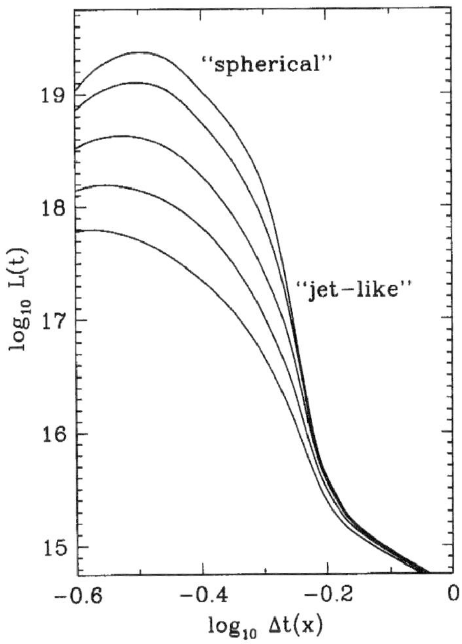

Fig. 39.2. Light curves constructed by summing the quantity $p_i(1 + \beta_i \cos\phi)/(1 - \beta_i \cos\phi)^2$ over the grid. The five values of the angle ϕ between the observer and the local velocity vector of a fluid element are (going to smaller emission values) 0°, 3°, 6°, 10°, and 15°. For the "face-on" viewer (inclination = 0°), there is a mild change in the decay slope, i.e., break, between the regimes of "spherical" and "jetlike" expansion.

are calculated by summing the emission from all points within the computational domain that lie, in a given time step, on 100 $y-z$ planes which straddle the position of the blob and advance toward the observer at c. One sees a slight break in the "light curve" for the $\phi = 0°$ observer at $\Delta \log t(x) \approx -0.3$ due to the transition from spherical to jetlike expansion. Although our model is too simple to allow any detailed quantitative comparison with observations at this point, it is worth noting that many of the observed afterglow transitions are also smooth (e.g., Fig. 38.2 of Stanek *et al.* 1999; Fig. 38.1 of Harrison *et al.* 1999), rather than the abrupt breaks evident in some of the semi-analytical models.

39.3 Conclusion

The calculations we present are the first 3-D relativistic hydrodynamical calculations of GRB jet evolution pertinent to the afterglow phase that do not enforce any special symmetry (e.g., spherical or axial). We find that (i) the CSM gas does not accrete onto the advancing blob, but rather is shunted aside by the bow shock, (ii) the decay light curve steepens roughly when one first "sees" the edge of the jet $\gamma^{-1} \approx \theta$, with this effect being strongest for "face-on" observers (confirming previous studies), and (iii) the rate of decrease of the x-component of momentum $\langle \gamma v_x \rangle$ is well-characterized by a simple model in which the cross sectional area of the blob increases quadratically with laboratory time (or distance). The primary impetus for the built-in assumption of accretion of matter in previous studies was the influential work of Blandford & McKee (1976 = BM76) in which spherical relativistic expansion was considered. Accretion of gas onto the relativistically expanding shell is obviously justified for spherical expansion, but subsequent GRB workers applied the results to the case of the GRB jet, in which a thin wedge of material propagates through a low density medium. Here however material tends to be pushed around the jet, rather than to accrete. In summary, there is nothing in our results to suggest that BM76, Rhoads (1999), and Sari *et al.* (1999) are not internally consistent, rather it appears that the subsequent application of their results to afterglow evolution may have been inappropriate.

Acknowledgements

We are grateful to Luca Del Zanna, Pawan Kumar, Andrew MacFadyen, Jay Norris, Craig Wheeler, and Weiqun Zhang for helpful discussions.

References

Berger, E., Kulkarni, S. R., & Frail, D. A. 2003, *Astrophys. J.*, **590**, 379–385.
Blandford, R. D., & McKee, C. F. 1976, *The Phys. of Fluids*, **19**, 1130–1138.
Cannizzo, J. K., Gehrels, N., & Vishniac, E. T. 2004, *Astrophys. J.*, **601**, (Jan. 20 issue) in press.
Del Zanna, L., & Bucciantini, N. 2002, *Astr. & Ap.*, **390**, 1177–1186.
Frail, D. A., Kulkarni, S. R., Sari, R., Djorgovski, S. G., Bloom, J. S., Galama, T. J., Reichart, D. E., Berger, E., Harrison, F. A., Price, P. A., Yost, S. A., Diercks, A., Goodrich, R. W., & Chaffee, F. 2001, *Astrophys. J.*, **562**, L55–L58.
Harrison, F. A., Bloom, J. S., Frail, D. A., Sari, R., Kulkarni, S. R., Djorgovski, S. G., Axelrod, T., Mould, J., Schmidt, B. P., Wieringa, M. H., Wark, R. M., Subrahmanyan, R., McConnell, D., McCarthy, P. J., Schaefer, B. E., McMahon, R. G., Markze, R. O., Firth, E., Soffitta, P., & Amati, L. 1999, *Astrophys. J.*, **523**, L121–L124.
Panaitescu, A., & Kumar, P. 2001a, *Astrophys. J.*, **554**, 667–677.
Panaitescu, A., & Kumar, P. 2001b, *Astrophys. J.*, **560**, L49–L53.
Panaitescu, A., & Kumar, P. 2002, *Astrophys. J.*, **571**, 779–789.
Panaitescu, A., Mészáros, P., & Rees, M. J. 1998, *Astrophys. J.*, **503**, 314–324.
Piran, T., Kumar, P., Panaitescu, A., & Piro, L. 2001, *Astrophys. J.*, **560**, L167–L169.
Rhoads, J. E. 1999, *Astrophys. J.*, **525**, 737–749.
Sari, R., Piran., T., & Halpern, J. P. 1999, *Astrophys. J.*, **519**, L17–L20.
Stanek, K. Z., Garnavich, P. M., Kaluzny, J., Pych, W., & Thompson, I. 1999, *Astrophys. J.*, **522**, L39–L42.
Zhang, W., Woosley, S. E., & Heger, A. 2003, astro-ph/0308389.
Zhang, W., Woosley, S. E., & MacFadyen, A. I. 2003, *Astrophys. J.*, **586**, 356–371.

40

The first direct supernova/GRB connection: GRB 030329 / SN 2003dh

T. Matheson

Harvard-Smithsonian Center for Astrophysics 60 Garden Street, Cambridge, MA 02138

Abstract

Observations of gamma-ray burst (GRB) afterglows have yielded tantalizing hints that supernovae (SNe) and GRBs are related. The case had been circumstantial, though, relying on irregularities in the light curve or the colors of the afterglow. I will present observations of the optical afterglow of GRB 030329. The early spectra show a power-law continuum, consistent with other GRB afterglows. After approximately one week, broad peaks in the spectrum developed that were remarkably similar to those seen in the spectra of the peculiar Type Ic SN 1998bw. This is the first direct, spectroscopic confirmation that at least some GRBs arise from SNe.

40.1 Introduction

The mechanism that produces gamma-ray bursts (GRBs) has been the subject of considerable speculation during the four decades since their discovery (see Mészáros 2002 for a recent review of the theories of GRBs). Optical afterglows (e.g., GRB 970228: Groot *et al.* 1997; van Paradijs *et al.* 1997) opened a new window on the field (see, e.g., van Paradijs, Kouveliotou, & Wijers 2000). Subsequent studies of other bursts yielded the redshifts of several GRBs (e.g., GRB 970508: Metzger *et al.* 1997), providing definitive evidence for their cosmological origin.

Models that invoked supernovae (SNe) to explain GRBs were proposed from the very beginning (e.g., Colgate 1968; Woosley 1993; Woosley & MacFadyen 1999). A strong hint was provided by GRB 980425. In this case, no traditional GRB optical afterglow was seen, but a supernova, SN 1998bw, was found in the error box of the GRB (Galama *et al.* 1998a). The SN was classified as a Type Ic (Patat & Piemonte 1998), but it was unusual, with high expansion velocities (Patat *et al.* 2001). Other SNe with high expansion velocities (and usually large luminosity as well) such as SN 1997ef and SN 2002ap are sometimes referred to as "hypernovae" (see, e.g., Iwamoto *et al.* 1998, 2000).

The redshift of a typical GRB is $z \approx 1$, implying that a supernova component underlying an optical afterglow would be difficult to detect. At $z \approx 1$, even a bright core-collapse event would peak at $R > 23$ mag. Nevertheless, late-time deviations from the power-law decline typically observed for optical afterglows have been seen and these bumps in the light curves have been interpreted as evidence for supernovae (for a recent summary, see Bloom 2003). Perhaps the best evidence that classical, long-duration gamma-ray bursts are generated by core-collapse supernovae was provided by GRB 011121. It was at $z = 0.36$, so the supernova component would have been relatively bright. A bump in the light curve was observed both from the ground and with *HST* (Garnavich *et al.* 2003a; Bloom *et al.* 2002). The color changes in the light curve of GRB 011121 were also consistent with a supernova (designated SN 2001ke), but a spectrum obtained by Garnavich *et al.* (2003a) during the time that the bump was apparent did not show any features that could be definitively identified as originating from a supernova. The detection of a clear spectroscopic supernova signature was for the first time reported for the GRB 030329 by Matheson *et al.* (2003a, 2003b), Garnavich *et al.* (2003b, 2003c), Chornock *et al.* (2003), and Stanek *et al.* (2003a). Hjorth *et al.* (2003) also presented spectroscopic data obtained with the VLT. In addition, Kawabata *et al.* (2003) obtained a spectrum of SN 2003dh with the Subaru telescope.

The extremely bright GRB 030329 was detected by instruments aboard *HETE II* at 11:37:14.67 (UT is used throughout this paper) on 2003 March 29 (Vanderspek *et al.* 2003). Due to the brightness of the afterglow, observations of the optical transient (OT) were extensive, making it most likely the best-observed afterglow so far. From the moment the low redshift of 0.1685 for the GRB 030329 was announced (Greiner *et al.* 2003), we started organizing a campaign of spectroscopic and photometric follow-up of the afterglow and later the possible associated supernova. Stanek *et al.* (2003a) reported the first results of this campaign, namely a clear spectroscopic detection of a SN 1998bw-like supernova in the early spectra, designated SN 2003dh (Garnavich *et al.* 2003c). In this paper, I describe the evidence for the supernova in the spectroscopy during the first two months. For a more complete discussion, see Matheson *et al.* (2003c).

40.2 Spectra

The brightness of the OT allowed us to observe the OT each of the 12 nights between March 30 and April 10 UT, mostly with the MMT 6.5-m, but also with the Magellan 6.5-m, Lick Observatory 3-m, LCO du Pont 2.5-m, and FLWO 1.5-m telescopes. This provided a unique opportunity to look for spectroscopic evolution over many nights. The early spectra of the OT of GRB 030329 (top of Figure 40.1) consist of

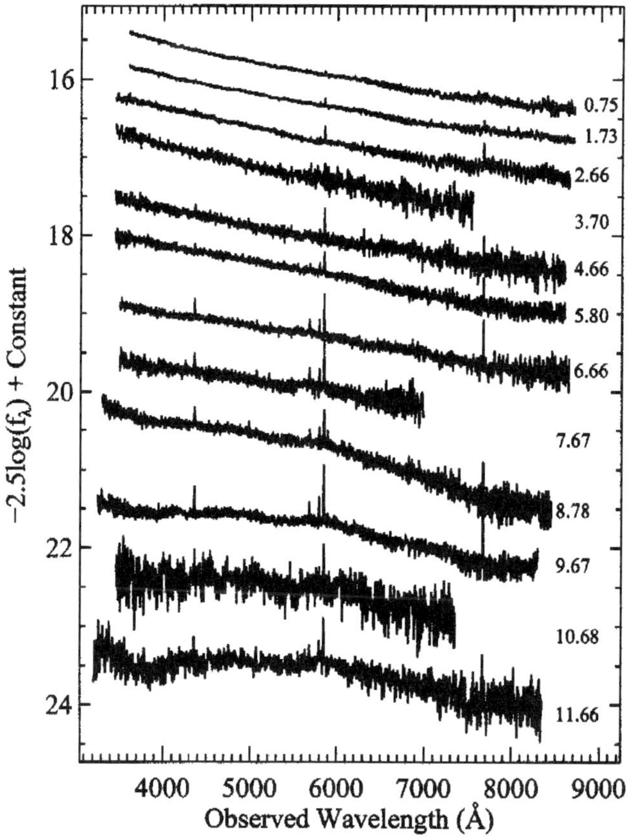

Fig. 40.1. Evolution of the GRB 030329/SN 2003dh spectrum, from March 30.23 UT (0.75 days after the burst), to April 10.14 UT (11.66 days after the burst). The early spectra consist of a power-law continuum with narrow emission lines originating from H II regions in the host galaxy at $z = 0.1685$. Spectra taken after $\Delta T = 6.66$ days show the development of broad peaks characteristic of a supernova.

a power-law continuum typical of GRB afterglows, with narrow emission features identifiable as Hα, [O III] $\lambda\lambda$4959, 5007, Hβ, and [O II] λ3727 at $z = 0.1685$ (Greiner *et al.* 2003; Caldwell *et al.* 2003) probably from H II regions in the host galaxy.

Beginning at $\Delta T = 7.67$ days, our spectra deviated from the pure power-law continuum. Broad peaks in flux, characteristic of a supernova, appeared. The broad bumps are seen at approximately 5000 Å and 4200 Å (rest frame). At that time, the spectrum of GRB 030329 looked similar to that of the peculiar Type Ic SN 1998bw a week before maximum light (Patat *et al.* 2001) superposed on a typical afterglow continuum. Over the next few days the SN features became more prominent as the afterglow faded and the SN brightened toward maximum.

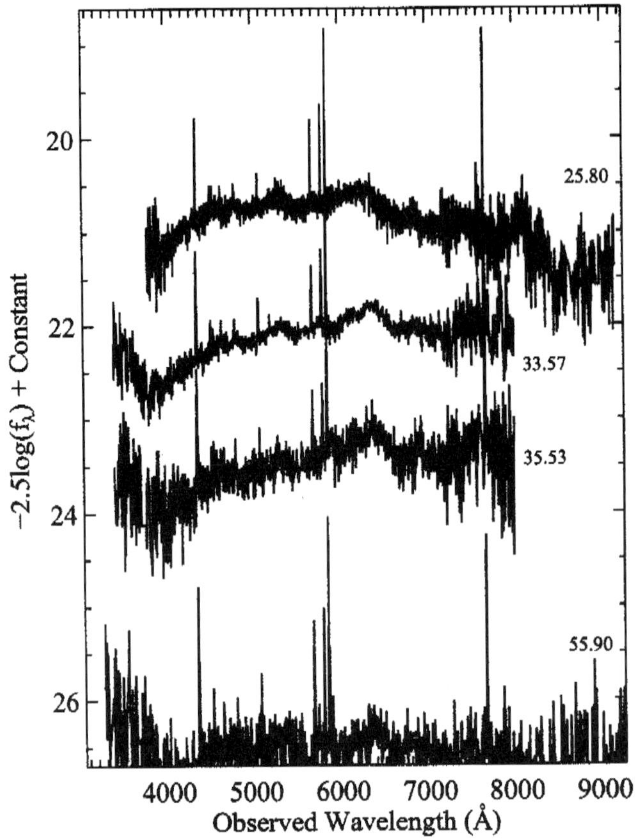

Fig. 40.2. Evolution of the GRB 030329/SN 2003dh spectrum, from April 24.28 UT (25.8 days after the burst), to May 24.38 (55.9 days after the burst). The power-law contribution decreases and the spectra become more red as the SN component begins to dominate, although the upturn at blue wavelengths may still be the power law. The broad features of a supernova are readily apparent, and the overall spectrum continues to resemble that of SN 1998bw several days after maximum.

Later spectra obtained on April 24.28, May 2.05, May 4.01, and May 24.38 continue to show the characteristics of a supernova. As the power-law continuum of the GRB afterglow fades, the supernova spectrum rises, becoming the dominant component of the overall spectrum (Figure 40.2).

40.3 Separating the GRB from the supernova

To explore the nature of the supernova underlying the OT, we modeled the spectrum as the sum of a power-law continuum and a peculiar Type Ic SN. Specifically, we chose for comparison SN 1998bw (Patat *et al.* 2001), SN 1997ef (Iwamoto *et al.*

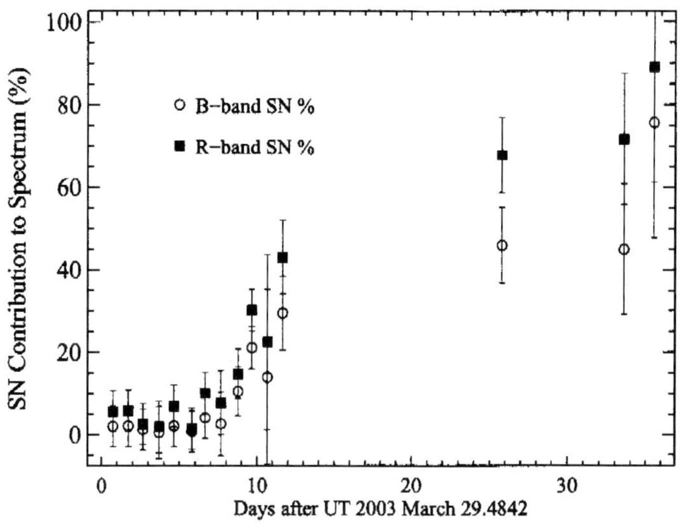

Fig. 40.3. Relative contribution of a supernova spectrum to the GRB 030329/SN 2003dh afterglow as a function of time in the *B* (*open circles*) and *R* (*filled squares*) bands. Using the technique described in the text, we derive a best fit to the afterglow spectrum at each epoch with the fiducial power-law continuum and the closest match from our set of peculiar SNe Ic. We then synthesize the relative *B*-band and *R*-band contributions. There is some scatter for the early epochs due to noise in the spectra, but a clear deviation is evident starting at $\Delta T = 7.67$ days, with a subsequent rapid increase in the fraction of the overall spectrum contributed by the SN. Errors are estimated from the scatter when the SN component is close to zero ($\Delta T < 6$ days) and from the scale of the error in the least-squares minimization.

2000), and SN 2002ap (using our own as yet unpublished spectra, but see, e.g., Kinugasa *et al.* 2002; Foley *et al.* 2003). We had 62 spectra of these three SNe, spanning the epochs of seven days before maximum to several weeks past. For the power-law continuum, we chose to use one of our early spectra to represent the afterglow of the GRB. The spectrum at time $\Delta T = 5.80$ days was of high signal-to-noise ratio (S/N), and suffers from little fringing at the red end. Therefore, we smoothed this spectrum to provide the fiducial power-law continuum of the OT for our model.

To find the best match with a supernova spectrum, we compared each spectrum of the afterglow with the sum of the fiducial continuum and a spectrum of one of the SNe in the sample. We performed a least-squares fit, allowing the fraction of continuum and SN to vary, finding the best combination of continuum and SN for each of the SN spectra. The minimum least-squares deviation within this set was then taken as the best SN match for that epoch of OT observation. Figure 39.3 shows the relative contribution to the OT spectrum by the underlying SN in the *B* and *R* bands as a function of ΔT.

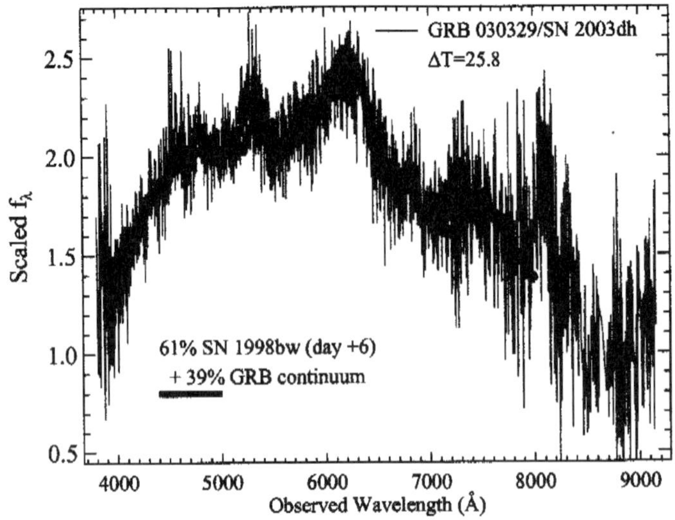

Fig. 40.4. Observed spectrum (*thin line*) of the GRB 030329/SN 2003dh afterglow at $\Delta T = 25.8$ days. The model spectrum (*thick line*) consists of 39% continuum and 61% SN 1998bw from 6 days after maximum.

Within the uncertainties of our fit, the SN fraction is consistent with zero for the first few days after the GRB. At $\Delta T = 7.67$ days, the SN begins to appear in the spectrum, without strong evidence for a supernova component before this. When the fit indicates the presence of a supernova, the best match is almost always SN 1998bw. The only exceptions to this are from nights when the spectrum of the OT are extremely noisy, implying that less weight should be given to those results. The least-squares deviation for the spectra that do not match SN 1998bw is also much larger.

Our best spectrum (i.e., with the highest S/N) from this time when the SN features begin to appear is at $\Delta T = 9.67$ days. For that epoch, our best fit is 74% continuum and 26% SN 1998bw (at day -6 relative to SN *B*-band maximum). The next$-$best fit is SN 1998bw at day -7. Using a different early epoch to define the reference continuum does not alter these results significantly. It causes slight changes in the relative percentages, but the same SN spectrum still produces the best fit, albeit with a larger least-squares deviation.

The SN fraction contributing to the total spectrum increases steadily with time. By $\Delta T = 25.8$ days, the SN fraction is $\sim$61%, with the best-fit SN being SN 1998bw at day $+6$ (Figure 40.4). The SN percentage at $\Delta T = 33.6$ days is still about 63%, but the best match is now SN 1998bw at day $+13$. The rest-frame time difference between $\Delta T = 9.67$ days and $\Delta T = 25.8$ days is 13.8 days ($z = 0.1685$). For the best-fit SN spectra from those epochs, SN 1998bw at day -6 and SN 1998bw at day $+6$ respectively, the time difference is 12 days. The rest-frame time difference between $\Delta T = 25.8$ days and $\Delta T = 33.6$ days is 6.7 days, with a time difference between the best-fit spectra for those epochs of 7 days. The spectral evolution determined from these fits indicates that SN 2003dh follows SN 1998bw closely, and it is not as similar to SN 1997ef or SN 2002ap. The analysis by Kawabata *et al.* (2003) of their May 10 spectrum gives a phase for the spectrum of SN 2003dh that is consistent with our dates, although they do consider SN 1997ef as a viable alternative to SN 1998bw as a match for the SN component in the afterglow.

The spectra of SN 1998bw (and other highly energetic SNe) are not simple to interpret. The high expansion velocities result in many overlapping lines so that identification of specific line features is problematic for the early phases of spectral evolution (see, e.g., Iwamoto *et al.* 1998; Stathakis *et al.* 2000; Nakamura *et al.* 2001; Patat *et al.* 2001). This includes spectra up to two weeks after maximum, approximately the same epochs covered by our spectra of SN 2003dh. In fact, as Iwamoto *et al.* (1998) showed, the spectra at these phases do not show line features. The peaks in the spectra are due to gaps in opacity, not individual spectral lines. Detailed modeling of the spectra can reveal some aspects of the composition of the ejecta (Mazzali *et al.* 2003).

If the $\Delta T = 9.67$ days spectrum for the afterglow does match SN 1998bw at day -6, then limits can be placed on the timing of the supernova explosion relative to the GRB. The rest-frame time for $\Delta T = 9.67$ days is 8.2 days, implying that the time of the GRB would correspond to $\sim$14 days before maximum for the SN. The rise times of SNe Ic are not well determined, especially for the small subset of peculiar ones. Stritzinger *et al.* (2002) found the rise time of the Type Ib/c SN 1999ex was $\sim$18 days (in the B band), while Richmond *et al.* (1996) reported a rise time of $\sim$12 days (in the V band) for the Type Ic SN 1994I. A rise time of $\sim$14 days for SN 2003dh is certainly a reasonable number. It also makes it extremely unlikely that the SN exploded significantly earlier or later than the time of the GRB, most likely within ± 2 days of the GRB itself.

The totality of data contained in this paper allows us to attempt to decompose the light curve of the OT into the supernova and the afterglow (power-law) component. From the spectral decomposition procedure described above, we have the fraction of light in the BR-bands for both components at various times, assuming that the spectrum of the afterglow did not evolve since $\Delta T = 5.64$ days. As we find that the spectral evolution is remarkably close to that of SN 1998bw, we model the R-band supernova component with the V-band light curve of SN 1998bw (Galama *et al.* 1998a, b) stretched by $(1 + z) = 1.1685$ and shifted in magnitude to obtain a good fit. The afterglow component is fit by using the early points starting at $\Delta T = 5.64$ days with late points obtained via the spectral decomposition. This can be done in both in the B and in the R-band and leads to consistent results, indicating that our assumption of the afterglow not evolving in color at later times is indeed valid.

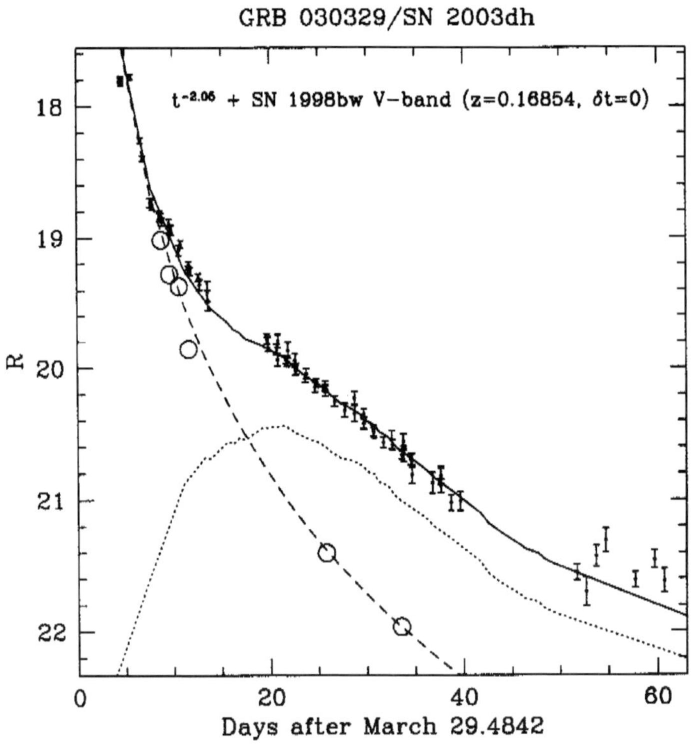

Fig. 40.5. Decomposition of the OT R-band light curve into the supernova (*dotted line*) and the power-law continuum (*dashed line*). As the light curve model for the supernova, we took the V-band light curve of SN 1998bw (Galama *et al.* 1998a, b) stretched by $(1 + z) = 1.1685$ and shifted in magnitude. The resulting supernova light curve peaks at an apparent magnitude of $m_R = 20.4$. No offset in time has been applied between the GRB and the supernova. To constrain the continuum, information from the spectral decomposition was used (*big open circles*).

The result of the decomposition of the OT R-band light curve into the supernova and the power-law continuum is shown in Figure 40.5. The overall fit is remarkably good, given the assumptions (such as using the stretched V-band light curve of SN 1998bw as a proxy for the SN 2003dh R-band light curve). No time offset between the supernova and the GRB was applied, and given how good the fit is, we decided not to explore time offset as an additional parameter. Introducing such an additional parameter would most likely result in a somewhat better fit (indeed, we find that to be the case for $\delta t \approx -2$ days), but this could easily be an artifact with no physical significance, purely due to small differences between SN 1998bw and SN 2003dh. At this point the assumption that the GRB and the SN happened at the same time seems most natural.

40.4 Summary

The spectroscopy of the optical afterglow of GRB 030329, as first shown by Stanek *et al.* (2003a), provided direct evidence that at least some of the long-burst GRBs are related to core-collapse SNe. We have shown with a larger set of data that the SN component is similar to SN 1998bw, an unusual Type Ic SN. It is not clear yet whether all long-burst GRBs arise from SNe. Catching another GRB at a redshift this low is unlikely, but large telescopes may be able to discern SNe in some of the relatively nearby bursts. With this one example, though, we now have solid evidence that some GRBs and SNe have the same progenitors.

Acknowledgments

Kris Stanek and Peter Garnavich were very supportive and equal colleagues in the research described here. I would like to thank the many observers who sacrificed their time to observe this GRB (see Matheson *et al* 2003c for a full list).

References

Bloom, J. S., 2003, in *Gamma-Ray Bursts in the Afterglow Era*, ed. M. Feroci *et al.* (San Francisco: ASP), 1.
Bloom, J. S., *et al.*, 2002, *Astrophys. J.*, **572**, L45.
Caldwell, N., Garnavich, P., Holland, S., Matheson, T., & Stanek, K. Z., 2003, *GCN Circ.* 2053.
Chornock, R., Foley, R. J., Filippenko, A. V., Papenkova, M., & Weisz, D., 2003, *GCN Circ.* 2131.
Colgate, S. A., 1968, *Canadian J. Phys.*, **46**, 476.
Foley, R. J., *et al.*, 2003, *Pub. Astron. Soc. Pac.*, in press (astro-ph/0307136).
Galama, T. J., *et al.*, 1998a, *Nature*, **395**, 670.
Galama, T. J., *et al.*, 1998b, *Astrophys. J.*, **497**, L13.
Garnavich, P. M., *et al.*, 2003a, *Astrophys. J.*, **582**, 924.
Garnavich, P., *et al.*, 2003b, *IAU Circ.* 8108.
Garnavich, P., Matheson, T., Olszewski, E. W., Harding, P., & Stanek, K. Z., 2003c, *IAU Circ.* 8114.
Greiner, J., *et al.*, 2003, *GCN Circ.* 2020.
Groot, P. J., *et al.*, 1997, *IAU Circ.* 6584.
Hjorth, J., *et al.*, 2003, *Nature*, **423**, 847.
Iwamoto, K., *et al.*, 1998, *Nature*, **395**, 672.
Iwamoto, K., *et al.*, 2000, *Astrophys. J.*, **534**, 660.
Kawabata, K. S., *et al.*, 2003, *Astrophys. J.*l, **593**, L19.
Kinugasa, K., *et al.*, 2002, *Astrophys. J.*, **577**, L97.
Matheson, T., *et al.*, 2003a, *GCN Circ.* 2107.
Matheson, T., *et al.*, 2003b, *GCN Circ.* 2120.
Matheson, T., *et al.*, 2003c, *Astrophys. J.*, in press (astro-ph/0307435).
Mazzali, P. A., *et al.*, 2003, *Astrophys. J.*, submitted (astro-ph/0309555).
Mészáros, P., 2002, *Ann. Rev. Astron. Astrophys.*, **40**, 137.

Metzger, M. R., *et al.*, 1997, *Nature*, **387**, 878.
Nakamura, T., Mazzali, P. A., Nomoto, K., & Iwamoto, K., 2001, *Astrophys. J.*, **550**, 991.
Patat, F., & Piemonte A., 1998, *IAU Circ.* 6918.
Patat, F., *et al.*, 2001, *Astrophys. J.*, **555**, 900.
Richmond, M. W., *et al.*, 1996, *Astron. J.*, **111**, 327.
Stanek, K. Z., *et al.*, 2003a, *Astrophys. J.*, **591**, L17.
Stathakis, R. A., *et al.*, 2000, *Mon. Not. R. Astr. Soc.*, **314**, 807.
Stritzinger, M., *et al.*, 2002, *Astron. J.*, **124**, 2100.
Vanderspek, R., *et al.*, 2003, *GCN Circ.* 1997.
van Paradijs, J., Kouveliotou, C., & Wijers, R. A. M. J., 2000, *Ann. Rev. Astron. Astrophys.*, **38**, 379.
van Paradijs, J., *et al.*, 1997, *Nature*, **386**, 686.
Woosley, S. E., 1993, *Astrophys. J.*, **405**, 273.
Woosley, S. E., & MacFadyen, A. I., 1999, *Astron. Astrophys.*, **138**, 499.

41

Gamma-Ray Burst environment and energetics

A. Panaitescu and P. Kumar

Dept. of Astronomy, University of Texas at Austin, TX 78712

Abstract

There are currently a few cases where a supernova was associated with a Gamma-Ray Burst, proving that GRBs arise from the death of massive stars. Other lines of evidence supporting this conclusion are the spatial location of bursts in the host galaxy, the detection of multiple high velocity absorption lines in GRB 021004, and of X-ray emission lines and edges for a few afterglows. Massive stars drive powerful winds, shaping the circumstellar medium up to tens of parsecs. Modeling of the broadband afterglow emission with a relativistic fireball interacting with the circumburst medium, yields estimations of its particle density. The resulting values, ranging from 0.1 cm^{-3} to 50 cm^{-3}, are consistent with the density of the wind from a Wolf-Rayet star at the typical distance ($0.1 \div 1$ pc) where the afterglow is expected to occur. The r^{-2} density profile expected around a massive star is consistent with the results of afterglow modeling in a majority of cases; nevertheless there are a few afterglows for which a homogeneous medium accommodates much better the sharpness of the optical light-curve break. Afterglow modeling also shows that the kinetic energy of GRB jets spans the range 10^{50} and 3×10^{51} ergs, i.e. slightly less than that of the supernova ejecta. The burst γ-ray energy output, corrected for collimation, has a similar range.

41.1 Observational evidence linking GRBs with massive stars

The association of a Gamma-Ray Burst with a supernova is the best observational evidence we have for a link between GRBs and the collapse of a stellar core. The first discovered association is that of supernova 1998bw and GRB 980425 (Galama *et al.* 1998). The supernova was found less than a day after the burst, within its 8 arcmin error circle provided by the Wide Field Camera onboard the BeppoSAX satellite. Galama *et al.* (1998) estimate that the probability of observing by chance a supernova within a time window of 10 days around the burst epoch and in the error boxes of any of the 13 bursts localized by BeppoSAX prior to GRB 980425 is about 10^{-4}. Thus, the association of SN 1998bw with GRB 980425 is unlikely to be incidental.

The temporal structure, duration ($\sim$20 seconds), and spectrum of GRB 980425 (peak energy $\sim$150 keV) are typical for a cosmological burst; however the supernova redshift ($z = 0.0085$) and 20 keV$-$1 MeV fluence (4×10^{-6} erg/cm^{-2}) imply a burst γ-ray output of $\sim 10^{48}$ ergs (if isotropic), i.e. 3–6 orders of magnitude smaller than for a cosmological burst. For a GRB rate of $0.1 \div 0.5$ Gpc^{-1} yr^{-1} (Schmidt 2001, Stern *et al.* 2002), the probability of observing a burst during 15 months of observations (from January 1997 until April 1998) and occurring within 40 Mpc is about 10^{-4}. Collimation of the GRB outflow and an observer location outside the initial jet opening (which would explain the low luminosity of GRB 980425) increase the probability of observing a burst like GRB 980425 to less than 1%, which suggests that this is a new type of burst, with a significantly higher occurrence rate than that of cosmological bursts. The non-detection of *X*-ray transient emission from GRB 980425 is also peculiar, given that such an afterglow has been detected by BeppoSAX for more than 90% of bursts.

The spectrum of the supernova 1998bw lacks the hydrogen lines characteristic of Type II, the Si II characteristic of Type Ia, and the strong He I 5876 which characterizes Type Ib. The spectrum of SN 1998bw is somewhat similar to that of the Type Ic SN 1994I (Galama *et al.* 1998), but has less pronounced spectral features. SN 1998bw is unusual in other ways. Its peak brightness, $m_B = 14$, corresponds to a luminosity $L_B = 10^{43}$ ergs, thus SN 1998bw was as bright as a Type Ia and about four times brighter than Type Ib/c. SN 1998bw was also unusually bright at radio frequencies, where the emission exhibits two maxima, 12 and 30–40 days after the burst (Kulkarni *et al.* 1998). SN 1998bw is the only other case of a Type Ic for which *X*-ray emission has been detected (Pian 1999). The peculiarities of both the SN 1998bw and the GRB 980425 may further strengthen their association.

Close monitoring of the spectral evolution of the optical afterglow of GRB 030329 ($z = 0.17$) has shown the slow emergence of broad peaks after the first week, characteristic of a supernova (Stanek *et al.* 2003). The spectral evolution and light-curve of SN 2003dh obtained after subtracting the power-law afterglow continuum, resemble those of SN 1998bw (Matheson et al, these proceedings; see also Matheson *et al.* 2003, Stanek *et al.* 2003).

There are two other GRBs for which a supernova association is indicated by the flattening or brightening of the optical afterglow emission and by the shape of the spectrum. The optical light-curve of the GRB afterglow 980728 ($z = 0.7$) exhibits a flattening after 10 days, during which the optical spectrum departs from the power-law expected for a GRB afterglow and bears some resemblance that of the SN

1998bw (Richart 1999, Galama *et al.* 2000), after correction for redshift. A similar situation occurred for GRB 0101211 ($z = 0.36$), whose associated SN 2001ke peaks at about 15 days, fades faster and has a bluer spectrum than SN 1998bw, but reached the same peak luminosity (Garnavich *et al.* 2003). There are other GRB afterglows whose optical light-curves exhibited a brightening or flattening at 10–30 days after the burst, such as GRB 980326 (Bloom *et al.* 1999) and GRB 9912108 (Castro-Tirado *et al.* 2001), but lacking any spectral information to strengthen the association of the excess emission to a supernova.

That GRBs are related to the death of massive stars is consistent with the distribution of GRB locations in the host galaxy. As shown by Bloom *et al.* (2002), about half of the 20 well-localized bursts occur within the half-light radius of the host. More quantitatively, a K-S test shows that the distribution of burst localizations is consistent with an exponential disk. Furthermore, the distribution of burst localizations is inconsistent (K-S probability 10^{-3}) with the expectations for a delayed merging of compact stellar remnant (BH–NS, NS–NS) for a variety of models for the galactic potential (although calculations of the distribution of the GRB–host center separation expected in binary merger models depend on the merging time, which is subject to uncertainties in the binary separation at the end of the common envelope phase). Further indication that GRBs are associated with massive stars is that most or all bursts occur in actively star-forming galaxies (Djorgovski *et al.* 2001).

Another line of evidence for the GRBs–massive star connection is provided by the possible detection of the signatures of a clumpy medium (as expected around Wolf-Rayet stars) in the optical emission of some afterglows. Salamanca *et al.* (2002) have identified four C IV absorption components separated by about 3000 km/s in the optical spectrum of the GRB afterglow 021004. They note that this velocity dispersion is too large for a cluster of galaxies, and that, if the lines arise in a supernova shell around the burst, the narrowness of the lines require that the supernova has occurred many weeks before the burst, to allow the shell to cool, but then it would be unlikely for the observer's line of sight to intersect four filaments. Another possibility, suggested by Mirabal *et al.* (2002), is that the blueshifted absorption components arise in the wind of a WC star. Schaefer *et al.* (2003 and these proceedings) argue that the observed high velocity dispersion could be due to the resonant scattering of burst and early afterglow UV emission by the clumps or shells in the WR wind. Depending on the GRB spectrum, a significant acceleration may occur up to a distance of 1 pc. The existence of clumps within 1 pc of the burst could also explain the variability ("bumps") seen at 0.1, 1, and 3 days in the optical flux ("bumps") of the 021004 afterglow. Significant variability has also been observed in the GRB afterglows 970508 and 030329 during the first days after the burst.

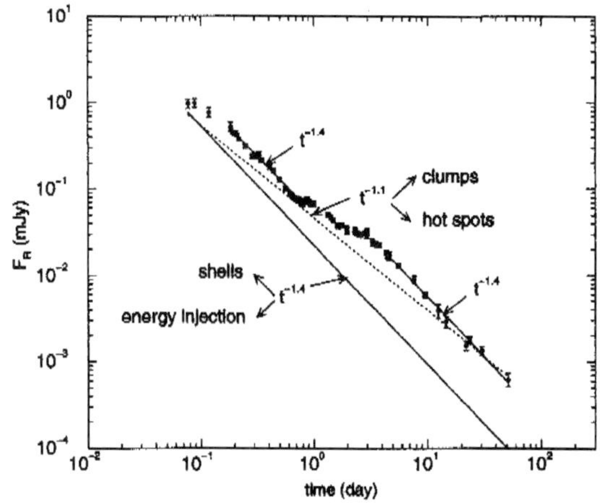

Fig. 41.1. Fluctuations in the optical emission of GRB afterglow 021004. Four possible sources of variability are indicated. Anisotropies in the angular distribution of the fireball kinetic energy ("hot spots") or clumps in the circumburst medium produce transient brightenings which fade, the underlying flux ($\propto t^{-1.1}$) resurfacing between fluctuations. A stratified circumburst medium ("shells") or an energy injection preserve the afterglow decay ($\propto t^{-1.4}$) but reset the flux level, as appears to be the case here. A clumpy or a stratified medium are consistent with the detection of multiple absorption components seen in the spectrum of this afterglow.

However, inhomogeneities in the circumburst medium (CBM) are not the only source of afterglow variability. A stratified medium consisting of shells of various densities, an anisotropic distribution of the GRB ejecta kinetic energy, or injection of energy in the GRB remnant could also lead to fluctuating light-curves. Distinguishing among these possibilities may be possible with detailed calculations of the fluctuations (Heyl & Perna 2003, Nakar *et al.* 2003) and observations at other frequencies (radio and X-ray). At a simpler, qualitative level, some discrimination among the above scenarios can be obtained by analyzing the decay of the optical bumps (Figure 41.1), provided that the behavior of the underlying optical flux can be inferred from the spectrum and from X-ray observations.

A supernova explosion may also be required to account for the medium rich in iron (or lighter elements) producing the X-ray lines or edges observed in some GRBs and afterglows. Amati *et al.* (2000) report the detection of a 3.8 ± 0.3 keV transient Fe absorption edge in the spectrum of GRB 990705 ($z = 0.86$), corresponding to an optical depth $\tau \sim 1$. The photoionization of the Fe-rich medium during the first 10 seconds of the burst, after which the absorption feature disappeared, places the absorber at $D \sim 0.1$ pc from the burst, implying a 10 yr delay between the SN and the GRB. The hydrogen column density $N_H = 10^{22}$ cm^{-2} inferred by fitting the X-ray continuum, implies an ejecta mass $M \sim 10 M_{\odot}$, a relative abundance

$A_{Fe} = Fe/\,Fe_{\odot}. = 75$, and an iron mass $M_{Fe} \sim 1\ M_{\odot}$. A smaller ejecta mass, $M = 0.5 A_{Fe}^{-1} M_{\odot}$, located closer to the burst, $D \sim 3 \times 10^{15}$ cm, has been inferred by Piro *et al.* (1999) from the intensity and variability timescale of the 3.5 keV Fe K_{α} emission line observed during the first day in the *X*-ray afterglow of GRB 970508 ($z = 0.84$). A 3.5 keV iron K_{α} emission line, and 4.4 keV iron and 1.7 keV sulfur recombination edges have been detected in the afterglow of GRB 991216 ($z = 1.0$), at 3σ to 4.7σ significance levels (Piro *et al.* 2000). More recently, K_{α} transient emission lines from highly ionized Mg, Si, S, Ar, Ca have been reported for the GRB afterglow 011211 (Reeves *et al.* 2002). For a burst redshift $z = 2.14$, all lines appear blueshifted by about the same velocity of 26,000 km/s. Due to the lack of Fe emission lines, the authors favor a thermal emission model, for which they infer an abundance of the light metals that is 10 times the solar value, and an Fe abundance not exceeding solar value. The duration of the line emission (50 ksec), can be used to constrain the location of the line emitting region (10^{15} cm), which implies a short delay (few days) between the SN and the GRB, consistent with the lack of an over-solar iron abundance.

Finally, that we did not detect an optical afterglow for more than half of the localized GRBs can be considered circumstantial evidence for a GRB – massive star association, as such stars are expected to reside in giant molecular clouds, thus the GRB afterglow may be heavily extincted. As noted by Lazzati *et al.* (2002), an average extinction of only 2 magnitudes suffices to explain the fraction of optically dark GRBs, given the depth of the searches. Dust sublimation by the UV flash accompanying the GRBs or from the early afterglow could clear more than 10 pcs, depending on the UV output and grain size (Waxman & Draine 2000), thus, for a hydrogen density of $10^4\ \mathrm{cm}^{-3}$, only clouds more massive than $10^5\ M_{\odot}$ can "hide" the optical afterglow. However, the discovery of an optically dim afterglow lacking the signature of significant dust reddening (Berger *et al.* 2002) and that optically dark GRBs are on average 5 times dimmer in the *X*-rays than the bursts with optical transients (de Pasquale *et al.* 2003), indicate that the dark bursts may be intrinsically dim in the optical, with dust extinction playing a smaller role.

41.2 GRB environment

If the GRB progenitors are massive stars that undergo core collapse, then the environment surrounding the GRB is either the unperturbed wind blown by the progenitor or that resulting from the interaction between the stellar material expelled at different stages and/or the circumstellar medium. For a steady wind, the particle density in the immediate vicinity of the progenitor is $n(r) = \dot{M}/\,(4\pi m_p v r^2) = 3 \times 10^{35}\ A_*/\,r^2$, where $\dot{M} \sim 10^{-5}\ M_{\odot}/$ yr is the mass-loss rate, $v \sim 1000$ km/s is the wind velocity, and A_* is the $\dot{M}/4\pi m_{\mathrm{p}} v$ factor normalized to its value expected for the given canonical values of $\dot{M}$ and v, which are characteristic for a Wolf-Rayet star. The radius at which the afterglow emission is released can be calculated from energetic considerations. Taking into account relativistic effects, it can be shown that, for an r^{-2} density profile of the CBM, $r_{aglow}(t) = 0.14\ E_{53} A_*^{1/2} t_d^{1/2}\ pc$, where E_{53} is the fireball kinetic energy measured in units of 10^{53} ergs and t_d is the observer time in days ($z = 1$ was assumed). Thus one obtains the wind density at the location of the GRB remnant is $n(t) = 1\ E_{53}^{-1} A_*^2 t_d^{-1}\ \mathrm{cm}^{-3}$. Typical afterglow observations are made at a fraction of a day until few tens of days, i.e. when the fireball radius is between $0.1 A_*^{-1/2}$ pc and $1 A_*^{-1/2}$ pc and the CBM density between $3A_*^2\ \mathrm{cm}^{-3}$ and $0.03\ \mathrm{cm}^{-3}$ for $E_{53} \sim 1$.

This straightforward implication of the association of GRBs with massive stars can be tested by constraining the CBM from the afterglow emission. The external particle density determines the flux at the peak of the afterglow synchrotron spectrum, the self-absorption and the cooling frequency. These quantities also depend on E and on two microphysical parameters quantifying the typical post-shock electron energy and the magnetic field. By combining radio, optical and X-ray observations it is possible, at least in principle, to constrain the above three spectral characteristics, as well as the characteristic frequency corresponding to the typical electron energy (which is also the peak frequency if electrons are cooling radiatively slower than adiabatically). Thus, one can obtain four observational constraints for four unknowns, one of which is the external particle density.

In practice, such an exercise can be carried out only for GRB 970508 for which all four spectral characteristics can be determined observationally. Using a homogeneous CBM, Wijers & Galama (1999) infer $E_{53} = 4$ and $n = 0.03\ \mathrm{cm}^{-3}$ from $t = 12$ days observations. The resulting CBM density is very close to that expected for $A_* = 1$. For other afterglows, a reliable determination of the CBM density can be obtained by fitting the radio, optical, and X-ray afterglow data with the emission expected from a relativistic fireball decelerated by the interaction with the CBM (e.g. Chevalier & Li 2000, Panaitescu & Kumar 2002). In those cases where a break has been observed in the afterglow optical light-curve, which is considered a signature of a collimated outflow, one can determine the initial jet opening from the time when the break is seen, thus the jet energy can also be calculated. To some extent, the shape of the break resulting when the moderately relativistic jet begins to expand sideways provides an additional constraint on the CBM density.

With the aid of such fits, we found best-fit values for the wind A_* parameter ranging from 0.2 to 3.5, for 7 out of 10 cases being below 1, which are consistent with a Wolf-Rayet wind. If the stratification of the CBM were ignored and a homogeneous environment were assumed, the resulting densities are between $0.1\ \mathrm{cm}^{-3}$ and $50\ \mathrm{cm}^{-3}$ (Figure 41.2), again roughly consistent with what is expected for the density of a Wolf-Rayet wind ($A_* \sim 1$) at the fireball radius corresponding to an observer time of days.

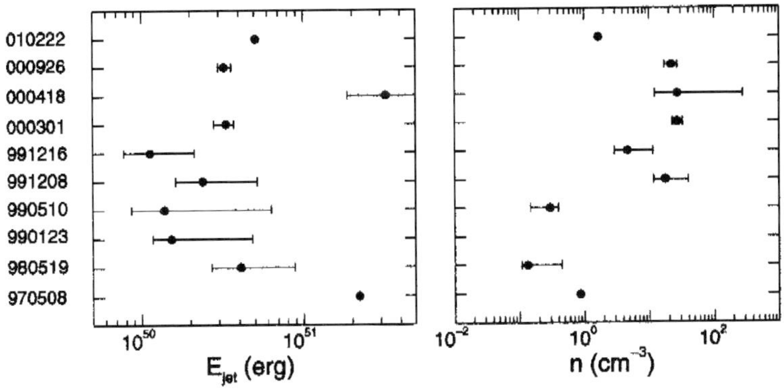

Fig. 41.2. Jet energy and CBM density (assumed homogeneous) inferred from modeling the broadband emission of 10 GRB afterglows whose optical light-curves decay exhibit a break (except GRB 000418), which allows the determination of the initial jet opening. For GRB 990123, we obtain $n < 10^{-2}$ cm^{-3}, which may be due to a lack of radio measurements. Error bars indicate 1σ uncertainties, with exception of GRBs 970508 and 010222, for which the best fits obtained are very poor.

The reason for discussing a homogeneous medium is that, in some cases, it provides a better description than a wind-like CBM of the optical light-curve breaks. In comparison with a homogeneous medium, a decreasing density leads to a slower deceleration of the fireball, so that it takes a longer time for the jet edge to become visible to the observer, especially when the jet lateral spreading is taken into account. Consequently, the steepening of the afterglow fall-off due to the jet collimation is smoother for a wind CBM than for a homogeneous medium (Kumar & Panaitescu 2000). A wind CBM yields significantly poor fits than a homogeneous medium for GRB 990123, GRB 990510, and GRB 000926; for the other afterglows, the quality of the fits being comparable.

Given that the circumstellar medium impedes the free expansion of the WR wind, it is worth investigating if, in those cases where a uniform medium is required by the jet interpretation of the optical breaks, we actually observed the interaction of the fireball with the shocked part of the WR wind, which forms a bubble of uniform density. An analytical treatment of bubbles blown by stellar winds can be found in Castor *et al.* (1975). From energetic and dynamical considerations and taking into account the evaporation of swept-up interstellar medium into the bubble, it can be shown that the bubble temperature, density, and outer radius are $n_b = 0.05$ $(\dot{M}_{-5}v_3^2)^{6/35} n_0^{19/35} t_5^{-22/35}$ cm^{-3}, $T_b = 3 \times 10^6\ (\dot{M}_{-5}v_3^2)^{8/35}\ n_0^{2/35} t_5^{-6/35}$ K, $R_b = 8$ $(\dot{M}_{-5}v_3^2)^{1/5} n_0^{-1/5} t_5^{3/5}$ pc, where $\dot{M}_{-5}$ is the mass-loss rate in $10^{-5}\ M_\odot$/yr, v the wind velocity scaled to 1000 km/s, t is the bubble age scaled to 10^5 yr, and n_0 is the ambient particle density in cm^{-3}. The inner radius of the bubble, where the r^{-2} wind profile ends, can be estimated by equating the wind ram pressure, ρv^2, with the bubble thermal pressure, $n_b k T_b$: $R_w = 5\dot{M}_{-5}^{-3/10} v_3^{1/5} n_0^{-3/10} t_5^{2/5}$ pc. For the energy and densities inferred from afterglow modeling, the fireball radius after a fraction of a day is larger than 0.2 pc. Therefore, the entire afterglow emission arises within the uniform density bubble if $R_w < 0.1$ pc, which, for typical WR wind parameters requires an ambient density $n_a > 4 \times 10^4$ cm^{-3}. Such densities are characteristic for a dense cloud, and may lead to a significant extinction of the optical afterglow if the cloud is larger than 10 pc, as the burst cannot sublimate all the dust. For this ambient density, the bubble density is $n_b > 15$ cm^{-3}, which is compatible with that resulting from afterglow modeling. The wind mass up to the inner bubble radius is $\dot{M} R_w / v \sim 10^{-3} M_\odot$, thus the fireball is still relativistic after sweeping it up. The bubble termination shock is at $R_b = 1pc$.

The above results show that a WR bubble is a possible explanation for the homogeneous CBM required by some afterglows; however it is an over-simplified treatment of the medium structure around massive stars, as such stars do not expel winds at constant mass-loss and speed. As an example, a 30 $M_\odot$ O main sequence star loses up to 10^{-6} $M_\odot$/ yr at 3000 km/ s for about 4 Myr, followed by the red supergiant (RSG) phase with a mass-loss rate of 10^{-4} $M_\odot$/yr at 100 km/s, lasting for about 0.2 Myr. After most of the hydrogen envelope is lost, i.e. during the WR phase, the GRB progenitor expels about 10^{-5} $M_\odot$/yr at 1000 km/s for at least 0.2 Myr. Therefore, the WR wind interacts with the denser and slower RSG wind, whose density at $R_b = 1pc$ is of only 10 cm^{-3}, well below the n_a above required to produce a bubble with the right inner radius and density for GRB afterglows. If the GRB progenitor is more massive, then the wind blown during the luminous blue variable phase, preceding the WR stage, may carry 10^{-3} $M_\odot$/yr at only 10 km/s, thus its density at 1 pc is 10^3 cm^{-3}, i.e. closer, but still below n_a.

41.3 GRB energetics

As shown in Figure 41.2, the *kinetic* energies of collimated GRB outflows range from 10^{50} to 3×10^{51} ergs (1.5 dex), being less or comparable to that of the supernova ejecta, and about 3 orders of magnitude lower than the energy budget of all plausible GRB progenitors involving compact objects (Mészáros *et al.* 1999). From the observed GRB fluence, redshift, and jet break time t_b, Frail *et al.* (2001) have inferred the energy released in γ-rays. The first two quantities give the isotropice-quivalent GRB output ε_γ, while the last quantity determines the jet initial opening θ_{jet}, for a given GRB efficiency ε_γ and CBM density n: $\theta_{jet} \propto (n\varepsilon_\gamma/\varepsilon_\gamma)^{1/8} t_b^{3/8}$. The resulting burst energies, corrected for collimation, $E_\gamma \propto \varepsilon_\gamma \theta_{jet}^2$, range from 2×10^{50} to 2×10^{51} ergs (1.0 dex), for assumed $\varepsilon_\gamma = 0.2$ and $n = 1$ cm^{-3}. That the GRB output and the fireball kinetic energy are comparable implies that the γ-ray mechanism has a typical efficiency of 50%. Taking into account the various factors that determine the efficiency of internal shocks, this requirement appears rather hard to satisfy.

We note that the above methods constrain only the energy in the relativistic ejecta. Less relativistic ejecta could either surround the jet or lag behind it, only to catch up later as the jet is decelerated and refresh the forward shock. The former scenario has been invoked by Berger *et al.* (2003) to explain the brightening of the optical emission and the radio afterglow of GRB030329, while the later scenario is the explanation suggested by Fox *et al.* (2003) for the slow decay of the optical emission of the GRB afterglow 021004 during the first hour. In either scenario it is expected that emission from the less relativistic ejecta is at radio frequencies, thus these scenarios may explain the long-lived, shallow decay observed in some radio afterglows (991208, 991216, 000301, 000926, 010222). An analysis based on the general properties of the radio and optical light-curves of these afterglows (Panaitescu & Kumar 2003) has shown that the first scenario, involving an outflow with angular structure, cannot explain simultaneously the slow decaying radio emission and the steep fall-off observed in the optical. The second scenario, where the radio emission arises in the reverse shock crossing the incoming ejecta and the optical emission in the forward shock energizing the CBM, is more promising. We found that, in order to accommodate the observations, the energy carried by the less relativistic ejecta is at most equal to that initially existing in the GRB fireball, thus the energy budget of GRBs remains basically unchanged.

References

Amati, L. *et al.* 2000, Science, 290, 953
Berger, E. *et al.* 2002, ApJ, 581, 981
Berger, E. *et al.* 2003, Nature, in press (astro-ph/0308187)
Bloom, J. *et al.* 1999, Nature, 401, 453
Castor, J., McCray, R. & Weaver, R. 1975, ApJ, 200, L107
Castro-Tirado, A. *et al.* 2001, A&A, 370, 398
Chevalier, R. & Li, Z. 2000, ApJ, 536, 195
Djorgovski, S. *et al.* 2001, astro-ph/0107535
Fox, D. *et al.* 2003, Nature, 422, 284
Frail, D. *et al.* 2001, ApJ, 562, L55
Galama, T. *et al.* 1998, Nature, 395, 670
Galama, T. *et al.* 2001, ApJ, 536, 185
Garnavich, P. 2003, ApJ, 582, 924
Heyl, J. & Perna, R. 2003, ApJ, 586, L13
Kulkarni, S. *et al.* 1998, Nature, 395, 663
Kumar, P. & Panaitescu, A. 2000, ApJ, 541, L9
Lazzati, D., Covino, S. & Ghisellini, G. 2002, MNRAS, 330, 583
Matheson, T. *et al.* 2003, ApJ, in press (astro-ph/0307453)
Mészáros, P., Rees, M. & Wijers, R. 1999, New Astronomy, 4, 303
Mirabal, N. *et al.* 2002, GCN #1618
Nakar, E., Piran, T. & Granot, J. 2003, New Astronomy, 8, 495

Panaitescu, A. & Kumar, P. 2002, ApJ, 571, 779
Panaitescu, A. & Kumar, P. 2003, MNRAS, submitted (astro-ph/0308273)
de Pasquale, M. *et al.* 2003, ApJ, 592, 1018
Pian, E. 1999, astro-ph/9910236
Piro, L. *et al.* 1999, ApJ, 514, L73
Piro, L. *et al.* 2000, Science, 290, 955
Reeves, J. *et al.* 2002, Nature, 416, 512
Reichart, D. 1999, ApJ, 521, L111
Salamanca, I. *et al.* 2002, GCN #1611
Schaefer, B. *et al.* 2003, ApJ, 588, 387
Schmidt, M. 2001, ApJ, 552, 36
Stanek, K. 2003, ApJ, 591, L17
Stern, B., Tikhomirova, Y. & Svensson, R. 2002, ApJ, 573, 75
Wijers, R. & Galama, T. 1999, ApJ, 523, 177

Part VII

Conference Summary

42

Three-dimensional explosions

J. C. Wheeler

Department of Astronomy University of Texas at Austin

42.1 Introduction

This conference was packed with interesting and relevant developments regarding the three-dimensional nature of both thermonuclear and core-collapse supernovae. Before summarizing those presentations, I would like to summarize some of the developments regarding rotation and magnetic fields that were on my mind during the conference.

42.2 Dynamo theory and saturation fields

There has been a major breakthrough in the conceptual understanding of astrophysical dynamos in the last few years. In traditional mean field dynamo theory, the turbulent velocity field that drives the "alpha" portion of the $\alpha - \Omega$ dynamo was specified and held fixed. A weakness of the original theory was that the turbulent velocity field cannot be constant. The buildup of small scale magnetic field tends to inhibit turbulence, cutting off the dynamo process for both small and large scale fields. Since the small scale field tended to grow faster than the large scale field, it appeared that the growth of the large scale field would be suppressed (Kulsrud & Anderson 1992; Gruzinov & Diamond 1994). In these theories, the magnetic field energy cascades to smaller length scales where it is ultimately dissipated at the resistive scale. Large scale fields tend to build up slowly, if at all.

The solution to this problem has been the recognition (Blackman & Field 2000; Vishniac & Cho 2001; Field & Blackman 2002; Blackman & Brandenburg, 2002; Blackman & Field 2002; Kleeorin *et al.* 2002) that the magnetic helicity, $\mathbf{H} = \mathbf{A} \cdot \mathbf{B}$ is conserved in ideal MHD and that this conservation had not been treated explicitly in mean field dynamo theory. Incorporation of this principle leads to an "inverse cascade" of helical field energy to large scales that is simultaneous with the cascade of helical field energy from the driving scale to the dissipation scale. Basically, the large scale helical field and inverse cascade must exist with opposite magnetic helicity to that of the field cascading to small scale. The result (Blackman & Brandenburg 2002) is the rapid growth of large scale field in a kinematic phase (prior to significant back-reaction) to a strength where the field on both large and small scales is nearly in equipartition with the turbulent energy density. At that point, the back reaction sets in and there tends to be a slower growth to saturation at field strengths that can actually somewhat exceed the turbulent energy density. It may be that the early, fast, kinematic growth is the only phase that is important for astrophysical dynamos, especially in situations that have open boundaries so that field can escape (Brandenburg, Blackman & Sarson 2003; Blackman & Tan 2003) and that are very dynamic. The collapse ambience is clearly one of those situations.

Another possibly important insight is that the rapid kinematic phase can lead to magnetic helicity currents (Vishniac & Cho 2001). It is possible that these magnetic helicity currents can transport power out of the system in twisting, propagating magnetic fields. This is clearly reminiscent of jets or winds, but the physics is rather different than any that has been previously explored in driving jets or winds. This physics needs to be explored in the context of supernovae and gamma-ray bursts.

This new work on dynamo theory has not changed one basic aspect and that is the level of the saturation fields. It remains true that the saturation fields will be of order $v_a \sim \lambda\Omega$ or $B^2 \sim 4\pi\rho\lambda^2\Omega^2$ where the characteristic wavelength, $\lambda \lesssim r$, for quasi-spherical situations. For a proto-neutron star this yields a field of order 10^{15} to 10^{16} G. For collapse to form a black hole, the velocities will be Keplerian and the associated, dynamo-driven, predominantly toroidal field will have a strength of order $B \sim 10^{16}\mathrm{G}\,\rho_{10}^{1/2}$ assuming motion, including the Alfvén speed, near the speed of light near the Schwarzschild radius and a characteristic density of order 10^{10} g cm^{-3} (MacFadyen & Woosley 1999). Fields this large could affect both the dynamics and the microphysics in the black hole-formation problem. Because of the nearly Keplerian motion in the black hole case, the fields generated will be much closer to pressure equipartition than in the neutron star case, and hence, perhaps, even more likely to have a direct dynamical effect. The associated MHD power in the black hole case would be roughly 10^{52}–10^{53} erg s^{-1}.

42.3 Possible effects of large magnetic fields

A. Equation of state

Fields of order 10^{15} to 10^{16} G are far above the QED limit, $B_{QED} = 4 \times 10^{13}$ G, so quantum effects may become important. The calculations of Akiyama *et al.* (2003) predict regions $\sim 10^6$ to 10^7 cm where the electron Fermi energy is less than the first Landau level after about 100 ms (see the contribution in these proceedings by Akiyama *et al.*). In such conditions, electron motions will be quantized, with the electron component of the pressure being strongly anisotropic. This pressure anisotropy is likely to be balanced by the $\mathbf{j} \times \mathbf{B}$ force of induced magnetization (Blandford & Hernquist 1982), but in the absence of such isotropy, pressure anisotropy of order 10^{-4} and hence velocity anisotropy of order 10^{-2} might be induced. The electron pressure will be reduced compared to calculations that ignore quantization, but it is not clear that will make a significant difference to the dynamics.

For $B > B_{QED}$, the electrons can only flow along the field lines, that is $\mathbf{j} \parallel \mathbf{B}$. On the other hand, classic MHD includes currents only implicitly and assumes that the current is always normal to the field, $\mathbf{j} \perp \mathbf{B}$. The result is a manifest contradiction, as pointed out to me by Dave Meier. The resolution to this might be non-local currents, ion currents (which would require flows of only 10^{-6} cm s^{-1}), or most interestingly, but unlikely, a field that saturates at the QED limit. These issues are worth more thought.

B. Neutrino transport

Fields of order 10^{15} to 10^{16} G that will characterize both neutron star and black hole formation may affect neutrino transport. With a large magnetic field, direct $\nu - \gamma$ interaction is possible mediated by W and Z bosons. This would allow neutrino Cerenkov radiation, $\nu \rightarrow \nu + \gamma$, and would enhance plasmon decay, $\gamma \rightarrow \nu + \nu$ (Konar 1997).

In addition, processes like $\nu \rightarrow \nu + e^+ + e^-$ would no longer be kinematically forbidden. In that case, closed magnetic flux loops can trap pairs. The energy in pairs would grow exponentially to the point where annhilation cooling would balance pair creation. Thompson & Duncan (1993) estimated that an energy as much as $E_{\rm pair} \sim 10^{50}$ erg could be trapped in this way. This is not enough energy to cause a robust explosion, but it is enough energy to drive the dynamics of core collapse in a substantially different way, perhaps by inducing anisotropic flow if the flux loops are themselves distributed anisotropically.

With substantial magnetic fields, the cross section for inverse beta decay, $\nu_e + n \rightarrow p + e^-$, would become dependent on neutrino momentum, especially for asymmetric field distributions, which would be the norm (Lai & Qian 1998; Bhattacharya & Pal 2003; Ando 2003).

All these processes and more should be considered quantitatively in core collapse to form neutron stars and black holes.

42.4 Core collapse MHD and jet formation

A. Magnetic helicity currents

It is not at all proven that the large magnetic fields expected in core collapse generate jets, but there are a number of clues pointing in that direction. For the more traditional situation in which collapse leads to the formation of a neutron star, the premise is that there is a rapid formation of a strong magnetic field with $B \sim 100\ B_{QED} << (4\pi P)^{1/2}$, that is much above the QED limit, but less than equipartition with the ambient pressure. This field is expected to be primarily toroidal (simulations give ~80%; Hawley Gammie & Balbus 1996), but turbulent, with a maximum around the proto-neutron star surface, a location well within the standing shock. The expected MHD power, $\sim 10^{52}$ erg s^{-1}, would be delivered in some form beneath that shock and could help to reinvigorate it, or to provide entirely unique, jet-like dynamics in which the shock no longer played a key role. In this highly magnetized environment, there will be hoop stresses, gradients in magnetic pressure and perhaps in the electron pressure. These anisotropic components will be weak compared to the total pressure, but they will be non-radial and anisotropic.

As an example of the possibly relevant physics, Vishniac & Cho (2001) argue that along with conservation of magnetic helicity, $\mathbf{H} = \mathbf{A} \cdot \mathbf{B}$, and the inverse cascade of magnetic field energy to large scales, one will get a current of magnetic helicity that can be crudely represented by

$$j^H \sim B^2 \lambda v, \tag{42.1}$$

where the characteristic length, λ, might be comparable to a pressure scale height, $l_P = (d \ln P/dr)^{-1}$, and $v \sim v_a \sim l_P \Omega$. The energy flux associated with this magnetic helicity current is $J_H/\lambda \sim B^2 v_a$, and so with $B^2 \sim \rho l_P^2\, \Omega^2$ the associated power is:

$$L = r^2 B^2 v_a \sim B^2 r^2 l_P \Omega \sim \rho r^5 \Omega^3 \left(\frac{l_P}{r}\right)^3. \tag{42.2}$$

Note that the next-to-last expression on the RHS is essentially just the characteristic Blandford-Payne luminosity (Blandford & Payne 1982); however, in this case the field is not externally given, but provided by the dynamo process so that the final expression on the RHS is given entirely in terms of local, internal quantities. The implication is that this amount of power is available in an axial, helical field without twisting an external field. Again, while this analysis has superficial resemblance to other jet mechanisms, it involves rather different physics and is self-contained. Whether this truly provides a jet remains to be seen. A first example of driving a polar flow with the MRI is given by Hawley & Balbus (2002).

Note that this process of creating a large scale field with an MRI-driven dynamo with its promise of naturally driving axial, helical flows does not require an equipartition field. As pointed out by Wheeler *et al.* (2002), the field does not have to have equipartition strength and hence to be directly dynamically important in order to be critical to the process of core collapse. The field only has to be significantly strong to catalyze the conversion of the free energy of differential rotation of the neutron star into jet energy. As long as this catalytic function is operative, the rotational energy should be pumped into axial flow energy until there is no more differential rotation. For the case of stellar collapse, this would seem to imply that, given enough rotational energy in the neutron star, this machine will work until there is a successful explosion. Even if the core collapses directly into a black hole, or does so after some fall-back delay, the basic physics outlined here, including magnetic helicity currents and their associated power should also pertain to black hole formation.

B. Poleward slip instability

Another interesting bit of physics that may pertain to core collapse is the poleward slip instability. This is analogous to wrapping a rubber band around the equator of a ball and then sliding it upward. For the case of a magnetized plasma, a toroidal field is absolutely unstable to this effect in the absence of rotation (Spruit & Ballegooijen 1982). The case with differential rotation has been considered by, among others, Chanmugam (1979). In that case the axisymmetric ($m = 0$) mode still appears to be unstable, but this case is a bit tricky because the absence of a sufficient condition for stability as derived by Chanmugam does not necessarily imply a necessary and sufficient condition for instability. The interesting behavior, in any case, is not merely the linear instability, but the non-linear dynamics. This does not seem to have been explored at all in the literature.

As a crude way of examining this, let us assume that the pressure gradient balances gravity to first order and look at the acceleration resulting from the hoop stress and centrifugal potential, assuming conservation of angular momentum of the matter associated with a flux tube. The result is

$$a \sim -\frac{v_a^2}{r} + \frac{R^4 \Omega_{eq}^2}{r^3}, \tag{42.3}$$

where r is the cylindrical radius, R the value on the equator, and Ω_{eq} the value of the angular velocity on the equator. For the case of interest, the saturation field condition is that $v_a \sim R\Omega_{eq}$, so that these terms nearly cancel on the equator. This is a caution, at least, that care must be taken to take all the forces into account self-consistently. The issue of what happens as the field starts to slip toward the pole seems to depend on the behavior of the Alfvén velocity, and hence the magnetic field and entrained density, as the flux tube moves.

Note that the poleward slip instability, whatever its ultimate non-linear behavior, should not depend on whether the field is continuously connected around the body (literally like a rubber band) or whether it is turbulent and discontinuous. This is because, for instance, the hoop stress is a local property of a field with a mean radius of curvature. Williams (2003 and this conference) has argued that even a tangled field with $\langle B \rangle \sim 0$, but $\langle B^2 \rangle^{1/2} \neq 0$, will act like a viscoelastic fluid (see also Ogilvie 2001) and, in particular, exert a hoop stress.

The conjecture is that the ultimate non-linear behavior is for the field to accumulate near the pole where it reaches approximate equipartition, $B \sim (8\pi P)^{1/2}$, and hence becomes dynamically significant. Again, this suggests activity at the pole that is reminiscent of a jet. Yet again, this remains to be seen.

One interesting aspect of the poleward slip instability is that it would seem to pertain directly to neutron stars that have a strong density gradient at the surface, essentially a hard surface, but it should *not* work for black holes, where there is no surface to support the poleward slip. Whether or not this makes any difference in the jet formation in neutron stars versus black holes is an interesting question.

42.5 Summary of contributions

42.5.1 Asymmetry rules

The conference began with excellent summaries of the new and growing sample of supernova spectropolarimetry by Lifan Wang and Alex Filippenko. It is this data that has driven the new conviction that core collapse supernovae are essentially universially asymmetric and that the asymmetry is driven by the engine of core collapse itself. With this new conviction, disparate data on otherwise isolated events like the Crab nebula with its pulsar and jet, Cas A, and SN 1987A, begin to make sense in a large picture of fundamentally asymmetric supernovae. Roger Chevalier discussed our evolving knowledge of supernova remnants and pulsar wind nebulae. Rob Fesen described observations on the morphology of supernovae remnants, especially Cas A. Bob Kirshner summarized the imaging spectroscopy on SN 1987A. Doug Swartz described the new data on supernova remnants available from the Chandra Observatory. Vikram Dwarkadas showed that his multidimensional simulations of supernova ejecta colliding with previously expelled wind material are rife with Rayleigh-Taylor instabilities and look remarkably like the observations.

One of the lessons that comes through from this work is that neither Type Ia nor the zoo of core collapse supernovae are spherically symmetric. Peter Höflich and the fully-represented Oklahoma mafia – David Branch, R. C. Thomas, Dan Kasen, and Eric Lenz, with Eddie Baron kibbutzing from the audience – outlined the various ways in which polarization could be induced in supernova spectra. Among these are: an intrinsically asymmetric shape, blocking of part of the photosphere by some off-center distribution of matter, and an off-center energy source. All of these may contribute in various supernovae or even for a single supernova, depending on circumstances.

42.5.2 *Type Ia*

As mentioned in my introduction, one of the goals of the study of Type Ia supernova research for decades has been to obtain direct observational evidence that Type Ia arise in binary systems, as widely accepted on circumstantial grounds. This conference may have revealed some of the first evidence in this direction. Lifan Wang, Peter Höflich, and Dan Kasen discussed the observations and interpretations of polarization data from Type Ia supernovae, particularly the "normal" event SN 2001el that shows remarkable departure from symmetry in the form of a highly polarized high-velocity component to the Ca II IR triplet. After the conference, Gerardy *et al.* (2003) submitted a paper arguing that a similar high-velocity Ca feature in SN 2003du might arise in a hydrogen-rich circumstellar medium. The data have not yet revealed definate proof, but tantalizing suggestions that the asymmetry may be connected to a disk or binary companion, the existence of which would be proof that a binary system was needed.

Mario Hamuy added a dramatic new development in this area with his discussion of SN 2002ic, an event that shows familiar Type Ia features, but also strong hydrogen emission lines similar to those from Type IIn. A substantial amount of hydrogen, of order a solar mass, must be involved. After the conference, Wang *et al.* (2003) submitted a paper based on VLT spectropolarimetry observations that showed that the hydrogen envelope is substantially polarized and probably arrayed in a large, dense, clumpy disk-like way. SN 2002ic is very similar both near maximum light and 200 days later to SN 1997cy and SN 1999E, both classified as Type IIn. This raises the issue of whether or not at least some of these events previously classified as SN IIn are hydrogen-surrounded Type Ia. These events are rather rare, so it cannot be true that all Type Ia erupt in this configuration, but it is also clear that Hamuy has provided us with a stimulating new avenue of exploration of the nature of Type Ia and their binary configuration.

As a complement to this, Don Winget described the work that he and his group are doing with asteroseismology to probe the inner composition of white dwarfs. Sumner Starrfield and S. C. Yoon gave very thought-provoking summaries of their work that gives new insights into the possible configurations of white dwarf accretion and growth that could lead to Type Ia explosions. Jim Truran and Andy Howell both provided insights into how the diversity of Type Ia supernovae may arise.

There has been amazing progress on understanding and simulating the combustion physics associated with Type Ia thermonuclear explosions, as summarized by Alexei Khokhlov, Elaine Oran, Vadim Gamezo, Eli Livne, and Peter Höflich. In particular Gamezo illustrated the state of the art with a three-dimensional simulation of a detonation that starts deep in the fingers of unburned carbon and oxygen that survive at the end of the phase of subsonic, turbulent, deflagration. Fundamental understanding of the deflagration/detonation transition in this "unconfined" problem may be just around the corner.

We also had summaries of the dramatic application of Type Ia supernovae to cosmology and the prospects for probing the "dark energy" from Brian Schmidt and Saul Perlmutter. The astounding discovery of the acceleration of the Universe did not depend on any deep understanding of the physics of the explosion, nor on the evidence for asymmetry being revealed by spectropolarimetry. As we try to measure the effective acceleration as a function of space and time, effectively the equation of state of the dark energy, systematic effects must be mastered at an unprecedented level of precision. This will require a greater physical understanding and an understanding of the origin of the asymmetries that may give a dependence of the luminosity on the angle of observation. If all Type Ia are basically alike, then such angle-dependent effects will average out in a large sample, but if, for instance, the cause of the asymmetry varies with redshift because the underlying cause of the asymmetry does, then great care will be required to make the appropriate analysis of the high-redshift observations.

42.5.3 Core collapse

To emphasize, the lesson that emerges strongly from recent studies of the polarization of supernovae and related issues is that core collapse supernovae are always asymmetric, and frequently, but not universally, bi-polar. I must emphasize that this is a hard won conclusion, with heroic observational work by Lifan Wang and by Doug Leonard, Alex Filippenko and their colleagues.

In terms of giving credit, Stirling Colgate revealed the true father of modern supernova research: Scratchy Serapkin. Anatoly Serapkin was the head of the Soviet delegation to the Geneva talks aimed at the Limited Test Ban Treaty to abandon space, atmospheric, and underwater nuclear talks in 1963. Colgate was one of the representatives on the U.S. side with the self-appointed goal of convincing both sides that we needed to understand the astrophysical "background" to avoid confusing a natural event with a bomb test. The yet-to-be famous Vela satellites played a role in these discussions. Colgate said supernovae might be confused with a test. Scratchy, not a scientist himself, fixed him with a steely glare and inquired, "Who knows how supernovae work?" Colgate realized what thin ground he, and the U.S. delegation, were on, returned to Livermore and made the case to Edward Teller that understanding supernovae must become a primary goal of the lab. The rest is history.

At this conference, new perspectives on the mechanisms of core collapse, neutrino transport, rotation, and magnetic fields were given by Stirling Colgate, Adam Burrows, Thiery Foglizzo, Dave Meier, Dong Lai, and Peter Williams. The impact of asymmetries on the dynamics and on the question of neutron star versus black hole formation were also discussed. Issues of nucleosynthesis were discussed by John Cowan, Keichi Maeda, and Raph Hix.

Mario Hamuy also spoke about the large range in apparent kinetic energy of the explosions of Type II supernovae. He raised the question of whether or not the distribution of energy is a continuum, or is more complex, implying, perhaps, different physical processes. An example would be neutron star versus black hole formation. These questions must also be posed for Type Ic supernovae. We need to determine if the events labeled "hypernovae" by Maeda and his collaborators are truely special, or part of a continuum. Given the evidence for strong asymmetries, there will be line-of-sight effects early on. Asymmetric flows are also apt to alter the systematics of gamma-ray deposition in later phases, and hence the luminosity and slope of the radioactive tail. Alejandro Clocchiatti reported some old, but still quite relevant, data on Type Ic events that may help to resolve such issues.

Norbert Langer argued that the variations in single star and binary evolution make it unlikely that all massive stars will undergo the same rotational history. Thus, it is improbable to have all iron cores evolve with the same rapid rotation prior to collapse as may be required for MHD jet models of supernovae and of GRBs. Rotation remains a key parameter in stellar evolution studies and, as for collapse dynamics, it is unlikely that rotation will be present in the absence of magnetic fields. Both rotation and magnetic fields must be included self-consistently from the proto-stellar phase to collapse before we really understand this issue. It may not be appropriate, for instance, to evolve a model star a considerable amount and then add the effects of magnetic viscosity once some amount of shear has developed.

42.6 Gamma-Ray bursts

We ended the conference with a stimulating session on GRBs and their possible connection to supernovae. Tom Matheson brought the exciting observations by his team of supernova SN 2003dh associated with GRB 030329. This supernova apparently closely resembled SN 1998bw despite the fact that GRB 030329 was a "normal," if exceptionally close, GRB and that GRB 980425, if associated with SN 1998bw, was rather odd. Given the liklihood of asymmetries and large variations in energy, it is not at all clear why these two supernovae should be so similar.

Don Lamb summarized work on the X-ray rich bursts discovered by HETE, concluding that the gamma-ray component might be more collimated than previously thought and hence that the rate of explosion of GRBs might rival that of SN Ic. Edo Berger, on the other hand, presented radio data on SN Ic that apparently showed that rather few Type Ic could be associated with GRBs. Alin Panaitescu outlined his work with Pawan Kumar that was more consistent with the "standard" value of collimation, but consistent for some bursts with magnetic fields that were quite large in the early, reverse shock phase, and that then decayed with time. Brad Schaefer illustrated how use of the variability/luminosity and spectral lag/luminosity relations are both self-consistent and potentially useful tools to provide distance estimates to GRBs and hence to use GRBs as independent cosmological probes.

42.7 Conclusions and charge

With SN 2003dh, we have incontrovertible evidence that at least some GRBs are associated with spectrally identifiable supernova explosions. This still leaves open a raft of fascinating questions. What is the machine of the explosion? My bet is that it involves rotation and magnetic fields. With the results of Panaitescu and Kumar and of Coburn & Boggs (2003), there are new suggestions that the magnetic field may not just be required to produce synchrotron radiation, but may be dynamically important in producing the burst. If (long) GRBs are routinely associated with the collapse of massive stars, how does the burst and associated magnetic field get out of the star? Another important question, touched on above, is what fraction of Type Ic supernovae make GRBs? If GRBs come from massive stars, why do we not see evidence for winds and shells, dense circumstellar media, in every burst? Must all (again long) GRBs be associated with the formation of black holes, or can some be associated with neutron stars? Do we expect the distribution of rotation rates of stars to vary over the history of the Universe, and if so, should there be some impact on the rate of production of GRBs with redshift?

These are important questions, but there are fundamental issues lying at the core of all of them. That leads to my charge to attendees of the meeting and to readers of these proceedings. *Go thee forth and think about rotation and magnetic fields!*

Acknowledgements

I again express my gratitude to Peter Höflich and Pawan Kumar for the work they did on this meeting and to all my colleagues who attended and said embarrassingly nice things. This work was supported in part by NSF AST-0098644 and by NASA NAG5-10766.

References

Ando, S. 2003, *Phys Rev D*, **68**, 63002

Akiyama, S. Wheeler, J. C., Meier, D. & Lichtenstadt, I. 2003, *Astrophysical Journal*, **584**, 954

Bhattacharya, K. & Pal, P. B. 2003, hep-ph/0209053

Blackman E. G. & Brandenburg, A. 2002, *Astrophysical Journal*, **579**, 359

Blackman, E. G. & Field, G. B. 2000, *Astrophysical Journal*, **534**, 984

Blackman, E. G. & Field, G. B. 2002, *Phys. Rev. Lett.*, **89**, 265007

Blackman E. G. & Tan, J. 2003, in "*Proceedings of the International Workshop on Magnetic Fields and Star Formation: Theory vs. Observation*," in press (astro-ph/0306424)

Blandford, R. D. & Hernquist, L. 1982, *Journal of Phys. C.*, **15**, 6233

Blandford, R. D. & Payne, D. G. 1982, *Monthly Notices of the Royal Astronomical Society*, **199**, 833

Brandenburg, A., Blackman E. G. & Sarson, G. R. 2003, *Adv. Space Res.*, **32**, 1835 (astro-ph/03005374)

Chanmugam, G. 1979, *Monthly Notices of the Royal Astronomical Society*, **187**, 769

Coburn, W. & Boggs, S. E. 2003, *Nature*, **423**, 415

Field, G. B. & Blackman E. G. 2002, *Astrophysical Journal*, **572**, 685

Gerardy, C. L., Höflich, P. Quimby, R., Wang, L., Wheeler, J. C., Fesen, R. A., Marion, G. H., Nomoto, K. & Schaefer, B. E. 2003, *Astrophysical Journal*, **607**, 391 (astro-ph/0309639)

Gruzinov, A. V. & Diamond, P. H. 1994, *Phys. Rev. Lett.*, **72**, 1651

Hawley, J. F. & Balbus, S. A. 2002, *Astrophysical Journal*, **573**, 738

Hawley, J. F., Gammie, C. F. & Balbus, S. A. 1996, *Astrophysical Journal*, **464**, 690

Kleeorin, N. I., Moss, D., Rogachevskii, I. & Sokoloff, D. 2002, *Astronomy & Astrophysics*, **387**, 453

Konar, S. 1997, PhD. Thesis

Kulsrud, R. M. & Anderson, S. W. 1992, *Astrophysical Journal*, **396**, 606

Lai, D. & Qian, Y.-Z. 1998, *Astrophysical Journal*, **505**, 844

MacFadyen, A. & Woosley, S. E. 1999, *Astrophysical Journal*, **524**, 262

Ogilvie, G. I. 2001, *Monthly Notices of the Royal Astronomical Society*, **325**, 231

Spruit, H. C. & van Ballegooijen, A. A. 1982, *Astronomy & Astrophysics*, **106**, 58

Thompson, C. & Duncan, R. C. 1993, *Astrophysical Journal*, **408**, 194

Vishniac, E. T. & Cho, J. 2001, *Astrophysical Journal*, **550**, 752

Williams, P. T., 2003, IAOC Workshop "*Galactic Star Formation Across the Stellar Mass Spectrum*," *ASP Conference Series*," ed. J. M. De Buizer, in press (astro-ph/0206230)